WITHDRAWN
UTSA Libraries

P9-AFQ-439

The University of Texas
at San Antonio
UTSA Libraries

Ecological Studies, Vol. 71

Analysis and Synthesis

Edited by

W. D. Billings, Durham, USA
F. Golley, Athens, USA
O. L. Lange, Würzburg, FRG
J. S. Olson, Oak Ridge, USA
H. Remmert, Marburg, FRG

Ecological Studies

Volume 58
Ecology of Biological Invasions of North America and Hawaii
Edited by H.A. Mooney and J.A. Drake
1986. X, 320 p., 25 figures. cloth
ISBN 3-540-96289-1

Volume 59
Acid Deposition and the Acidification of Soils and Waters
By J.O. Reuss and D.W. Johnson
1986. VIII, 120 p., 37 figures. cloth
ISBN 3-540-96290-5

Volume 60
Amazonian Rain Forests
Edited by C.F. Jordan
1987. X, 133 p., 55 figures. cloth
ISBN 3-540-96397-9

Volume 61
Potentials and Limitations of Ecosystem Analysis
Edited by E.-D. Schulze and H. Zwölfer
1987. XII, 435 p., 141 figures. cloth
ISBN 3-540-17138-X

Volume 62
Frost Survival of Plants
By A. Sakai and W. Larcher
1987. XI, 321 p., 200 figures. cloth
ISBN 3-540-17332-3

Volume 63
Long-Term Forest Dynamics of the Temperate Zone
By Paul A. Delcourt and Hazel R. Delcourt
1987. XIV, 450 p., 90 figures.
33 maps. cloth
ISBN 3-540-96495-9

Volume 64
Landscape Heterogeneity and Disturbance
Edited by Monica Goigel Turner
1987. XI, 241 p., 56 figures. cloth
ISBN 3-540-96497-5

Volume 65
The Community Ecology of Sea Otters
Edited by G.R. VanBlaricom and J.A. Estes
1987. XVI, 247 p., 71 figures. cloth
ISBN 3-540-18090-7

Volume 66
Forest Hydrology and Ecology at Coweeta
Edited by W.T. Swank and D.A. Crossley, Jr.
1987, XIV, 512 p., 151 figures. cloth
ISBN 0-387-96547-5

Volume 67
Concepts of Ecosystem Ecology
Edited by L.R. Pomeroy and J.J. Alberts
1988. XIV, 345 p., 94 figures. cloth
ISBN 0-387-96686-2

Volume 68
Stable Isotopes in Ecological Research
Edited by P.W. Rundel, J.R. Ehleringer and K.A. Nagy
1988. Approx. 570 p., 160 figures. cloth
ISBN 0-387-96712-5

Volume 69
Vertebrates in Complex Tropical Systems
Edited by M.L. Harmelin-Vivien and F. Bourlière
1988. Approx. 240 p., 17 figures. cloth
ISBN 0-387-96740-0

Volume 70
The Northern Forest Border in Canada and Alaska
By James A. Larsen
1988. Approx. 260 p., 73 figures. cloth
ISBN 0-387-96753-2

Volume 71
Tidal Flat Estuaries
Edited by J. Baretta and P. Ruardij
1988. XVIII, 353 p., 123 figures. cloth
ISBN 3-540-19323-5

Tidal Flat Estuaries

Simulation and Analysis of the Ems Estuary

Edited by
J. Baretta and P. Ruardij

With a Foreword by
P. de Wolf

With 123 Figures

Springer-Verlag
Berlin Heidelberg New York
London Paris Tokyo

Dr. Job Baretta
Dr. Piet Ruardij

Netherlands Institute for Sea Research
P.O. Box 59
1790 AB Den Burg-Texel
The Netherlands

LIBRARY
The University of Texas
at San Antonio

ISBN 3-540-19323-5 Springer-Verlag Berlin Heidelberg New York
ISBN 0-387-19323-5 Springer-Verlag New York Berlin Heidelberg

Library of Congress Cataloging in Publication Data. Tidal flat estuaries: simulation and analysis of the Ems estuary/edited by J.W. Baretta and P. Ruardij. p. cm. – (Ecological studies: vol. 74) Includes index. 1. Estuarine ecology – Germany (West) – Ems River Estuary. 2. Tidal flat ecology – Germany (West) – Ems River Estuary. I. Baretta, J.W. (Job W.), 1944–. II. Ruardij, P. (Piet), 1949–. III. Series: Ecological studies; v. 74. QH149.T53 1988 574.5′26365′0943 – dc 19 88-23478

This work is subject to copyright. All rights are reserved, whether the whole or part of the material is concerned, specifically the rights of translation, reprinting, re-use of illustrations, recitation, broadcasting, reproduction on microfilms or in other ways, and storage in data banks. Duplication of this publication or parts thereof is only permitted under the provisions of the German Copyright Law of September 9, 1965, in its version of June 24, 1985, and a copyright fee must always be paid. Violations fall under the prosecution act of the German Copyright Law.

© Springer-Verlag Berlin Heidelberg 1988
Printed in Germany

The use of registered names, trademarks, etc. in this publication does not imply, even in the absence of a specific statement, that such names are exempt from the relevant protective laws and regulations and therefore free for general use.

Typesetting (media conversion), printing and binding: Brühlsche Universitätsdruckerei, Giessen
2131/3130-543210 – Printed on acid-free paper

This book is dedicated to our friend and colleague, HUUB SCHRÖDER, who will never read it, but who was our first and most enthusiastic convert to the joys of ecosystem modelling.

Foreword

Thanks to Vincent van Gogh's painting "The Potato Eaters", the Dutch are thought by many to use their huge potato crop for just that: eating. Though human consumption of potatoes in the Netherlands certainly is considerable, a large fraction of the crop is processed by the potato processing industry in the northeastern part of Groningen. This industry, now producing half a million tonnes of potato flour annually, started at the end of the 19th century. At that time the production was much smaller, but even so the first objections to the pollution of the canals of Groningen by process water and process waste date from 1910; already then the canals were anaerobic in the autumn. Around 1965 the amount of organic waste from the potato industry had risen to $20 \cdot 10^6$ inhabitant equivalents, roughly 50% of the total organic waste of the Netherlands, and the situation in the province of Groningen was virtually unbearable. Plans were made to pipeline the waste from the factories to the Ems-Dollard estuary and these plans were in the end causative for the research described in this book.

Originally it was expected that the existing knowledge of the hydrology, chemistry and biology of the estuary would be sufficient to calculate the oxygen saturations, following the introduction of given amounts of organic waste at selected sites in the estuary, but it was also realized that there was little or no knowledge available to predict the behaviour of the estuary as a whole under these circumstances. For this reason, when our research project (Biological Research Ems-Dollard Estuary, Dutch acronym BOEDE) started in 1973, from the very beginning we insisted upon the possibility of building a simulation model of the estuary; the actual implementation of the model, however, could only be realized at a later stage. Our modelling efforts are the subject of this book.

The project BOEDE has been contracted and financed by:
1. The Ministry of Transport and Public Works;
2. The Ministry of Education and Science;
3. The Ministry of Housing, Physical Planning and the Environment;
4. The Ministry of Agriculture and Fisheries.

The representatives of these ministries, as members of the (scientific) coordination committee, have been instrumental in concluding this project successfully by their willingness to give such long-term support

(14 years) to an undertaking that sometimes must have seemed rather nebulous to them. I hope that this book will justify their support. It must be noted, however, that the ideas and opinions expressed in this book are entirely the authors' responsibility.

The project was executed at the Netherlands Institute for Sea Research and the University of Groningen. Many thanks are due to the directors and staff of these institutes for availability of facilities, advice and opportunity for discussion. Thanks are also due to the Board of Directors of the Organization for Applied Scientific Research (TNO) who acted as employers for the personnel. The authors of this book have all been associated with the BOEDE project for a long time, most of them since the beginning, and at that time fresh from their universities. Job Baretta and Piet Ruardij deserve our gratitude, together with Wim Admiraal, for their efforts to make this book as it is.

I welcome this opportunity to note that in an earlier stage Huub Schröder (1945–1985) was a driving force behind the model and the planning of this book, as well as an industrious editor of earlier versions of the manuscripts. It is a sad thought that he was not to live to see the completion of a job that was dear to him. We all knew that he was suffering, for a number of years, from a wasting disease, but although this must have been a constant worry, he rarely spoke about it. After completion of his manuscript on anaerobic processes on 13th September 1985, he underwent major surgery (again) on 16th September, during which he died. We all lost an excellent colleague and a dear friend.

I recall with gratitude many of the often very heated discussions, more than once bordering on a real fight, but always conducted in the best scientific tradition, during the workshops in which we structured the model described here. In retrospect it even appears as if most authors did disagree about something at one time or another, but altogether I think that they will agree with my opinion that we had a great and interesting time.

My last remark, before wishing the book a broad readership, concerns the waste water problem that stood at the cradle of our efforts. This is in the process of being solved in the way it should be: by purification at the source. The industry is, at great expense, building purification plants, and the amount of waste that is brought into the Ems estuary declined to 40% of the original, while the cleaned effluent is piped to the estuary. But that, of course, is another story.

P. DE WOLF
BOEDE project coordinator
Organization for Applied Scientific Research
Division Technology for Society

Acknowledgements

The BOEDE ecosystem study, of which the Ems model is the final synthesis, has involved many people, too many to thank all by name for their assistence in completing this project. They include all the assistants and students, both at the biology department in Haren of the Groningen University and at the NIOZ in Texel, who did most of the experimental work.

The fieldwork, mainly done from the RV Ephyra, profited from the enthusiastic assistance of her crew, especially that of Cor Wisse. His unfailing good temper and his skill at improvising repairs to sampling quipment are gratefully acknowledged. More or less coincident with the shift in emphasis from fieldwork to simulation and analysis, Frank van Es left the group, but he pioneered many of the concepts on microbial ecology now incorporated in the model. We also thank Lucas Bouwman and Karin Romeyn for providing valuable contributions to the first concept of the model. Piet de Wolf created and the directors of NIOZ gave us the opportunity as well as the means for transforming a study into a model.

Volker Kartheus and Wulf Greve have introduced us to the workshop approach as well as given us the modelling software for our first workshop, while Norsk Data Nederland lent us the necessary computer hardware for that period. Wolfgang Ebenhöh taught us a viable approach to ecosystem modelling and solved many implementation problems. Mark Markofsky and Bill Silvert each led a modelling group to great effect during the first worshop. Of all the NIOZ scientists who have helped us, we especially acknowledge Bouwe Kuipers and Henk van der Veer for their assistance with epibenthos data and processes.

We thank Don Gordon and his group from the Marine Ecology Lab of BIO for the stimulating cooperation in building an ecosystem model of the Cumberland Basin, which refined our concept of the Ems model.

We thank Wim Admiraal for his editorial help on early versions of the manuscript. Anneke Bol-den Heijer typed the whole manuscript and Jeannet Schröder-ter Avest helped with the numerous revisions. Hans Malschaert and the graphics department of NIOZ produced the illustrations.

Ab Dral has helped us enormously by taking on the role of impartial reader who was not going to be fooled by high-tech jargon but wanted all explanations to be understandable to the general practitioner of marine ecology. We thank him for spending so much of his time on what must have seemed an endless task.

JOB BARETTA
PIET RUARDIJ

Contents

Part I: Description of the Ecosystem

1 Modelling the Ecosystem of the Ems Estuary.
J. W. BARETTA and P. RUARDIJ (With 3 Figures) 3
1.1 Introduction 3
1.2 Earlier Investigations of the Ems Estuary 7
1.3 Other Models of the Ems Estuary 8
1.4 Structure of the Ecosystem Model 9
1.5 A Glimpse of the Modelling Practice 11
1.6 Simulation Hard- and Software 12

2 The Abiotic Environment. V. N. DE JONGE (With 10 Figures) . 14
2.1 History 15
2.2 Topography and Morphology 15
2.3 Sediment Composition 16
2.4 Climate 18
2.5 Fresh Water Supply 19
2.6 Tide 20
2.7 Water Movement and Mixing 21
2.8 Sediment Transport and Accumulation 23
2.9 Nutrients 25

3 Biology. F. COLIJN, M. A. VAN ARKEL and A. STAM
(With 3 Figures) 28
3.1 General Description of the Estuary 28
3.2 The Pelagic System 30
3.3 The Benthic System 32
3.4 The Epibenthic System 33

4 General Features in the Model. W. ADMIRAAL, J. W. BARETTA,
F. COLIJN, V. N. DE JONGE, R. W. P. M. LAANE, P. RUARDIJ
and H. G. J. SCHRÖDER (With 18 Figures) 36
4.1 Universal Biological Processes 36
  4.1.1 Maintenance Processes 37
  4.1.2 Activity-Dependent Processes 37
  4.1.3 Bookkeeping Calculations 38

4.1.4 Modelling Food Uptake 39
4.1.5 Food Selection 40
4.1.6 Food Levels 41
4.2 Compartments and Boundary Conditions 42
4.2.1 Boundaries 42
4.2.2 The Dimensions of the Compartments 43
4.2.3 Boundary Conditions 50
4.3 Forcing Functions 53
4.3.1 Temperature 53
4.3.2 Irradiance 55
4.4 The Nature and Flux of Organic Matter 57
4.4.1 Partition of Organic Matter 58
4.4.2 Transport of Organic Matter 61
4.4.3 The Nutritional Value of Detrital Organic Matter
      (Detritus) 61

Part II: The Blueprint of the Model

5   The Construction of the Transport Submodel.
      P. RUARDIJ and J. W. BARETTA (With 3 Figures) 65
5.1 Introduction 65
5.2 Diffusive Transport 65
5.2.1 Determination of the Constants 66
5.3 Transport of Particulate Material 67
5.3.1 Description 68
5.3.2 Vertical Transport Processes: Sedimentation and
      Resuspension 70
5.3.3 Horizontal Transport Processes 71
5.3.4 Boundary Conditions 73
5.3.5 Calibration of the Model 74
5.3.6 Detrital Carbon in the Mud 76

6   The Construction of the Pelagic Submodel.
      J. W. BARETTA, W. ADMIRAAL, F. COLIJN, J. F. P. MALSCHAERT
      and P. RUARDIJ (With 3 Figures) 77
6.1 Introduction 77
6.2 Phytoplankton and Nutrients 79
6.2.1 State Variables 80
6.2.2 Irradiance and Photosynthesis 81
6.2.3 Nutrients 82
6.2.4 Assimilation 83
6.2.5 Loss Terms 84
6.3 Zooplankton 86
6.3.1 Microzooplankton 87
6.3.2 Mesozooplankton 89
6.3.3 Carnivorous Zooplankton 94

6.4 Pelagic Bacteria 96
   6.4.1 Biomass 97
   6.4.2 Uptake 97
   6.4.3 Loss Terms 99
6.5 Oxygen in the Water 101
   6.5.1 Physical and Chemical Aspects 102
   6.5.2 Sources and Sinks of Oxygen 103

7 The Construction of the Benthic Submodel.
W. ADMIRAAL, M. A. VAN ARKEL, J. W. BARETTA, F. COLIJN,
W. EBENHÖH, V. N. DE JONGE, A. KOP, P. RUARDIJ and
H. G. J. SCHRÖDER (With 15 Figures) 105
7.1 Introduction 105
7.2 Benthic Primary Producers 107
   7.2.1 Vertical Distribution of Biomass 107
   7.2.2 Effects of Light 109
   7.2.3 Specific Growth Rate and Primary Production 110
   7.2.4 Excretion and Respiration 112
   7.2.5 Suspension 113
7.3 Meiofauna 114
   7.3.1 Food Uptake 114
   7.3.2 Loss Factors 117
7.4 Macrobenthos 118
   7.4.1 Biology 119
   7.4.2 Model Description: Deposit Feeders 123
   7.4.3 Model Description: Suspension Feeders 127
7.5 The Spatial Structure of the Sediment and Vertical Transport 130
   7.5.1 Oxygen Demand and Sulphide Production 130
   7.5.2 Nondiffusive Transport Processes 135
   7.5.3 Diffusive Transport Processes 137
   7.5.4 The Transition of Organic Matter Between the Benthic
       Layers 145
7.6 Benthic Bacteria 146
   7.6.1 Distribution of Biomass 147
   7.6.2 Uptake of Labile Organic Carbon 148
   7.6.3 Detritus and Refractive Organic Matter 149
   7.6.4 Mortality and Excretion 150
   7.6.5 Respiration and Temperature Correction 151
   7.6.6 Conclusion 151

8 The Construction of the Epibenthic Submodel.
J. G. BARETTA-BEKKER and A. STAM (With 7 Figures) 153
8.1 The State Variables 153
8.2 The Faunal Components of the Epibenthos 154
8.3 Migration 155
   8.3.1 Redistribution 155
   8.3.2 Comfort Functions 157

8.3.3 Immigration 159
8.3.4 Emigration 161
8.4 Uptake of Food 162
8.4.1 Calculation of Available Food 162
8.4.2 Uptake 163
8.5 Loss Factors 164
8.5.1 Respiration and Excretion 164
8.5.2 Mortality 165
8.6 Correction for Temperature 165
8.7 The Role of Foraging Birds 165
8.7.1 Estimation of the Number of Birds in the Ems Estuary 166
8.7.2 Calculation of Available Food for Birds 166
8.7.3 Uptake by Birds 167

Part III: Results and Analysis

9   Running the Model. J. W. BARETTA 171
9.1 The Standard Run 171
9.2 The Initial Values of the State Variables 171
9.3 Validation of the Model 172
9.3.1 The Validation Sets 173

10   Results and Analysis of the Pelagic Submodel.
      J. W. BARETTA and W. ADMIRAAL (With 22 Figures) 175
10.1 Model Results and Validation of the Standing Stocks . . . 175
10.1.1 Primary Producers 175
10.1.2 Microzooplankton 180
10.1.3 Mesozooplankton 181
10.1.4 Carnivorous Zooplankton 181
10.1.5 Pelagic Bacteria 182
10.1.6 Oxygen 182
10.1.7 Nutrients 183
10.1.8 The Detrital State Variables (PLOC, PDET, PROC,
       PDROC) 183
10.2 Fluxes 186
10.2.1 Validated Fluxes 186
10.2.2 Nonvalidated Fluxes 189
10.3 Sensitivity Analysis 193
10.3.1 Degradation of Algal Production by Bacteria . . . 194
10.3.2 Growth Efficiency of Bacteria 197
10.3.3 Suspended Phytobenthos as a Food Source for
       Zooplankton 198
10.3.4 The Role of Microzooplankton in the Model . . . 200
10.3.5 Regulation of the Phytoplankton Succession 201
10.4 Conclusions 204

11   Results and Analysis of the Benthic Submodel.
     W. EBENHÖH (With 11 Figures) 205
11.1 The Seasonal Cycle of the State Variables 205
     11.1.1 Total and Anaerobic Bacterial Biomass 205
     11.1.2 Phytobenthos 206
     11.1.3 Meiobenthos 207
     11.1.4 Macrobenthos 207
     11.1.5 Organic Matter 209
     11.1.6 Sulphide and Pyrite 209
     11.1.7 The Location of the Sulphide Horizon 211
11.2 The Fluxes from and to the State Variables 212
     11.2.1 Validated Fluxes 212
11.3 Unvalidated Fluxes 214
     11.3.1 Diet of Meiofauna 214
     11.3.2 Grazing and Predation in the Benthos 215
     11.3.3 Carbon Budgets of Organisms 215
     11.3.4 Net Production and Productivity 216
     11.3.5 Dynamics of Refractory Organic Carbon in the
            Sediment 217
11.4 Sensitivity Analysis of the Benthic Subsystem 218
     11.4.1 Reduced Detritus 218
     11.4.2 Bacterial Excretion and Efficiency 220
     11.4.3 Phytobenthos Regulation 221
     11.4.4 The Role of Meiobenthos in the Benthic Community 224
     11.4.5 The Activity Pattern of Macrobenthos 225
11.5 Conclusions 227

12   Results and Analysis of the Epibenthos Submodel.
     J. G. BARETTA-BEKKER (With 8 Figures) 228
12.1 Model Results 228
     12.1.1 The Standing Stocks in the Five Compartments . . 228
     12.1.2 Internal Fluxes and Productivity 230
     12.1.3 Carbon Fluxes to the Epibenthic Submodel 231
12.2 Sensitivity Analysis 233
     12.2.1 Varying the Immigrating Biomass 233
     12.2.2 Model Runs, Partially or Wholly Excluding Epibenthos 234
12.3 Conclusions 239

13   Results and Validation of the Transport Submodel.
     P. RUARDIJ (With 13 Figures) 240
13.1 Introduction 240
13.2 Transport of Dissolved Compounds 240
     13.2.1 Salinity 240
     13.2.2 Dissolved Organic Carbon 241
13.3 Transport of Particulate Compounds 243
     13.3.1 Suspended Sediment 243
     13.3.2 Particulate Organic Carbon 245
     13.3.3 Suspended Matter 246

13.4 Flows and Annual Budgets 246
        13.4.1 Suspended Sediment Transport. 246
        13.4.2 Organic Carbon Transport 247
13.5 Sensitivity Analysis. 252
        13.5.1 Behaviour of Particulate Matter During Transport . 252
        13.5.2 The Role of Imported POC in the System 255
        13.5.3 The Role of Silt in the Estuary 256
13.6 Conclusions 258

14   Model Applications and Limitations.
        J. W. BARETTA and P. RUARDIJ (With 7 Figures) 259
14.1 Introduction 259
14.2 Limitations of the Model 259
        14.2.1 Category I: "Physiological" Parameters 260
        14.2.2 Category II: "Structural" Parameters 260
        14.2.3 Category III: "Dimensional" Parameters 261
14.3 Applications of the Model. 262
        14.3.1 Case 1: Reduced Organic Loading 262
        14.3.2 Case 2: Applying the Ems Model to the Severn
                Estuary 266
14.4 Conclusions 268

15   References 269

Appendix A: Parameter Descriptions and Values 282
A-1: Parameters of the Physical Submodel 282
A-2: Parameters of the Pelagic Submodel 285
A-3: Parameters of the Benthic Submodel 294
A-4: Parameters of the Epibenthic Submodel 303

Appendix B: The FORTRAN Listing of the Submodels 308
B-1: Subroutine BAHBOE 308
B-2: Subroutine SUBPEL 313
B-3: Subroutine SUBBEN 321
B-4: Subroutine SUBEPI 328

Appendix C: Data Files 333
C-1: Data of the Physical Model 333
C-2: Data of the Pelagic Model 336
C-3: Data of the Benthic Model 340
C-4: Data of the Epibenthic Model 343
C-5: The Initial Values of the State Variables 344

Subject Index 345

Contributors

W. Admiraal, National Institute for Public Health and Environmental Hygiene, P.O. Box 1, 3720 BA Bilthoven, The Netherlands

M. A. van Arkel, Netherlands Institute for Sea Research, P.O. Box 59, 1790 AB Den Burg, Texel, The Netherlands

J. W. Baretta, Netherlands Institute for Sea Research, P.O. Box 59, 1790 AB Den Burg, Texel, The Netherlands

J. G. Baretta-Bekker, Netherlands Institute for Sea Research, P.O. Box 59, 1790 AB Den Burg, Texel, The Netherlands

F. Colijn, Rijkswaterstaat, Tidal Waters Division, AOB, Van Alkemadelaan 400, 2579 AT 's Gravenhage, The Netherlands

W. Ebenhöh, Universität Oldenburg, Fachbereich Mathematik, Postfach 2503, D-2900 Oldenburg, Federal Republic of Germany

V. N. de Jonge, Rijkswaterstaat, Tidal Waters Division/AOB, P.O. Box 207, 9750 AE Haren, The Netherlands

A. Kop, Netherlands Institute for Sea Research, P.O. Box 59, 1790 AB Den Burg, Texel, The Netherlands

R. W. P. M. Laane, Rijkswaterstaat, Tidal Waters Division/AOC, Van Alkemadelaan 400, 2579 AT 's Gravenhage, The Netherlands

J. F. P. Malschaert, Netherlands Institute for Sea Research, P.O. Box 59, 1790 AB Den Burg, Texel, The Netherlands

P. Ruardij, Netherlands Institute for Sea Research, P.O. Box 59, 1790 AB Den Burg, Texel, The Netherlands

H. G. J. Schröder, deceased 16-9-1985, Netherlands Institute for Sea Research, P.O. Box 59, 1790 AB Den Burg, Texel, The Netherlands

A. Stam, Netherlands Institute for Sea Research, P.O. Box 59, 1790 AB Den Burg, Texel, The Netherlands

Part I: Description of the Ecosystem

1 Modelling the Ecosystem of the Ems Estuary

J. W. BARETTA and P. RUARDIJ

1.1 Introduction

The subject of this book is the modelling of an estuarine ecosystem that is not always completely covered with water: a tidal flat estuary. These tidal flats make it a rather special instance of a marine ecosystem, since this feature exposes large parts of the benthic system to direct sunlight and makes it accessible to non-diving predators, such as waders.

When we travel through the marine environment along a depth gradient coming from the deep ocean into oceanic – shelf – estuarine systems we find along with the depth gradient a number of other gradients that determine the functioning of the system. Generally, with decreasing depth the primary producers are coming closer to the benthic consumers and, in the case of estuarine systems with tidal flats, there appears a second group of producers, the phytobenthos. Where in deeper waters the benthic system is dependent on the fallout from the pelagic, in systems where tides keep the watercolumn more or less mixed, the benthos can get at the pelagic primary production directly, and conversely, the pelagic can directly profit from nutrient regeneration from the sediments.

When we proceed further into tidal flat estuaries, the exposure of the benthic system to direct sunlight may lead to a shift from a system dominated by pelagic production to a system dominated by benthic production.

Models representing different stages along this gradient have been made. Steele's (1974) North Sea model is a deep-sea model in that it is a purely planktonic model. Crucial in this model is that all primary production has to be grazed, because the detrital food web is neglected. This neglect is surprising, since Riley et al. (1949), whose model of the NW Atlantic gave a very good fit with available data, pointed out that a substantial fraction of primary production is not grazed and thus enters the detrital food web. In this respect Riley et al. were far ahead of their time, since the notion that detritus may be important in the marine food web, apparently has not been adopted by other modellers for over 25 years.

Kremer and Nixon (1978) published a model of Narragansett Bay. This model represents a coastal marine ecosystem, having a well-mixed shallow water column, which is not very turbid. There are no tidal flats and benthic primary production was left out of the model. The model emphasized the exchange between the spatial compartments and across the boundaries of the system, thus assuring realistic transport rates of plankton and dissolved substances. The biological groups that are modelled with mechanistic detail are phytoplankton, zooplankton and nutrients. Since the benthic variables were given as time series of biomas-

Table 1.1. The state variables in the biological submodels with their names as used in the model code. The single letter or figure following the model code name is also used in the model code (Appendix B) to denote the state variable. The last column lists the state variables as food sources for other state variables

Submodel	Description	Name	Code	Food sources
PELAGIC	Particulate refractory organic carbon (mg C · m⁻³)	PROC	A	—
	Dissolved refractory organic carbon (mg C · m⁻³)	PDROC	B	—
	Particulate organic detritus (mg C · m⁻³)	PDET	C	—
	Labile organic carbon (mg C · m⁻³)	PLOC	D	—
	Microzooplankton (mg C · m⁻³)	PMIC	E	PBAC PLOC PDET PDIA BDIA PFLAG PMIC
	Flagellate phytoplankton (mg C · m⁻³)	PFLAG	F	(Light) (PO4)
	Pelagic diatoms (mg C · m⁻³)	PDIA	G	(Light) (PO4) (SILIC) BDIA PDIA
	Mesozooplankton (mg C · m⁻³)	PCOP	H	PFLAG PDIA PDIA BDIA PCOP
	Carnivorous zooplankton (Ctenophora) (mg C · m⁻³)	CARN	I	PCOP PBLAR PMIC
	Benthic larvae (mg C · m⁻³)	PBLAR	J	—
	Pelagic bacteria (mg C · m⁻³)	PBAC	M	PROC PDET PLOC
	Noncarbon state variables:			
	Silicate (μmol · m⁻³)	SILIC	L	—
	Phosphate (μmol · m⁻³)	PO4	K	—
	Pelagic silt (mg · m⁻³)	SILT	—	—
	Silt on the channel bed (fluid mud) (mg · m⁻³)	CSILT	—	—
	Salinity (S)	SALT	—	—
	Oxygen (g · m⁻³)	OX	O	—
EPIBENTHOS	Birds (numbers)	BIRDS	X	BBBM BSF EMES EMAC
	Mesoepibenthos (mg C · m⁻²)	EMES	Y	BDIA BMEI BDET BBAC EMES
	Macroepibenthos (mg C · m⁻²)	EMAC	Z	BMEI BBBM BSF EMAC EMES PCOP
BENTHOS Aerobic layer	Particulate refractory organic carbon (mg C · m⁻²)	BROC	1	—
	Particulate organic detritus (mg C · m⁻²)	BDET	2	—

Variable (units)	Code	No.					
Labile organic carbon (mgC·m⁻²)	BLOC	3	–				
Diatoms (mgC·m⁻²)	BDIA	4	(Light)				
Meiobenthos (mgC·m⁻²)	BMEI	5	BDIA	BMEI	BBAC	ABAC	
Macrobenthos (deposit feeders) (mgC·m⁻²)	BBBM	6	BDIA	BMEI	BBAC	ABAC	
Macrobenthos (suspension feeders) (mgC·m⁻²)	BSF	7	BDIA	PDIA	PFLAG	PBLAR	PMIC
Bacteria (mgC·m⁻²)	BBAC	8	BLOC	BDET	BROC		
Noncarbon state variable:							
Depth of the sulphide horizon (cm)	BAL	–					
Anaerobic layer							
Particulate organic detritus (mgC·m⁻²)	ADET	2A	–				
Labile organic carbon (mgC·m⁻²)	ALOC	3A	–				
Bacteria in anaerobic layer (mgC·m⁻²)	ABAC	8A	ALOC	ADET	(AROC)		
Benthic sulphides (mgC·m⁻²)	BSUL	–					
Pyrites (mgC·m⁻²)	BPYR	–					

ses, the interaction between the benthic and planktonic system cannot be analyzed. Since this system was viewed as being plankton-based, however, this may
not matter very much.

The next model along our oceanic-estuarine gradient is GEMBASE, the General Ecosystem Model of the Bristol Channel and Severn Estuary, developed in
the 1970s by a group at IMER, led by Longhurst and reported partially (Radford
and Uncles, 1980; Radford, 1981).

GEMBASE must have been very difficult to formulate since it spans almost
the whole range of the oceanic-estuarine gradient, extending from the Celtic Sea
to the mud flats of the Severn estuary. The model is much more extensive than
the Narragansett Bay model but the only considered transfer of organic matter
into the benthic system is through the feeding activity of filterfeeders. Once the
particulate organic matter is in the benthic, however, the model provides pathways for utilization, which is an important extension and improvement to other
models, because here for the first time, an attempt is made to model both pelagic
and benthic processes dynamically.

If we move even further along the oceanic-estuarine gradient, the present Ems
model picks up where GEMBASE stops. GEMBASE in its most riverward compartment has a tidal-flat surface of about 5% of the total compartment surface
area. The Ems model in its outermost compartment has a tidal flat surface of 46%
of the total compartment surface, and this percentage increases inward to a maximum of 88% in the innermost compartment.

The Ems model, representing a temperate, turbid-tidal flat estuary is, as far
as we know, the first attempt to model the major benthic processes as fully as has
been customary for the pelagic. The model contains a large number of state variables (Table 1.1). Both in the pelagic and the benthic systems four main biological
groups have been incorporated: producers (phytoplankton and phytobenthos),
consumers (zooplankton, carnivores, meiobenthos and macrobenthos, epibenthic
organisms), decomposers (benthic and pelagic bacteria and also anaerobic bacteria in the benthos) and detrital carbon, which appears in the model in three different forms, according to its turnover time.

Detritus in this model occupies a central place in both the pelagic and the
benthic system, indicating the importance we ascribe to the detrital food web.
This does not imply that the energy fluxes through the detrital food web are larger
than those through the grazing/predation interactions because, even though the
concentrations of detrital material are large, a large fraction of this material is
rather refractory and thus has a very low turnover. The sources of the detritus
pool are many and diverse but all detritus has one thing in common: it consists
of the remains of organisms. This implies that detritus from all trophic levels contributes to it. Detritus is also a potential food source for consumers from different
trophic levels.

The scope of our model is to describe the functioning of the Ems ecosystem
in terms of what is known of it by research of our own and of others, in order
to be able to explain the probable causality of its functioning.

The first objective of our research project therefore was to identify and quantify the major components in the Ems estuary. The target groups at first were:
phytobenthos, pelagic and benthic bacteria, macrobenthos, zooplankton and the

epibenthos. The primary aim in studying these groups originally was to detect and quantify any effects of waste water discharges on them. It soon became apparent that a more systems-oriented approach was required in order to understand the functioning of the Ems ecosystem and that, moreover, the original number of components to be studied should be extended to include micro- and meiobenthos, phytoplankton and detritus in all its manifestations. Directing our efforts towards building a mathematical model of the system was considered to be the most promising way of integrating our knowledge of the different components into an understanding of the structure and function of the ecosystem.

A secondary objective was to gain the ability to quantify and explain the impact of man-made disturbances by waste water discharges, dredging and other activities.

The results of this long-term research project have been disseminated in a large number of publications, partially in our own series *"BOEDE publicaties en verslagen"* (BOEDE, 1985), partially in international publications. Thanks to this research and to the fact that there was an earlier data base to build upon, the Ems estuary now is one of the most extensively studied estuaries in the world, which made it ideally suitable for mathematical modelling. The Ems ecosystem simulation model, the subject of this book, is the integration and synthesis of all the separately published and unpublished research results.

1.2 Earlier Investigations of the Ems Estuary

One of the oldest Dutch scientific publications on the Ems estuary (Fig. 1.1) is a description of the physical geography of the Dollard by Stratingh and Venema (1855). This book contains a number of descriptions of different aspects of the area from the time before waste water discharges started.

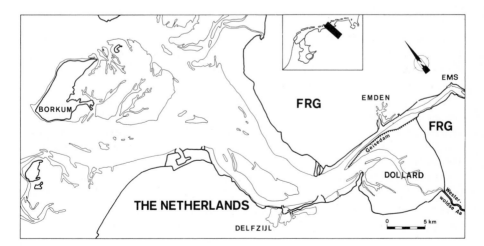

Fig. 1.1. The location of the Ems estuary

Lohmeyer made a survey of the fish fauna of the Ems estuary in 1907. Hydrographical and biological survey work has been done by Kühl and Mann (1968) in the early 1950s. Their hydrographical work has been very useful to us in providing baseline values for a number of parameters such as suspended sediment. Their biological work was of a qualitative nature and does not allow for quantification of the biota, but the species composition of the zooplankton seems not to have changed with the possible exception of Appendicularians. These have been caught only rarely during our investigations but were common in the early 1950s.

Dutch scientists from a number of disciplines have studied different aspects of the Ems estuary from 1954–1956 (Voorthuysen and Kuenen, 1960). These studies encompassed sedimentology, hydrography, morphology as well as benthic flora and fauna components. The results, especially the sedimentological and hydrographical ones formed the basis of our research. In the 1960s and early 1970s the German Forschungsstelle on Norderney and now and then the Netherlands Institute for Sea Research did hydrographical, biological and chemical work in the area. Some aspects of the Ems estuary thus were well known at the start of our project in 1973, but the ecology of the system was still virgin territory.

1.3 Other Models of the Ems Estuary

Dorrestein and Otto (1960) published a 12-compartment, one-dimensional diffusive water transport model for the Ems, using exchange coefficients between the compartments which were derived from salinity measurements and fresh water discharges into the estuary. Eggink (1965) used this model to estimate the biological oxygen demand (BOD) in the estuary which would be caused by a proposed new waste water discharge in the central region of the Ems estuary.

More recently (Anonymus, 1984), a one-dimensional hydrodynamic model with several grid sizes has been developed by Delft Hydraulics Laboratory which gives a much better spatial resolution than the 12-compartment one, but otherwise provides no new information.

Another approach was taken by the German Bundesanstalt für Wasserbau which developed hydraulic models of the Ems estuary for use in connection with the proposed new harbour on the German side of the Dollard, which would necessitate the rerouting of the Ems river.

A recent refinement to the Dorrestein and Otto model has been made by Helder and Ruardij (1982) who modelled the consequences of different fresh water discharge levels to the exchange coefficients and the resulting salinity distributions in the estuary. An extended version of this transport model has been used by Van Es and Ruardij (1982) to calculate BOD distributions in the Dollard resulting from different waste water discharge scenarios. This model forms the basis of the transport submodel (Chap. 5) in our ecosystem model. All these models are confined to calculating water levels and salinity distributions. There have been no earlier attempts at constructing an ecosystem model for the Ems estuary.

1.4 Structure of the Ecosystem Model

By its very nature a tidal-flat estuary contains a wide variety of environmental conditions, both temporally and spatially. The spatial variety in the Ems estuary is generally found along its longitudinal axis. Therefore we divided the estuary along this longitudinal axis into five compartments (Fig. 1.2). These compartments exchange water, dissolved and particulate matter through transport processes modelled in the transport model. All biological processes are modelled in the five compartments in the same way; since these processes are modified by the morphology of the compartments, the end result of the biological processes may be strongly different for each compartment. An example may clarify this: the level of a tidal flat relative to mean sea level in combination with the tidal range determines the duration of exposure of its surface to sunlight and hence the potential benthic primary production. Differences between elevation of the tidal flats in the compartments thus will lead to differences in benthic primary production, other things being equal.

Next to the transport model there are three biological subsystems, modelled as three separate submodels: the pelagic, the benthic and the epibenthic. Each of these has connections to the other two, but has much more and stronger connections internally and for this reason is treated as a unit. The whole model thus consists of four submodels representing the four subsystems (Fig. 1.3). Each of the biological submodels contains a number of state variables (Table 1.1). Each state variable represents a functional biological group consisting of a varying number of species that have a similar function in the system.

The pelagic state variables are subject to transport processes as modelled in the transport submodel. Two different modes of transport are incorporated. The first one, diffusive and advective transport, operates on planktonic state variables and on dissolved and suspended substances. The second one, suspended matter transport, operates on nonliving particulate material such as silt and detritus

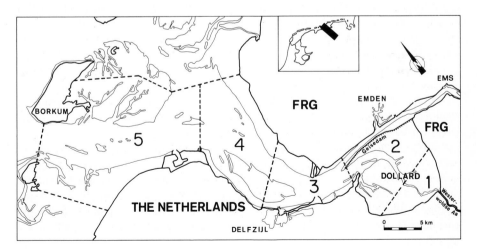

Fig. 1.2. The model compartments in the Ems estuary

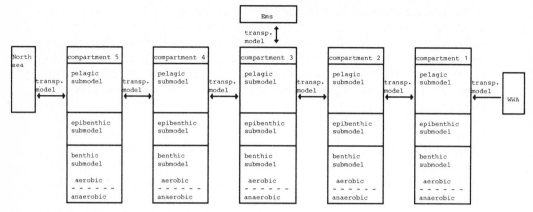

Fig. 1.3. Schematic diagram of the model structure

which is transported against the concentration gradient along the longitudinal axis of the estuary.

The epibenthic state variables can actively migrate through the compartments of the system, feed mainly on benthic organisms and their migration is influenced both by the abiotic conditions and the available food.

The benthic system is a two-layered system: it contains both an aerobic and an anaerobic layer. The thickness of these layers varies dynamically through the variation in the biological and chemical oxygen demand and the supply of oxygen from benthic algae and the atmosphere. Organic matter is transported by processes such as bioturbation and sediment accretion from the aerobic to the anaerobic layer, and vice versa. The benthic state variables that are susceptible to this transport have been defined in both layers.

The model is constructed as a carbon flow model, and organic carbon with a unit of mgC is the central entity. A large part of the carbon in the model is produced in the system by primary production, both benthic and pelagic. Another major source of carbon is allochtonous carbon, transported into the system from the fresh water sources and the sea.

No attempt has been made to give a strict trophic status to the state variables: instead the state variables for the fauna encompass within themselves a number of trophic levels ranging from bacterivorous to carnivorous. As Table 1.1 shows, feeding relationships in the model are structured as a food web, each consumer/predator having a number of food sources.

All the biological state variables perform the functions normal to living organisms: they eat, respire, excrete, reproduce, grow and die. In all these processes, except for growth and reproduction, organic carbon is transformed into a different state. In the case of transformation by respiration, the organic carbon is degraded/oxidized into CO_2 and thus lost from the system. This implies that respiration rates are very important parameters, since the values of these parameters determine the size of these losses. As an independent check on the sum of all respiratory losses in the model the oxygen dynamics have been included.

Organic carbon not only occurs in the guise of "living" state variables but also in several "nonliving", detrital state variables. The complexity of the detrital pathways has been approached in the model by separating detrital organic carbon into a number of state variables representing the different degrees of degradability. The detrital components occur both in the benthic and the pelagic system. In the pelagic an additional separation is made between dissolved and particulate organic detrital carbon.

1.5 A Glimpse of the Modelling Practice

The Ems ecosystem model had to reflect the accumulated knowledge of the whole BOEDE group, the published as well as the, as yet, unpublished parts.

To accomplish this, we opted for a workshop approach, intensive 5-day sessions during which the structure and content of the model were defined in plenary session. Then, the participants were divided into smaller groups that each constructed one of the submodels. These groups consisted of the scientists involved in research on this part of the ecosystem plus a modeller to assist in translating the formulated concepts and approaches into formulations and, later, into computer code. As soon as a submodel was up and running it was added to the other submodels and run again together with the other submodels. The goal of each workshop was to have a complete model running by the end of the workshop and to be able to do a preliminary analysis of the results in plenary session on the last day of the workshop. The latter objective was never fully realized.

In all, we needed four of these workshops, spaced over a period of 3 years, until the model arrived at its final structure and content, which will be presented in this book. During the last three workshops the goal of each submodel group was to redefine and restructure their submodel in order to resolve discrepancies between model behaviour and system behaviour as deduced from field and laboratory experiments. The periods between the workshops were used to implement and correct the changes to the model imposed upon it during the workshops. These workshops were somewhat like Great Leaps Forward: they resulted in great changes and improvements in the structure of the model, but they usually left a mathematical chaos behind, which the modellers pain-stakingly had to sort out afterwards. The advantages and disadvantages of developing a simulation model using such a workshop approach have been very well described by Holling (1978) in his book *Adaptive Environmental Assessment and Management*.

For us, the main advantage of building a model by a group effort was the higher degree of involvement of the individual scientists. Earlier efforts by one of us to construct a model after extracting information from the relevant scientists one by one (Ruardij, 1981) had shown that such an approach ended in the resulting model being viewed as strictly the responsibility of the modeller alone. This attitude severely restricted the amount of feedback the modeller, and hence the model, received.

1.6 Simulation Hard- and Software

The model has been developed in Fortran-77 on a Norsk Data 100 mini-computer, equipped with Tektronix graphics terminals and screen-editor capabilities on Tandberg alphanumeric terminals.

A simulation software package based on ideas developed by Holling and his group in Canada (Holling, 1975), taken up by Greve and Kartheus at the Biologische Anstalt Helgoland and modified and greatly extended by Ruardij at the Netherlands Institute for Sea Research (NIOZ), has been used as a meta-language for the model. A full description of this software package, which is called BAH-BOE is in preparation. Since the availability of this package has been instrumental in making a success of our approach we will shortly outline its main features.

A. Simulation facilities
1. Automatic compiling, linking and loading of the model
2. Definition of model run parameters
   a. Time step size
   b. Duration of the model run
   c. Output interval of the numerical/graphical output
   d. Precision of integration
   e. Selection of the year(s) to be simulated
      The large BOEDE database, resulting from 10 years of field work in the Ems estuary, made it desirable to be able to run the model for specific years at will. Data, such as boundary conditions and forcing function time series from the particular period to be simulated are selected by BAHBOE from the data base.
3. Facilities for changing parameter values and rerunning the model without recompiling. This facility turned out to be a mixed blessing: on the one hand, it saved a lot of time by avoiding recompiling, but on the other hand, it encouraged a lot of scientific diddling with parameter values.
4. Error checking of the parameter names. If a user wants to make a new run, the package checks if he only changed parameter values and not at the same time parameter names or added names. This avoids assigning the wrong values to parameters during initialization.
5. An integration routine, whose numerical precision is not the primary concern, but which is fast and able to handle numerical discontinuities. To fulfill these requirements, a rectangular integration is implemented, automatically changing the time step as the rates of change require.
B. Output facilities
   Numerical and graphical output on plotter, graphical terminals or line-printer.
   The user is able to generate graphics from his model output by using simple mnemonic commands and menu choices, calling the output variables by name. The output can be presented in different ways.
C. Facilities which were absolutely essential for a workshop approach were:
1. The use of a "standard" model. Every submodel, which is being improved, runs together with the other already approved submodels from the standard model. These submodels are automatically loaded together with the

submodel being tested. The standard model consists of the latest versions of all submodels approved together as the latest "best version" of the whole model.

2. The ability to simultaneously work on all submodels.

D. Analysis facilities

To fully incorporate our knowledge of the system and make use of our extensive data set, we had to build a complex ecosystem model, containing many state and other variables. To maintain a good grip on how the model was behaving as compared to the field data it was felt to be necessary to design a validation procedure. This procedure should provide us with an objective measure of the correspondence between field data and model results. The validation had to meet the following requirements:

1. Graphical output of simulation results and field data in the same plot.

2. Numerical output, giving the similarity between the field data and model results.

The software reads the field data for use in the validation procedure from a separate validation data base. The data entry for this data base has been made virtually "scientist-proof", to preclude ridiculous "validation" values resulting from input errors. The similarity procedure we used (Stroo, 1986) is derived from econometric and operation research literature. This procedure compares simulation results with field measurements. The comparison is based on three criteria: level, slope and residual variance and assigns a weighted value between zero (very bad indeed) and 10 (suspiciously perfect). Moreover, it also produces an overall validation score for all variables together, thus enabling quick comparisons between different model runs.

2 The Abiotic Environment

V. N. DE JONGE

Estuaries can be defined in many ways. A generally accepted and biologically useful definition of an estuary is "a semi-enclosed and coastal body of water which has a free connection with the open sea and within which sea water is measurably diluted with fresh water derived from land drainage" (Cameron and Pritchard, 1963).

The Ems estuary is part of the 600-km-long European Wadden Sea (Fig. 2.1). This shallow sea consists of a series of tidal basins protected from direct North Sea wave action by a string of barrier islands. The islands are separated by tidal inlets. The tidal basins are partly separated from each other by high tidal flats al-

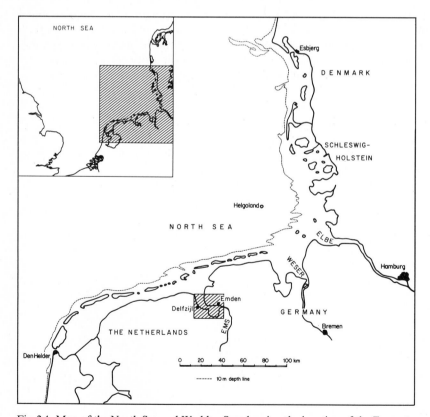

Fig. 2.1. Map of the North Sea and Wadden Sea showing the location of the Ems estuary

lowing water exchange only during high tide. A number of rivers drain into the Wadden Sea, the river Ems being one of them.

The surface of the Ems estuary (Fig. 1.1) is approximately 500 km^2, including a fresh water tidal area in the Ems of about 37 km^2.

2.1 History

The Wadden Sea and its estuaries originate from the last two glaciations. River valleys were formed during the glacial periods, while in the interglacial periods the sea transgressed and the river valleys became inundated. When large quantities of sediment were supplied and the tidal range was small, the valleys could completely fill in so that deltas developed. However, when sediment supply was less the river valleys drowned, giving rise to so-called coastal plain estuaries, which widen in a seaward direction.

The final shape of the Ems estuary was formed during the Pleistocene (Jelgersma, 1960; De Smet, 1960; Voorthuijsen, 1960) by the combined effects of sediment supply, a high tidal range and presumably storm surges (Russell, 1967). Originally the estuary had a funnel shape, with the river Ems more or less in its present place. A number of storm floods in the 14th and 15th century changed the original shape drastically and enlarged the estuary considerably (Stratingh and Venema, 1855). At the same time accretion began in the Dollard area (Wiggers, 1960; De Smet and Wiggers, 1960), followed by land reclamation from the sea. This process continued until the present day resulting in the Dollard as it is now.

2.2 Topography and Morphology

The upper part of the Ems estuary between Pogum and the mouth of the Dollard has an area of 100 km^2. It consists of the upper part proper and a shallow bay, the Dollard separated by a low guide dam, the Geisedamm (Fig. 1.1). During high tide some water exchange is possible by a number of perforations in the dam. The main channel is canalized and heavily dredged but it has tidal-flat embankments. Tidal flats also cover 85% of the Dollard area. Consequently the mean water depth in the Dollard is low (1.2 m). The volume of water in the upper estuary at mean sea level is approximately $120 \cdot 10^6$ m^3.

The middle part of the estuary, downstream of the Dollard (Fig. 1.1) has a funnel shape. This region extends to Eemshaven where the estuary joins the Wadden Sea. Most flats lie along the shore, but a large tidal flat divides the estuary longitudinally into two parts creating two channels. The main channel is on the east side. The total surface of this section is 155 km^2 of which 36% comprises tidal flats. The mean depth of this section increases gradually in a seaward direction, with the average water depth being 3.5 m. The water volume of this area at mean sea level is approximately $550 \cdot 10^6$ m^3.

The most seaward region of the estuary is the Wadden Sea part of the estuary. This region is situated between Eemshaven and the islands Rottumeroog and

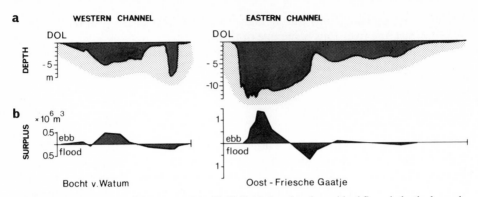

Fig. 2.2. Cross-section of the estuary near Delfzijl (a) showing the residual flows in both channels (b). *DOL* = Dutch Ordinance Level

Borkum (Fig. 1.1). The boundary with the North Sea is formed by these two islands and the tidal inlet between them. The high tidal flats between the islands and the mainland form the hydraulic boundaries between the estuary and the adjacent tidal basins. The area occupies 215 km^2 of which 46% consists of tidal flats. It has two channels separated by a series of shoals. The water volume of this region of the estuary is approximately $770 \cdot 10^6$ m^3 at mean sea level.

The morphology of the tidal flats, channels and gullies is complex, not only in their spatial distribution but also in their geometry. In Fig. 2.2 an example is given of a cross-section of the middle part of the estuary. The bottom profile shows two channels separated by a shoal, with considerable differences in the residual flow profiles.

The morphology of the Ems estuary is not static. Slow changes are not only induced by natural processes such as sedimentation, meandering of channels and gullies, but also by human activities such as land reclamation, building of harbours, dredging and sand mining. An example of natural morphological changes near the tidal inlet is given in Fig. 2.3 (Samu, 1979). Samu concluded that these morphological changes were cyclic and occurred with a periodicity of some 25 years. Similar morphological changes were initiated by dredging activities and sand mining (in Randzelgat). The effects of these activities were noticeable within 1 year (De Jonge, 1983).

2.3 Sediment Composition

The sediments in the outer estuary are mainly composed of sand with median grain sizes between 95 and 155 µm. The clay content (grain size <2 µm) varies, depending on the degree of exposure to currents and waves, between 0.3 to 3.5%; nearshore the clay content is higher. In the middle part of the estuary the clay content on the embankments increases to values of 9 to 18% with an accompanying decrease in the median grain size to values of 16–75 µm, while the sediments on the tidal flats are sandy with a clay content from 0.1 to 5.5% and a median grain size from 105–150 µm. In the Dollard clear gradients in the clay content are pres-

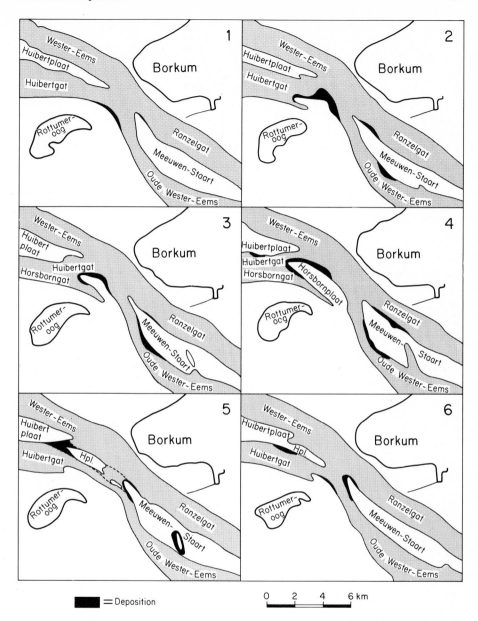

Fig. 2.3. Periodic changes in the morphology of the Ems estuary near the tidal inlet. The cycle of the changes is known for over a century and determined to be approx. 25 years (after Samu, 1979)

ent. In the central part it is less than 5% and increases towards the shore (Maschhaupt, 1948; Wiggers, 1960) to 35% near the salt marshes.

Thus, the main constituent by weight of every sediment is sand, even at the
very soft and clayey margins of the Dollard. Wiggers (1960) observed that over
centuries the composition of the sediments in the Dollard was conservative.

The soils of the whole Ems watershed (De Smet, 1960) and also the sediments
in the Ems estuary itself contain peat, both as layers, produced in situ and now
buried in the sediment as well as individual particles eroded from these layers.
Apart from this peat other organic matter occurs. The major part of this organic
material is present as aggregates of sand, clay and organic material (Meadows
and Anderson, 1968; Frankel and Mead, 1973; Eglington and Barnes, 1978). The
organic matter content of the sediments (Maschhaupt, 1948; Wiggers, 1960)
closely follows its clay content. On average 6.5–7.0% of the clay fraction consists
of organic carbon.

2.4 Climate

Temperature. The climate of northwestern Europe, in which the Ems estuary is
situated, is temperate with mild winters and cool summers (Fig. 2.4 a). The yearly
temperature curves differ slightly in each part of the estuary.

In summer, the air temperature in the upper region of the estuary is a few degrees higher than near the barrier islands while in winter the opposite is the case.

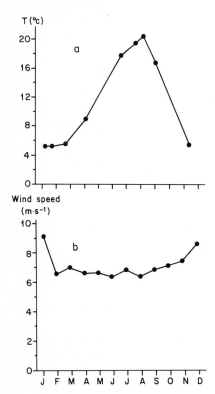

Fig. 2.4. Monthly means of water temperature near
Emshaven (a) and wind speed at Borkum for the period 1951 to 1960 (b). The values are irrespective of
wind direction (Deutsches Wetteramt, Bremen,
FRG)

This is due to the moderating influence of the sea, which rapidly diminishes inland, causing differences in the temperatures along the axis of the estuary.

Wind. There is a clear seasonal cycle in wind speed (Fig. 2.4 b), with lowest monthly averages in summer and highest values in winter. The wind is predominantly westerly, especially the high winds. The average wind speed is highest near the barrier islands. Because of the friction between the wind and the land surface the average wind speed decreases towards the upper estuary.

The impact of wind on the estuary is threefold. Depending on the direction and the wind speed it may raise or lower the water level and thus influence the mixing and flushing rates. Secondly, wave action modifies the sediments of the tidal flats and of the shallow subtidal. Thirdly the wind generates drift currents which also modify the mixing and flushing rates and thus the water exchange with the Wadden Sea.

2.5 Fresh Water Supply

Fresh water enters the Ems estuary by different sources of which the most important is the river Ems with a watershed of 12,650 km^2 (Hinrich, 1974). The discharge is very variable, ranging from a high of 390 m$^3 \cdot$ s^{-1} to a low of 25 m$^3 \cdot$ s^{-1}.

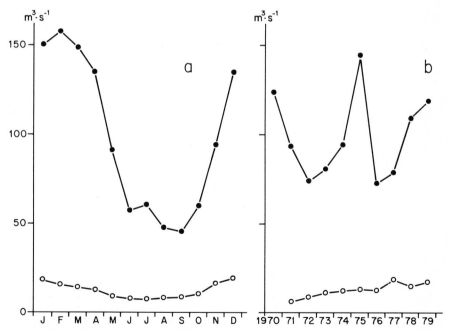

Fig. 2.5. Monthly mean (a) and annual mean (b) discharge of the river Ems (●) and the Westerwoldsche Aa (o) for the period 1970–1979

The second fresh water source is the Westerwoldsche Aa. Being part of the canal system of the northern provinces, the Westerwoldsche Aa has no well-defined watershed. The water discharge is roughly 10% of that of the river Ems (Fig. 2.5).

Another fresh water source is the precipitation in the estuary. The annual precipitation is 72.5 cm near the barrier islands and 74 cm at Emden (Dorrestein, 1960). The precipitation in the estuary is approximately balanced by evaporation (KNMI, 1972). Precipitation on the tidal flats during emergence can temporarily lower the salinity of the interstital water and may act as a stress factor on the benthic communities.

2.6 Tide

The tide in the Ems estuary is dominated by the semi-diurnal lunar tide and has a mean period of 12 h 25'. The tidal curve is asymmetric. At Delfzijl, the mean period of the flood is 5 h 45', which is 55 min shorter than the ebb period. Hence, the flood currents are stronger than the ebb currents (Fig. 2.6). In addition, there is a significant diurnal inequality in the tidal amplitude.

The tidal amplitude along the axis of the estuary increases upstream. In 1971 the mean annual amplitude was 2.20 m near Borkum, 2.79 m in Delfzijl and 3.03 m at Emden. In the southeastern part of the Dollard the mean tidal amplitude is somewhat higher than at Emden.

The annual average tidal amplitude is not constant but has increased by ca. 5% in the period from 1960–1980. This increase is possibly caused by heavy dredging over the last 2 decades (De Jonge, 1983).

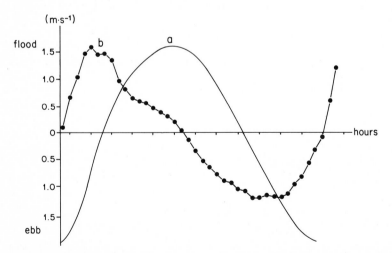

Fig. 2.6. Tidal amplitude (a) in meters at Delfzijl (Dutch tide tables, Staatsuitgeverij, Den Haag, The Netherlands). Tidal variation in current velocity (b) in $m \cdot s^{-1}$ in the middle section of the Ems estuary (9–5–1978)

2.7 Water Movement and Mixing

The rise and fall of the water level in the North Sea, caused by the passing tidal wave (vertical tide), generates the horizontal tidal currents (horizontal tide) in the Ems estuary. This horizontal water movement is the dominant feature of the tide and it can be expressed in terms of the tidal prism (the water volume passing through a cross-section twice each tidal period). The tidal prism at the tidal inlet is approximately $900 \cdot 10^6 \mathrm{m}^3$. At the entrance of the Dollard the tidal prism is reduced to approximately $115 \cdot 10^6 \mathrm{m}^3$. The tidal excursion (the distance a water parcel is displaced during half a tidal period) also varies over the estuary, although less than the tidal prism. Near the tidal inlet the tidal excursion is approximately 17 km and in the Dollard it amounts to 12 km.

The water movements lead to typical circulation patterns and to mixing. In the Ems estuary the following mixing mechanisms are present:
1. Tide-induced water circulation
2. River-induced water circulation
3. Wind-driven water circulation or drift currents.

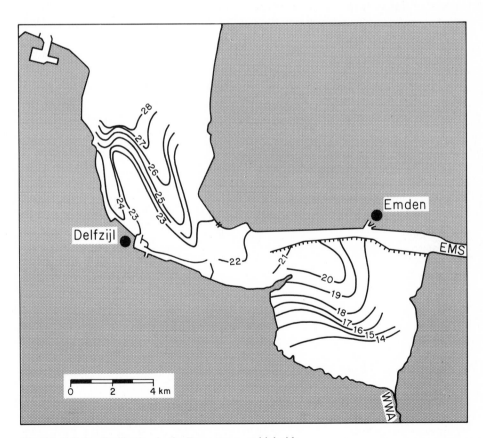

Fig. 2.7. Salinity distribution in the Ems estuary at high tide

During the flood period the water penetrates the estuary via the tidal inlet as a jet flow (Stommel and Farmer, 1952). However, after the turn of the tide the water runs from all directions straight towards the tidal inlet for as long as the tidal flats are submerged. Because the tidal excursion of the flood current in the channels is longer than that of the ebb current a residual water circulation arises. Residual circulation occurs between pairwise organized tidal channels as well as within one channel (cf. Fig. 2.2). They are caused by cross-sectional pressure gradients in the water and the Coriolis force (Zimmerman, 1976 a, b). Mean volumes of the residual circulations amount to 4–15% of the tidal volumes.

The river-induced water circulation is based on density differences between river water and seawater (Postma, 1967). When the river water meets the relatively stationary seawater its lower density causes it to flow further seaward on top of the denser seawater. The resulting pressure gradient can only be counteracted by an upstream displacement of the heavier seawater. This tends to create a well-stratified situation with horizontal isohalines. However, this stratification tendency is opposed by the process of mixing. When mixing is strong, as usually is the case in the Ems estuary, stratification is lacking and the salinity gradually increases from the river to the sea (Fig. 2.7). The shape and slope of the estuarine salinity gradient is continuously changing due to the river discharge. In summer when river discharge is low the salinity gradient is relatively steep and short. During the winter at high river discharge the slope decreases and the gradient extends over a larger part of the estuary.

The intensity of drift currents varies with wind speed and direction. The interference of drift currents with the tidal currents generates an increased dispersion

Table 2.1. Characteristics of water transport (in days) at high discharge (351 $m^3 \cdot s^{-1}$) and low discharge (34.1 $m^3 \cdot s^{-1}$) of fresh water for the Ems estuary as a whole and in the Dollard separately at high discharge (31.0 $m^3 \cdot s^{-1}$) and low discharge (5.1 $m^3 \cdot s^{-1}$) of Westerwoldsche Aa water. (Helder and Ruardij, 1982)

Time scale	time (days)	
	High discharge	Low discharge
Ems estuary		
Seawater:		
mean age	14.0	36.3
Ems water:		
mean age	19.3	64.7
flushing time	12.1	72.1
WWA water:		
mean age	12.9	32.4
Turnover time of basin:	17.7	35.7
Dollard		
Water entering at mouth:		
mean age	5.9	17.9
WWA water:		
mean age	8.5	27.1
flushing time	8.7	31.2
Turnover time of basin:	10.0	21.0

of water. In the Ems estuary drift currents play a role especially in the outer part and in the Dollard.

The various circulation processes are very complex, necessitating simplification to formulate model calculations. Dorrestein and Otto (1960) presented a simple one-dimensional transport model for the Ems estuary which was improved by Helder and Ruardij (1982). Some data on the mean age, flushing time and turnover time of the water mass in the estuary, derived from Helder and Ruardij (1982) are given in Table 2.1. Here, the mean age is the time that water parcels located in a certain region are present in the estuary since arrival from their source (viz. sea, Ems, Westerwoldsche Aa). The flushing time is the ratio between the total fresh water volume in the estuary and the discharge rate of fresh water. The turnover time is the time needed to decrease the volume of water or mass of dissolved constituent present at $t = 0$ in a region of the estuary to a fraction e^{-1}. Another characteristic often used is the residence time which is the average time a parcel of water needs to pass through a region.

2.8 Sediment Transport and Accumulation

The concentration gradient of suspended matter in the Ems estuary is typical for most coastal plain estuaries (Fig. 2.8), with low concentrations near the tidal inlet and high values in the inner part of the estuary.

The importance of the tide-induced sediment transport was first recognized by Postma (1954). Later, this was analyzed experimentally (Postma, 1961) as well as theoretically (Van Straaten and Kuenen, 1958; Groen, 1967; Postma, 1982).

Two factors contribute to this process. The first is the asymmetry of the tidal curve. The higher maximal current velocities during flood erode more bottom sediment than the ebb currents and therefore effectuate a net upstream transport of suspended matter. This is illustrated by the tidal fluctuations for the coarser fraction ($> 55 \mu m$) in Fig. 2.9. The finer suspended matter fraction ($< 55 \mu m$) does not show this phenomenon as clearly because of its lower settling velocity. Consequently the net transport of sand is larger than that of the finer fractions and larger close to the bottom (bed-load transport).

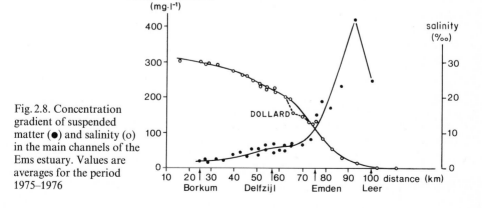

Fig. 2.8. Concentration gradient of suspended matter (●) and salinity (o) in the main channels of the Ems estuary. Values are averages for the period 1975–1976

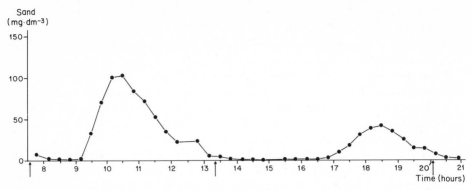

Fig. 2.9. Tidal variation in suspended matter (sand fraction: particles > 55 μm), in the middle section of the Ems estuary (9–5–1978). The *arrows* indicate the time of slack water

The second factor contributing to a net upstream transport of suspended matter is that the critical current velocity for erosion is higher than that for deposition of the same material (Postma, 1967). Thus particles deposited at a certain current velocity during flood are resuspended again during ebb at a later phase in the tidal cycle.

So far, only the net transport of suspended material has been considered. However, the net amount of transported material is extremely low in comparison to the amount that is deposited and resuspended. During each tidal period a considerable exchange of suspended material and sediment occurs between the tidal flats and the channels. The physical properties of particles also play a significant role in the transport process. Particles with a low density (e.g. colloids) behave approximately like water and hence can be exported to the sea while for heavier particles a net import can occur. Similarly the exchange rate of the relatively light mud aggregates is higher than that of the heavy sand grains.

Together with the sediments, also small organisms and organic matter are exchanged between the sediments and the water. This was observed by De Jonge (1985), who found a resemblance between the species composition of benthic diatoms on the tidal-flat sediments and those suspended in the water.

Deposition and resuspension of sediments is also affected by drift currents, wind waves and biological activity. Drift currents and wind waves increase the resuspension of sediments and lessen deposition. The higher the wind speed, the higher the wave energy, which results in an increased suspended matter concentration in the water and a higher degree of sorting (less fine fractions) of the sediments on the tidal flats. In high winds the highest values of suspended matter are found during high water near the margins of the tidal flats where resuspension is highest.

The effects of biological activity on resuspension and deposition of sediment and suspended matter are many and various. Organisms filtering water with suspended matter and producing faecal pellets or pseudofaeces can substantially contribute to the deposition of particles and aggregates (Verwey, 1952). Microorganisms such as benthic diatoms, bacteria, fungi, etc. can consolidate the sediment surface by forming a mat structure, which is often cemented to the heavier sand particles by adhesive excretion products (mucus, etc.). On the other hand, benthos and epibenthos (bottom fishes, shrimps, crabs, etc.) can increase the resuspension of sediment because they may actively suspend sediment or disturb surface mat structures. The latter generally leads to irregularities at the sediment surface increasing turbulent flow conditions (Jumars and Nowell, 1984) and consequently erosion.

The import of suspended matter by the two rivers is considerable and contributes to an accretion in the Dollard (and presumably also the rest of the estuary) of ca. 8 mm per year (Reenders and Van der Meulen, 1972). Nevertheless, the suspended matter in the water and the material deposited in the sediment of the estuary is of predominantly marine origin (Favejee, 1960; Salomons, 1975). This is in agreement with the theory (Postma 1982) that the import and sedimentation of marine particulate material predominates in the Ems estuary.

2.9 Nutrients

The rivers Ems, Westerwoldsche Aa and some small canals bring large amounts of nutrients into the estuary. These are partly in dissolved and partly in particulate form. Rainfall contributes only a small amount of nutrients to the estuary. Furthermore, large amounts of particulate nitrogen and phosphorus are transported from the Wadden Sea and the North Sea into the estuary. At present the particulate nitrogen from these sources cannot be quantified. The annual particulate phosphorus supply, calculated from data on particulate phosphorus in suspended matter, the accretion rate of the estuary and the mineral composition of

Table 2.2. Annual supply and concentration ranges of nutrients from the river Ems, the Westerwoldsche Aa and other sources

Source	Metric tonne y^{-1}	Data from	Nutrient component	Concentration range	
				Minimum ($\mu mol \cdot l^{-1}$)	Maximum ($\mu mol \cdot l^{-1}$)
River Ems					
Total nitrogen	18 500	a,b,c	Inorganic nitrogen	70	460
Total phosphorus	2 190	a,b	Orthophospate	3	20
Reactive silicate	13 700	a,b	Reactive silicate	40	240
Westerwoldsche Aa					
Total nitrogen	6 550	d,f	Inorganic nitrogen	430	5 000
Total phosphorus	890	d,f	Orthophosphate	3	100
Reactive silicate	1 625	a,d,e	Reactive silicate	80	315
Other sources					
Total nitrogen	2 780	d,f,g,h			
Total phosphorus	430	d,f,g,h			
Reactive silicate	>2 400	a,d,g			

Values calculated using: *a* BOEDE measurements; *b* river discharges (Wasser und Schiffahrtsamt Emden); *c* Preston (1978); *d* river discharges (Rijkswaterstaat); *e* Helder et al. (1983); *f* measurements Rijkswaterstaat; *g* Dankers et al. (1984); *h* precipitation (KNMI, 1972).

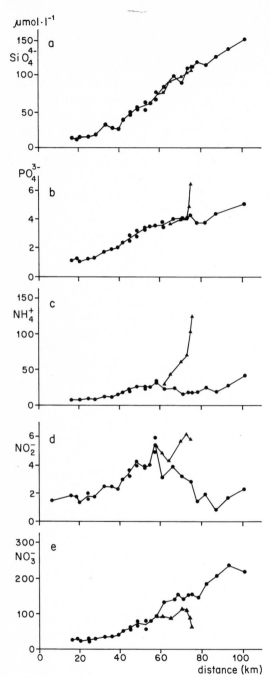

Fig. 2.10. Annual means of reactive silicate (a), phosphate (b), ammonia (c), nitrite (d) and nitrate (e) in the main channels of the Ems estuary. For meaning of distance cf. Fig. 2.8

the sediments, amounts to some 3500 tonnes P. This value agrees well with the estimate of De Jonge and Postma (1974) for the western part of the Dutch Wadden Sea. Calculated and estimated fresh water discharges of different forms of nitrogen, phosphorus and silicon are given in Table 2.2.

After arrival in the estuary the dissolved components are mixed and flushed in the same way as the water. This results in a concentration gradient (Fig. 2.10) with generally high values in the river water and low values near the tidal inlet. The particulate nutrients, however, are distributed according to the transport processes described for the suspended material.

Biologically, the most important areas are those where the dissolved nutrients reach low concentration levels so that nutrient limitation of primary production might occur, which is in the outer part of the estuary. On their way through the estuary dissolved nutrients are taken up by phytoplankton or are regenerated by mineralization. Particulate nutrients can pass into solution or dissolved nutrients can become particulate due to biological as well as physico-chemical processes. These processes together show a strong seasonal cycle. In the Dollard nutrient limitation is unlikely to occur as the concentrations are high throughout the year. However, in the outer regions seasonally low levels are reached. Dissolved silicate is often depleted in spring, indicating silicate limitation for diatoms. The role of nitrogen and phosphate is less evident because of rather large year to year variation. Occasionally nitrogen depletion occurs in early summer together with continuously high phosphate concentrations, but in more recent data sets phosphate depletion predominates. The distribution and uptake of phosphorus and silicate in the water column has been included in the model. Because it seemed unlikely that nitrogenous nutrients often become limiting to primary production, nitrogen dynamics are not included in the model.

As a considerable part of the total organic carbon production occurs by benthic diatoms, the nutrient levels in the pore water of the surface layers of the tidal flats are important. During low tide the concentration of reactive phosphate and reactive silicate in the thin water layer on the sediment and in the pore water of the surface layer of the tidal flats are usually higher than 1 $\mu mol \cdot l^{-1}$. Only in the outer region do the reactive silicate values drop below 1 $\mu mol \cdot l^{-1}$ during the phytobenthos bloom in spring. This indicates that growth limitation due to low nutrient values in this area can occur. In the other areas growth limitation due to low nutrient values probably never occurs.

The nitrogen cycle in the estuary was studied by Rutgers van der Loeff et al. (1981) and Helder (1983). Their data as well as the BOEDE data indicate that this nutrient does not regulate the benthic diatom growth through nutrient limitation. On the contrary, there are indications (Admiraal, 1977 c) that ammonia concentrations higher than 500 $\mu mol \cdot l^{-1}$, which occur in parts of the Dollard, inhibit diatom growth.

Since there were only scant indications that nutrient levels in the sediment might become limiting to phytobenthos production, the nutrient levels in the sediment have not been included in the model. The possible inhibition of phytobenthos production by high levels of ammonia, however, has been incorporated.

3 Biology

F. Colijn, M. A. van Arkel and A. Stam

3.1 General Description of the Estuary

As long ago as 1855 Stratingh and Venema gave a description of the Dollard which still holds at first glance. Their description runs as follows (translated from the original, loc. cit. p. 233): "Imagine: a vast, ochre, more or less sandy, silty plain, on the interface of land and water, which at ebb is present in the reeds and drains; smooth in the whole, but anyhow wrinkled by the untiring action of the waves; a sediment which is influenced by the seasons not else than through frost and its concurrent instability owing to thaw; for short, an inanimate plain, which is neither charming, nor graceful, nor wild, nor majestic, nor lofty, but of which the sight bores you- and you have a faithful picture of the tidal flats of the Dollard".

Of course at that time little was known about the biology of the estuary. The incomplete understanding of the authors, a physician and a member of the Provincial Government, is illustrated by their doubts about the character of the diatoms, whether they were plants or animals and about the origin of "blooming of the mud".

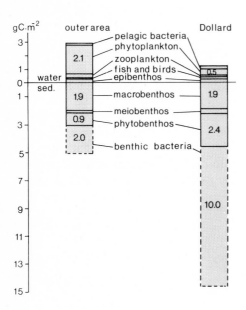

Fig. 3.1. Annual mean biomass of pelagic and benthic communities in the outer part of the estuary and the inner region, the Dollard

Nowadays most biologists working in the field look upon the estuary with a different eye. Closer inspection reveals an overwhelming richness of living organisms both in the water and in the sediment. The distribution of these organisms is controlled by a number of environmental factors which gradually change along the estuary (Chap. 2).

In the Dollard the system is dominated by the tidal flat communities which occupy more than half of the total area, whereas in the seaward part of the estuary the pelagic system dominates the biological activity. This difference is reflected in the relative biomass of pelagic and benthic organisms (Fig. 3.1) but even more in the primary production rate and the mineralization rate that differ greatly between the inner and outer parts of the estuary (Fig. 3.2).

The salt marshes fringing the Dollard are a special part of the ecosystem. They cover approximately 14 km^2 and are only occasionally flooded. On such occasions, mostly in winter, part of the vegetation may be washed off, importing detritus into the estuary. On the other hand, the salt marsh may function as a particle trap and thus cause an export of organic material from the estuary. There are indications that the marshes do not play a quantitatively important role in the ecosystem of the estuary (Dankers et al., 1984) and therefore they have not been included in the model.

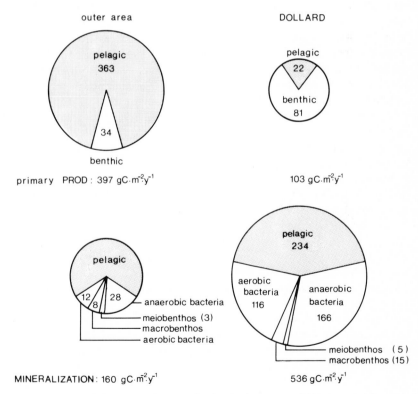

Fig. 3.2. Annual primary production and mineralization in the outer and inner part of the estuary

The biological entities for modelling purposes have been assigned to one of three subsystems: entities that live planktonically and also feed exclusively in the water column have been assigned to the pelagic subsystem.

Entities that live in or on the sediments and lead a sedentary life in the sense that they do not actively migrate, belong to the benthic subsystem.

Entities that feed on the benthos, often live in the sediments, but do migrate within the estuary or even over longer distances have been assigned to the epibenthic subsystem. This subsystem thus contains very mobile organisms such as birds, demersal fish, but also *Nereis*. The one thing these organisms have in common is that they live off the benthic system without being a permanent member of that community. If they manage to deplete a certain area of food, they can and do move to richer pastures, something the fully benthic (in our definition) organisms cannot do.

The next sections follow this criterion in describing the biological species included in the various subsystems.

3.2 The Pelagic System

Characteristic for estuarine waters is the large amount of detrital material composed of dead phytoplankton, plant debris, animal remains and dissolved organic compounds, mainly humic acids. The quality of this material as food for suspension feeders and planktonic organisms can vary from easily degradable and thus useful as food, to almost totally refractive. A partial mineralization is accomplished by bacterioplankton (Van Es and Laane, 1982). Both free-living bacteria and bacteria attached to particles occur. Their abundance increases towards the Dollard. Their numbers fluctuate, with maxima during phytoplankton blooms (Admiraal et al., 1985). Very high bacterial activity is also found near the outlet of the Westerwoldsche Aa during the waste water discharges in autumn. This temporarily causes low oxygen saturation values in the southern part of the Dollard or even complete anaerobiosis (Van Es et al., 1980).

The phytoplankton is composed of centric diatoms, flagellates, cyanobacteria and is often mixed with suspended pennate (bottom-dwelling) diatoms. The early spring bloom starts with diatoms (*Skeletonema costatum, Biddulphia aurita, Asterionella glacialis* and *A. kariana, Thalassionema nitzschioides, Thalassiosira* spp.) and is succeeded by a bloom of *Phaeocystis pouchetii*, a colony forming Haptophycean alga which can be very abundant (Colijn, 1983). Most species reach as far as the Dollard but there mixing with fresh water from the Westerwoldsche Aa and the river Ems results in populations of cyanobacteria (*Oscillatoria* spp.) and coccoid green algae (*Pediastrum* spp.). Some of the species in this area are indicative for polluted and eutroficated waters.

A further inspection of the pelagic system shows the presence of microzooplankton; occasionally large populations of tintinnid species occur. These animals feed on bacteria and algae, especially on single cells of *Phaeocystis pouchetii* and probably reduce such blooms directly (Admiraal and Venekamp, 1986). The microzooplankton was not studied properly but it undoubtedly provides a link between the bacterioplankton and the larger zooplankters.

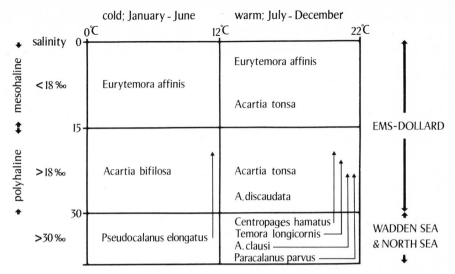

Fig. 3.3. Copepod distribution in the Ems estuary in relation to temperature and salinity

A step further in the food web we come across the mesozooplankton with about 30 taxa of holoplanktonic species. Among these the calanoid copepods are the most important with regard to numbers and biomass and as grazers of the phytoplankton and presumably of the microzooplankton. The copepod species are distributed in the estuary according to the salinity gradient, but also a seasonal succession mainly caused by temperature is observed (Baretta, 1980). A comprehensive view of the distributional patterns of the dominant species is shown in Fig. 3.3. The seasonal pattern of the mesozooplankton abundance shows a distinct spring peak in April. During the summer meroplankton (e.g. barnacle larvae, polychaeta larvae) sometimes contributes substantially to the biomass of the zooplankton. While salinity and temperature influence the distribution and succession of zooplankton, it is uncertain whether the distribution also depends on the suspended matter concentration, which may interfere with food selection of the copepods.

The group of the fishes consists of about 40 species, most of which are characteristic for shallow seas with extensive tidal areas. Earlier studies (Lohmeyer, 1907) indicate that at that time more species inhabited the estuary (the sturgeon and the Atlantic salmon have disappeared) and that the relative abundance of species has changed.

One of the predators on the fishes is the common seal (*Phoca vitulina*). The population in the estuary is an important part of the small population living in the Dutch Wadden Sea (Reijnders and Wolff, 1981). Their impact as a predator, however, is quantitatively negligible and not included in the model.

3.3 The Benthic System

The BOEDE group studied several aspects of the benthos: the micro- phytobenthos, the meiobenthos, the macrozoobenthos and the benthic bacteria.

It is now well established that bacteria do play an important quantitative role
in aquatic ecosystems both in the water and in the sediment. However, methods
to study benthic bacteria were scarce and insufficient at the start of our research.
Only recently suitable methods to count living cells have become available, such
as epifluorescence microscopy. This method has been used to estimate bacterial
biomass in the pelagic. To estimate the growth rate of pelagic bacteria the thymidine incorporation method has been employed. To measure the benthic aerobic
mineralization oxygen bell-jar experiments were conducted, and to determine anaerobic mineralization the sulphate reduction rates were measured. Both types of
measurements were needed because of the importance of anaerobic mineralization in these reduced sediments. Limited diffusion of oxygen into the sediments
causes oxygen depletion even at low mineralization rates. Therefore the depth of
the oxidized top layer ranges from 1 mm in muddy sediments in the Dollard exposed to waste water, to 10 cm in sandy sediments in the outer wave-exposed part
of the estuary.

The anaerobic mineralization sometimes accounts for more than 50% of the
total mineralization in the sediment (Schröder, unpublished). In most parts of the
estuary the mineralization in the sediment exceeds the benthic primary production (Van Es, 1982 a). This indicates that organic substrates from the pelagic are
transferred to the benthos by sedimentation (Colijn, 1983; Van Es, 1982 b).

Everywhere and at any time in the estuary brownish patches on the sediments
can be seen: this brown layer consists of the sediment inhabiting diatoms of which
millions of mostly motile cells can be found per cm^2. A large part of this diatom
flora belongs to the pennates, e.g. *Navicula*, *Nitzschia*, *Amphora* and *Diploneis*
spp. Their role in the estuarine food web was recently reviewed by Admiraal
(1984). In the sheltered parts of the estuary, especially in the Dollard, mat-form-
ing vegetations occur. Near the waste water outlet an impoverished flora is found
with few species, sometimes associated with e.g. *Oscillatoria* sp. or *Vaucheria* spp.
in addition to the abundant diatoms (Admiraal and Peletier, 1980 a). The sandy
sediments in the outer region contain a species-rich flora (Colijn and Dijkema,
1981). Occasionally other groups of microalgae were observed (*Cyanobacteria*),
but the diatoms as a whole form the main part of the benthic primary producers.
The annual primary production is correlated with the elevation of the tidal flats
above mean sea level which determines the effective photoperiod for the algae
during emersion.

A group which is easily observed with a dissecting microscope is the meiofauna; within this group the nematodes are most important. Other groups of
minor importance are the harpacticoid copepods, the oligochaeta, the ostracoda
and the turbellaria. Their biology is dealt with in detail by Bouwman (1983 b) and
references therein. The community structure of the meiofauna is regulated by the
estuarine gradients in environmental conditions (salinity, sediment structure, pollution) but also by biotic conditions such as food availability and food selection.
Two communities were distinguished, one in the outer region of the estuary and

one in the Dollard, both extending towards the middle reaches of the estuary. The characteristic species of the superficial sediment layers in the outer part of the estuary were *Viscosia rustica*, *Neochromadora trichophora*, *Monoposthia mirabilis* and *Odontophora rectangula*. In the middle region *Paracanthonchus caecus*, *Neochromadora poecilosoma*, *Atrochromadora microlaima* and *Sabatiera pulchra* were dominant, whereas in the southeastern polluted part of the Dollard *Eudiplogaster pararmatus* and *Dichromadora geophila* formed more than 90% of the nematode numbers. Besides the distribution patterns along the axis of the estuary also large differences in the vertical distribution were observed. Bacteria and diatoms are both important food sources though the stable part of the nematode fauna mainly feeds on bacteria while the part that fluctuates seasonally feeds on diatoms.

The most conspicuous faunal group on the tidal flats is the macrobenthos. Most species are hidden in the first 10 cm of sediment thus minimizing predation by birds (Hulscher, 1982) or by fishes (De Vlas, 1985). Yet they give themselves away by faeces production (some worms such as *Arenicola* and *Heteromastus*, and shellfish, such as *Macoma*) or by pits and small burrows in the sediment, e.g. *Cerastoderma*. The faecal deposits of *Arenicola* are certainly a very characteristic feature of the tidal flats. The macrobenthos consists of suspension feeders, deposit feeders and some omnivores/carnivores. The species composition resembles that of the Wadden Sea but at lower salinities the number of species decreases. In the Dollard the macrobenthos is dominated by 8 species only, whereas over 40 species have been found in the whole estuary. Important species are the Baltic tellin (*Macoma baltica*), lugworm (*Arenicola marina*), sand gaper (*Mya arenaria*) and the common cockle (*Cerastoderma edule*) which only reproduce once a year, which precludes a fast numerical response to an increase in the food supply. So the seasonal variation in their standing stocks is less pronounced than in the stocks of smaller benthic organisms mentioned before. Recruitment into the different macrobenthos species is highly variable, giving rise to considerable year to year differences in standing stocks.

The importance of the macrozoobenthos as a food source for fishes and birds is well established. Another important function of the macrobenthos is the permanent reworking of the sediment (bioturbation). The presence of burrows influences the permeability of the sediment and the consumption of sediment by deposit feeders (*Arenicola*) causes a vertical transport between the anaerobic and aerobic layer.

3.4 The Epibenthic System

The highly diverse group of the smaller epibenthic species includes: crabs, amphipods (*Corophium*, *Gammarus*), mysids, harpacticoid copepods and *Nereis diversicolor*, the common ragworm.

Amphipods form an important element in the epifauna of estuaries. The best known are the genera *Corophium* and *Gammarus*. *Corophium* sp. is a tube-building species and occurs in all intertidal areas in the Ems. The highest densities are found in very fine-grained sediments with high silt contents as occur in the Dollard and the middle part of the Ems estuary. The animal hardly occurs in the sandy sediments of the outer part.

The most abundant Mysidacea in the Ems estuary are *Neomysis integer*, *Mesopodopsis slabberi* and *Praunus flexuosus*. The first and second one occur near the bottom for at least a large part of the day, the third one occurs higher in the water column. All three species are euryhaline. *Neomysis integer* of those three penetrates farthest into the low salinity reaches of estuaries (Tattersall and Tattersall, 1951; Kinne, 1955).

A survey of the harpacticoids of the Ems estuary was made by Vaeremans (1977). She found 26 species, three of them new for the Netherlands. The greater part of the copepod fauna was found in the upper 2 cm of the sediment. Highest densities were found in the outer part of the estuary.

Nereis diversicolor is a bottom-living animal, burrowing in the sediment and constructing a network of tubes, ending at several places on the surface of the sediment. The animal lives at a depth of 20–30 cm, in severe winters down to 60 cm. *Nereis* is omnivorous (Hartmann-Schröder, 1971), with cannibalistic tendencies (Bogucki, 1953). *Nereis* lives in the high level mud flats in the marine reaches of the estuary, but is especially common in the mud flats along the brackish parts of the estuary (the Dollard). However, their numbers diminish again towards the waste water discharge point, where only occasional specimens are found in the lower part of the intertidal zone.

The most important member of the larger epibenthic species is the brown shrimp, *Crangon crangon*. On the tidal flats as well as in the channels the biomass of the shrimp mostly exceeds the biomass of the other members of the larger epibenthos, the flatfishes. In winter, brown shrimps migrate to the deeper channels where the water temperature is higher and in severe winters to sea, returning in spring when the water temperature increases. The brown shrimp is omnivorous. Almost everything that can be managed is eaten: detritus, worms, amphipods, schizopods, copepods, cyprid larvae of *Balanus*, snails, young mussels (Plagmann, 1939) and small flatfish (Bergman et al., 1976). Cannibalism has been observed among the older brown shrimps (Plagmann, 1939).

The other species of large epibenthos are the flatfishes, of which the most important representative is the plaice (*Pleuronectes platessa*). In the first years of this century it was already known that *Pleuronectes* spends the first 3 years of its life in shallow waters and migrates to deeper water when growing older. Heincke (1913) summarized this knowledge in what is called Heincke's law: size and age of plaice in the North Sea are inversely proportional with abundance and increase with the distance to the coast and the depth.

A small commercial fishery based on these fishes and on shrimps existed in the area, but nowadays only some shrimp fishing remains. Most fishes only spend part of their life in the estuary, using it as a nursery area, a feeding area or as a conduit to their spawning grounds. Therefore the composition of the fish fauna varies throughout the year and the standing stock fluctuates considerably.

The birds are also included in the epibenthic submodel. They are important as predators of the macrobenthos. About 50 different species occur in the estuary, some use it as breeding area, e.g. mallard (*Anas platyrhynchos*), others for feeding or wintering (teal, *Anas crecca*; oystercatcher, *Haematopus ostralegus*). Eighteen species are so numerous that they exceed the standard of the International Water-fowl Research Bureau (BOEDE, 1983) which means that the estuary is essential for the species.

4 General Features in the Model

W. Admiraal, J. W. Baretta, F. Colijn, V. N. de Jonge, R. W. P. M. Laane, P. Ruardij and H. G. J. Schröder†

4.1 Universal Biological Processes

The biological state variables in this concept are viewed as being nonstructured (not differentiated into size classes) populations. These state variables represent functional groups with the feeding mode being one of the main criteria for inclusion in a certain group.

In Fig. 4.1 the biological processes being modelled in all the state variables are given with the internal and external carbon flows. The information we use to model the state variables (P) is usually derived from studies at the population level in the case of groups of small organisms such as primary producers, decomposers and primary consumers. In such cases the experimental work and the modelling are at the same level of abstraction.

In modelling state variables, representing larger organisms such as macrobenthos, the experimental data usually refer to individual organisms, not to the population as a whole. Since we want to model whole populations, we have to scale

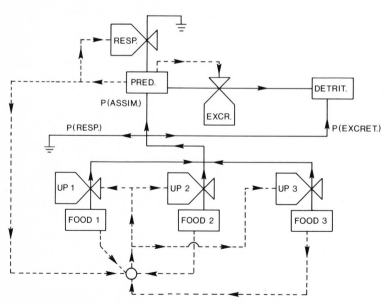

Fig. 4.1. Diagram of the biological processes in and around the state variable *P*, given in Forester notation. *Continuous* lines: carbon flows; *dashed* lines: feedback loops

the data from the organismal level to the population level. This will only be valid if a population is nothing but a number of identical individuals all performing identically. This may not be very realistic, but as long as we cannot model all individuals of the population we will have to make do with this approach.

In general, the larger and older the individuals in a functional group may become, the more diversity may be expected between the biological performance of individuals, and the more difficult it will be to model the functional group as a whole accurately.

The unit in which P is expressed is either $mgC \cdot m^{-3}$ or $mgC \cdot m^{-2}$ depending on whether the population to be modelled is pelagic or (epi)benthic.

The following sections deal with how the different processes in the state variables are formulated and how environmental conditions may influence them.

4.1.1 Maintenance Processes

a) Rest respiration: $rrsP = C_{rrs} \cdot P \cdot z(temp)$;
b) Rest excretion: $rexP = C_{rex} \cdot P \cdot z(temp)$.

Respiration and excretion processes (basal metabolism) and mortality are processes which are directly proportional to the biomass performing them. The rates of these processes differ strongly in the various groups. Mortality rates show the largest spread since mortality is coupled to the longevity of the group considered.

Basal metabolism and mortality are internal processes which are only coupled to the external world through their dependence on ambient temperature [$z(temp)$]. The rate constants (C_{rrs}, C_{rex}) of these processes have been defined at a specific temperature and are corrected for the prevailing temperature.

Mortality here is treated as a form of excretion: the state variable loses biomass through mortality, which enters the detrital pool.

4.1.2 Activity-Dependent Processes

a) Uptake: $UP = C_{uptake} \cdot z_{env} \cdot P \cdot z(temp)$;
b) Activity respiration: $arsP = C_{ars} \cdot UP$;
c) Activity excretion: $aexP = C_{aex} \cdot UP$.
The uptake has a temperature dependency [$z(temp)$] and is directly proportional to the biomass of P and to a function z_{env} which represents the interaction of P with its environment. This environmental function is fundamentally different for producers and consumers. For consumers the environmental function is a function in which food supply is a component:

$$z_{env} \approx z(FOOD).$$

The total amount of food in its turn consists of a number of different food sources. The number of food sources may be different for each P:

$$FOOD = \sum_{i=1}^{n} [\text{food source (i)}],$$

n = number of food sources.

For producers the environmental function is controlled by the chemical and physical environment. For example, diatoms in our model are controlled by light, the availability of phosphate, silicate and CO_2:

z_{env} = z (chemical, physical environment).

The uptake rate C_{uptake} is different for each P and inversely related to the average size of the organisms in P. It is largest for small organisms such as bacteria (5 d^{-1}) and smallest for large organisms such as benthic filterfeeders (0.05 d^{-1}).

The other two processes, activity respiration and activity excretion are similar in producers and consumers, and depend on the uptake UP. The rate constants C_{ars} and C_{aex} are different for each P and are closely related to the way the uptake is modelled. Details are given in Chapters 6–8.

4.1.3 Bookkeeping Calculations

These calculations are not really connected with the modelled populations, but are necessary to keep the carbon budget balanced, and to compute the changes in state variables due to predation, mortality, etc.

4.1.3.1 Impact of Grazing

$$fl(i)P = \frac{UP}{FOOD} \cdot \text{food source}(i) \quad \text{for} \quad i = 1, n$$

fl(i)P = carbon flux to P from food source (i).

Normally, not all the available food (FOOD) is taken in 1 day. The uptake UP thus is less than FOOD.

4.1.3.2 Balance Equations

Because carbon enters P in the form of uptake, on the one hand, but leaves P in one form or another because of respiration, excretion, etc. the biomass of P has to be recalculated for every time step (day) including all loss and gain terms:

$P_{t+\Delta t}$ = P_t + $\Delta t \cdot$ (UP-arsP-aexP-rrsP-rexP);
prey(i)$_{t+\Delta t}$ = prey(i)$_t$ – $\Delta t \cdot$ fl(i)
detritus$_{t+\Delta t}$ = detritus$_t$ + (rexP + aexP) $\cdot \Delta t$.

Δt in this model is 1 day and all rate constants therefore have the dimension d^{-1}.

In the first equation the net change in P is given. The net growth is the difference between the uptake and the different loss terms. Net growth thus also can be negative (in winter for example).

Grazing/predation losses are calculated in the next equation, and subtracted from the prey biomass. The amount grazed has already appeared on the plus side in the equation for food uptake.

Excretion products are added to the detritus pool in the last equation.

For primary producers, nutrients are treated as a food source. The dependence of primary producers on nutrients thus is treated in a similar way as the dependence of consumers on the available food.

4.1.4 Modelling Food Uptake

The process of food uptake is modelled differently for the various groups of consumers. Two different concepts may be distinguished which are related to the average size of the individual organism in P.

The first concept (Fig. 4.2 a) is to treat the organism as a black box: food goes in at some unspecified point, is assimilated with a specified efficiency and the remainder is respired and excreted in some form. This is the way the smaller groups (microorganisms, meiobenthos) and those groups where the main interest was not in their functioning but in their effect on other components of the system (epibenthos, planktonic carnivores) have been modelled.

The other concept (Fig. 4.2 b) has been to acknowledge the existence of a digestive tract in the organism and to specify in more detail what happens to the uptake on its way along the digestive tract. Where this concept was adopted, processes such as ingestion (uptake minus losses by sloppy feeding), assimilation efficiency and egestion have been taken into account. State variables where this approach has been taken, include the mesozooplankton and the macrobenthos.

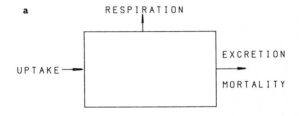

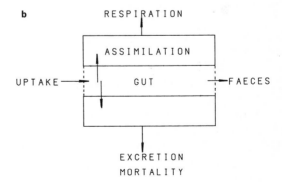

Fig. 4.2. a The state variable as a black box. b The state variable with a digestive tract

4.1.4.1 Availability of Food

For the determination of the potential uptake, the potential food sources for P are defined. A food source is defined by the product of the biomass of a state variable and an availability/accessibility parameter:

food source $(i) = a_i \cdot \text{prey}(i)$;

and $\quad \text{FOOD} = \sum_{i=1}^{n} \text{food source}(i)$;

so $\quad \text{FOOD} = \sum_{i=1}^{n} a_i \cdot \text{prey}(i)$.

Prey(i) is a state variable which may be eaten by P. The availability/accessibility parameters (a_i) reflect a number of quite different things:
1. The degree of digestibility of a food source. Diatoms are considered more edible than nondiatomaceous phytoplankton.
2. The relative size of a food source in relation to the size of the predator. Mesozooplankton is not able to consume the colonial forms of flagellates, but is able to consume individual flagellates.
3. The accessibility of a food source. Macrobenthic organisms, such as the lugworm (*Arenicola marina*), are only accessible to predation by flatfishes for a small part: when they expose their tail end to defaecate. Benthic filterfeeders expose only their siphons to surface-feeding fishes and are otherwise fairly safely located deep in the sediment.

The setting of the various a_i parameters is subject to a large uncertainty.

4.1.5 Food Selection

Each biological state variable P in the model represents a functional group which contains many species in reality. All these species may have different food preferences and food selection mechanisms.

The model cannot accommodate such detail, even if it was known who eats what, when and how. However, we can incorporate some of the consequences of the different species within a functional group having a different diet.

If a species cannot obtain its full ration because its preferred food is scarce, it will not realize its full potential growth rate. In the worst case it will die and disappear from the functional group. Other species with a different menu in the same functional group may be quite successful because their food is abundant. The species composition of the functional group thus will adapt to the composition of the available food. This will lead to increasing grazing pressure on the abundant food sources, since the components (species) within the functional group of grazers/predators feeding on the abundant food sources will grow at or near their potential growth rate and tend to dominate the group. In the model this effect is incorporated in the following way:

$$\text{FOOD} = \left\{ \sum_{i=1}^{n} [a_i \cdot \text{prey}(i)]^{\exp} \right\}^{1/\exp}.$$

a

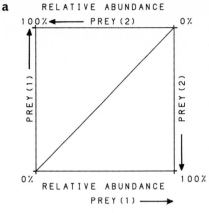

b

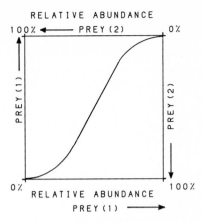

Fig. 4.3. The fractional uptake of prey relative to its abundance. Fractional uptake is proportional to prey abundance (exp=1) (a). Fractional uptake increases exponentially with relative prey abundance (exp=2) (b)

If exp=1 the grazing on prey(i) is proportional to its abundance. If exp>1 relatively more is eaten from abundant prey(i). This can be demonstrated by calculating the fraction frprey(i) which each prey(i) contributes to the total amount:

$$\text{frprey(i)} = \frac{a_i \cdot \text{prey(i)}^{\text{exp}}}{\text{FOOD}^{\text{exp}}}.$$

For abundant prey(i), an exp>1 leads to prey(i) being more heavily grazed than the less abundant food sources. As the relative abundance of any prey(i) increases, its contribution to the total food supply increases even more steeply. This is illustrated in Fig. 4.3, where the food supply for a state variable P, consisting of two sources, prey(1) and prey(2) is plotted as a function of their relative abundance, for exp = 1 (Fig. 4.3 a) and for exp = 2 (Fig. 4.3 b). A disadvantage of the use of exponents larger than 1 is that total FOOD is less than if calculated with exp = 1, because

$$\Sigma \, a_i \cdot \text{prey(i)} > (\Sigma \, a_i \cdot \text{prey(i)}^{\text{exp}})^{1/\text{exp}} \quad \text{for } \text{exp} > 1.$$

This side effect is unavoidable. It results in a slight reduction in the uptake of grazers, because they "see" less food than there really is.

4.1.6 Food Levels

At food levels distinctly below saturation the uptake by consumers is proportional to the concentration of available food. At food levels at or over the saturation level, the uptake reaches a plateau. This plateau might be viewed as the upper threshold to uptake, determined by the maximum uptake capacity (foodmax) of P. The biotic environment function z_{env} thus is some saturation function:

$$z_{\text{env}} \approx F \, (\text{foodmax, FOOD}).$$

These saturation functions are generally Michaelis-Menten functions:

$$F(\text{foodmax}, \text{FOOD}) = \frac{\text{FOOD} \cdot \text{foodmax}}{\text{FOOD} + \text{foodmax}}$$

or simply minimum functions:

$$F(\text{foodmax}, \text{FOOD}) = \text{Minimum}(\text{foodmax}, \text{FOOD}).$$

At low food levels there may be a lower threshold; if food levels fall below this threshold, uptake may cease totally. This lower threshold handling may be especially important in groups that depend on random encounters with food particles (meiobenthos, filter-feeding and deposit-feeding macrobenthos), because low food density means a large expenditure of energy for a small return in energy as food. In the model it is assumed that the functional groups cut their losses by shutting down feeding below certain food levels.

4.2 Compartments and Boundary Conditions

4.2.1 Boundaries

The boundaries of the Ems estuary to a large extent are defined by the coastline which is completely diked. The watershed of the Ems estuary is indicated by the elevated tidal flats between the island Rottumeroog and the Dutch coast on the western side and between the island Borkum and the German coast on the eastern side (see Fig. 1.1). As such they serve as boundaries to the model also. The boundary between the Ems estuary and the North Sea is taken to be the straight line between the islands Rottumeroog and Borkum.

From a geographical and a biological point of view the estuary can be separated into three parts: the Dollard, a middle part and an outer Wadden Sea part. To model the transport processes reasonably smoothly, the middle region was divided into two compartments, as was the Dollard. This latter division had the additional advantage that we could incorporate the anaerobic conditions in the inner compartment caused by the waste water discharges. In the model the river Ems enters the estuary not near Pogum as is the real situation but at the termination of the Geisedamm (Fig. 2.2). This is done because the water exchange over the Geisedamm is small. The result is that the complicating turbidity maximum is not included in the model.

The five compartments (Fig. 1.2) distinguished can be roughly characterized as follows.

1. Compartment 1, the innermost compartment of the estuary in the Dollard is characterized by very clayey sediments, extensive tidal flats and the discharge of large amounts of organic waste (approximately 30,000 t $C \cdot y^{-1}$, Van Es, 1982b). The mineralization of this material in the Dollard results in anaerobic conditions in the water and in the sediments during the discharge period, except for a few mm thick sediment layer on the intertidal flats. Due to this and the discharge of fresh water the diversity of the flora and the fauna is low compared to the other compartments.

Table 4.1. Parameters of the five compartments used in the calculation of all derived morphological parameters

Basic parameters	Compartment				
	1	2	3	4	5
Volume at high water vth (10^6 m^3)	15.0	136.00	235.00	525.0	1000.00
Volume at low water vtl (10^6 m^3)	2.2	36.00	120.00	275.0	635.00
Total surface areat (10^6 m^2)	12.5	78.00	48.50	109.0	215.00
Tidal flat surface areaf (10^6 m^2)	11.0	66.00	22.00	35.0	100.00
Tidal amplitude depbLH (m)	3.1	2.96	2.82	2.6	2.32

2. The main characteristic of compartment 2 in the Dollard is the extensive tidal flat area (cf. Table 4.1). The sediments vary from sandy to clayey.
3. The river Ems enters the estuary in compartment 3 and here the river water is mixed with the brackish Dollard water and the mid-estuary water. Consequently salinity is usually low here. The sediments are sandy except for the tidal flats along the dikes where the sediments are silty.
4. Compartment 4 is a transitional area between compartments 3 and 5. Sediments may be both of a sandy or a silty nature depending on their location relative to the coast. Due to the increasing influence of the coastal water salinities are higher than in the preceding compartments.
5. The most seaward compartment 5 is strongly influenced by the coastal water. The salinities are relatively high. The sediments are sandy because the area is exposed to wind waves, favouring the resuspension of fine material. The flora and fauna strongly resemble that of the western Wadden Sea.

4.2.2 The Dimensions of the Compartments

To use the tidally averaged situation in the model, while incorporating the effects of the tides correctly, the percentage of the tidal period that the tidal flats are submerged is calculated from the morphological parameters. Using the period of submergence, all the other parameters can be calculated by integration over a tidal period. For the purpose of the calculation the following simplifications have been made:
1. The tidal flats within a compartment are uniform: there are no height differences within a tidal flat or between different tidal flats within a compartment.
2. All channels and gullies within a compartment are treated as one large channel which has a constant depth over the length of the compartment (Fig. 4.4).
3. The water level varies with time according to a sinus function with a period of 12 h 25' (Fig. 4.5).
4. Tidal flats are those areas whose surfaces emerge at average low water level. This implies that tidal flats which emerge only at low water spring are taken to belong to the channel system.

   A consequence of averaging over a tidal cycle is that the way a morphological parameter will be used determines the manner in which it is calculated. If a mor-

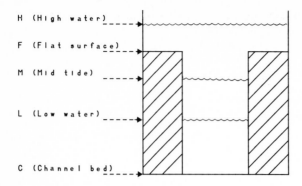

Fig. 4.4. Schematic cross-section of a model compartment

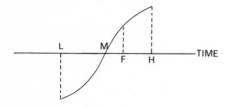

Fig. 4.5. Schematic representation of the water level during the tidal period from low to high water. The *letters* correspond to the tidal stages as indicated in Fig. 4.4

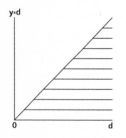
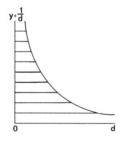

Fig. 4.6. The surfaces of the functions $Y = d$(epth) and $Y = 1/d$(epth)

phological parameter such as depth is used as a multiplier in the model the arithmetic mean is used. But when depth is to be used as a divisor, it is computed by the integration of 1/depth over the tidal cycle or over the relevant part of the tidal cycle. In other words, then the harmonic mean is used. If the integration is over a range of small depth values, taking the harmonic mean will result in a smaller value than when the arithmetic mean is used. This is illustrated by Fig. 4.6. The concave form of the 1/depth function leads to a lower average than the depth function. For the 1/depth function the smaller values are of relatively more importance.

In the next section we will develop the equations which are used to compute the morphological parameters.

All the equations used in the computation of the parameters will be given here. All the parameters except for some time parameters are compartment-dependent. The values of these parameters are given in Table 4.2.

Table 4.2. Derived morphological and hydrographical parameters of the model compartments. The parameters marked with * are used directly in the model

Parameter	Compartment				
	1	2	3	4	5
areac (10^6 m^2)	1.50	12.00	26.50	74.00	115.00
depbCL (m)	1.47	3.00	4.53	3.72	5.52
depbCH (m)	4.56	5.96	7.35	6.32	7.84
depbFH (m)	0.74	0.98	1.83	1.65	0.98
depbMF (m)	0.81	0.50	− 0.35	− 0.35	0.18
depbCF (m)	3.83	4.98	5.52	4.67	6.86
depbCM (m)	3.02	4.48	5.94	5.02	6.68
*depfa (m)	0.48	0.63	1.14	1.03	0.63
depcatLF (m)	2.39	3.75	4.88	4.05	6.02
depca (m)	see depbCM				
*depch (m)	2.59	4.23	5.77	4.85	6.58
*depfh (m)	0.25	0.31	0.45	0.42	0.30
*depchtLF (m)	2.17	3.66	4.86	4.03	5.99
*depchtFH (m)	4.30	5.60	6.61	5.65	7.48
*depth (m)	0.68	0.80	1.37	1.61	1.21
vCH (10^6 m^3)	6.85	71.50	195.00	467.00	902.00
vfh (10^6 m^3)	8.15	64.50	40.30	57.60	98.20
vbCF (10^6 m^3)	5.74	59.80	146.00	346.00	798.00
vtm (10^6 m^3)	4.53	53.80	167.00	383.00	768.00
*VOL (10^6 m^3)	6.26	70.10	172.30	392.00	797.00
tF (h)	1.09	0.69	− 0.60	− 0.54	0.31
*ptWET (−)	0.32	0.39	0.60	0.59	0.45
*pWAFL (−)	0.85	0.60	0.15	0.09	0.08

The data in Tables 4.1 and 4.2 are schematized in Fig. 4.7.

4.2.2.1 Basic Parameters

The total water volume in each compartment at high tide and at low tide, the total surface area, the surface areas of the tidal flats and the tidal amplitudes have been obtained directly from Rijkswaterstaat or calculated from maps or data from Rijkswaterstaat (Table 4.1). Starting from these variables all other variables are calculated.

4.2.2.2 Derived Parameters

The derived parameters (Box 4.1) are direct extensions from the definitions of mid-tide, the tidal amplitude, the tidal period and the surface areas.

By definition at low-tide all water in a compartment is in the channels. Combining this with vtl (see Table 4.1) the depth in the channel at low water (depbCL) can be calculated. Now that we have fixed the base level for the whole area in this way the other parameters follow quite naturally. By adding the tidal amplitude to the depth of low water, the depth at high water (depbCH) can be computed. Multiplying this variable with the total channel area we obtain the volume at high

---

Derived parameters

Half-tidal amplitude   : depbLM = depbMH = 0.5 · depbLH
Time from low to high water   : tLH = 0.5 · tidal period
Time at low water   : tL = 0.5 · tLH
Time at high water   : tH = 0.5 · tLH

Channel surface   : areac = areat-areaf
Channel depth at lw   : depbCL = vtl/areac
Channel depth at hw   : depbCH = depbCL + depbLH
Channel volume at hw   : vch = depbCH · areac
Volume over flat at hw   : vfh = vth-vch
Depth above flat at hw   : depbFH = vfh/areaf
Depth from mid-tide to flat level   : depbMF = depbMH-depbFH

Channel depth at mid-tide   : depbCM = depbCL + depbLM
Channel depth at flat level   : depbCF = depbMF + depbCM
Channel volume at flat level   : vbCF = depbCF · areac

Box 4.1.

water in the channels (vch). Now the volume over the flats (vfh) and the water depth on the flats (depbFH) at high water can be computed by subtracting vch from the total volume vth (input parameter) and dividing vfh by the flat area.

By subtracting a half-tidal amplitude depbFH from depbMH the height of the tidal flat relative to the mid-tide level (depbMF) can be computed.
Note that depbMF is negative in compartments 3 and 4 (Table 4.2). This indicates that the flats of these compartments are already submerged at mid-tide. The last three parameters in Box 4.1 are auxiliary parameters necessary for the computation of the arithmetic and harmonic means of depths and volumes.

4.2.2.3 Time Parameters

From depbMF the moment in the tidal cycle is computed when the flats are just flooding. Then the fraction of the time of one-half tidal period that the flats are submerged can be computed (Box 4.2). Obviously this parameter is also valid if expressed as a fraction of a day. In the computation we take mid tide as time = 0.

Time parameters

Moment of flooding of the flat with mid-tide as tM = 0:

$$tF = \text{Arcsin}\left[\frac{depbMF}{depbMH}\right] \cdot tLH/\pi$$

Fraction of a day that the flats are covered with water:

$$ptWET = \frac{tH - tF}{tLH}$$

Box 4.2.

Arithmetic mean depths

Average depth on flat

$$depfa = \left[\frac{depbMH}{tH - tF} \int\limits_{t=tM}^{t=tH} \sin(t\pi/tLH) \right] dt - depbMF$$

Average depth in channel during:

$$depcatLF = depbCM + \frac{depbLM}{(tL + tF)} \int\limits_{t=tL}^{t=tF} \sin(t\pi/tLH) dt$$

the time that the flats are dry

Box 4.3.

Harmonic mean depths

Harmonic mean depth in the channel from low water until the tide reaches the flat level:

$$\frac{1}{depchtCF} = \int\limits_{t=tL}^{t=tF} \frac{1}{depbCM + depbMH \cdot \sin(t\pi/tLH)} dt$$

Harmonic mean depth in the channel from the start of submergence of the flats until high water:

$$\frac{1}{depchtFH} = \int\limits_{t=tF}^{t=tH} \frac{1}{depbCM + depbMH \cdot \sin(t\pi/tLH)} dt$$

Harmonic mean depth in the channel:

$$\frac{1}{depch} = \frac{ptWET}{depchtFH} + \frac{(1 - ptWET)}{depchtLF}$$

Harmonic mean depth on the flat:

$$\frac{1}{depfh} = \int\limits_{t=tF}^{t=tH} \frac{1}{-depbMF + depbMH \cdot \sin(t\pi/tLH)} dt$$

Harmonic mean depth of a compartment:

$$\frac{1}{depth} = (1 - ptWET) \frac{1}{depchtCF} + \frac{ptWET}{areat} \left[\frac{areaf}{depfh} + \frac{areac}{depchtFH} \right]$$

Box 4.4.

4.2.2.4 Arithmetic Mean Depths

With the parameters derived before, the arithmetic mean water depth (Box 4.3) over the tidal flat is computed for the period that the tidal flat is flooded. Therefore we integrate the depth which increases according to a sinus from the moment tF to high water. We do not have to integrate for the time that the water is receding because the ebb period is assumed to follow the same sinus and hence will give the same answer.

In the same way the arithmetic mean depths in the channel are calculated. These depths are used in the pelagic submodel for the computation of the photic zone.

4.2.2.5 Harmonic Mean Depths

The harmonic mean depths are computed (Box 4.4) analogous to the arithmetic mean depths. The only difference is that in the next equations we divide by the actual depth and thus integrate over 1/sinus.

The harmonic mean channel depth is computed for two time intervals. Both depths are used in the pelagic submodel for the computation of the illuminated fraction of the water column. The harmonic mean depth for a whole compartment is used in the equation for computing the reaeration.

4.2.2.6 Volumes

Volumes

Volume at mid-tide:

vtm = depbCM · areac + MAX(0, (depbCM − depbCF) · areaf)

Arithmetic mean volume at mean sea level:

VOL = (1 − ptWET) · depcatLF · areac + ptWET ·
[(depfa + depbCF) · areac + depfa · areaf]

Box 4.5.

In the model VOL (Box 4.5) is used as the exchange volume between compartments.

4.2.2.7 Percentage Water on Flats

To determine what fraction of the average volume is on the tidal flat the integral of the ratio of the volume on the tidal flats to the total volume is computed over the time that the tidal flats are flooded. Both volumes are dependent on the tide:

Fractional volume over the flat during tFH:

$$pWAFL = \frac{areaf \cdot depfa}{VOL}$$

Box 4.6.

The parameter pWAFL (Box 4.6 and Table 4.2) for the two Dollard compartments is much larger than for the other compartments. This implies, since the sedimentation is directly related to the value of pWAFL, that sedimentation will be highest in the Dollard. The data in Tables 4.1 and 4.2 are schematized in Fig. 4.7.

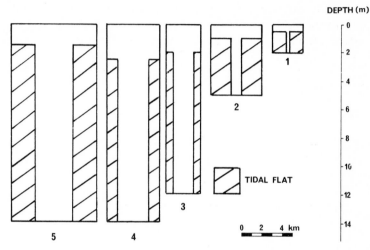

Fig. 4.7. Cross-sections of the five model compartments. Surfaces and depths drawn to scale

4.2.3 Boundary Conditions

The boundary conditions determine the quantity or concentration of state variables which are subject to transport at one of the boundaries of the estuary as given in Section 4.2.1. Of course boundary conditions are only given for the open boundaries across which transport takes place. These are the sea, the Ems and the Westerwoldsche Aa boundaries.

The boundary conditions may be given as a constant (e.g. salt concentration on the Westerwoldsche Aa = 0.3 S), as a calculation (e.g. oxygen in the Westerwoldsche Aa is calculated from relative saturation values and saturation concentration at the prevailing temperature) or as a time series (e.g. PO4 on the Ems and Westerwoldsche Aa).

The boundary conditions have to be defined for every time step of the model. The time series usually give the value of the boundary condition with intervals of a month. It is therefore necessary to interpolate between the time series values, which is done in an interpolation routine.

The boundary conditions for fresh water discharge from the Ems (QEMS in the model) and the Westerwoldsche Aa (QWWA) are given as monthly mean flows ($m^3 \cdot s^{-1}$) from 1977 to 1982.

The effects of the fluctuating fresh water discharges on the salinity (SAL) are relatively simple. The salinity distribution is directly governed by the fresh water discharges. The salinity values on the boundaries with the North Sea and the Ems depend on the discharge of the rivers, as described in Chapter 5.

The boundary concentrations of the suspended matter (EMSSUSM and WWASUSM) are given in Fig. 4.8. The values for the river Ems refer to the river area where the salinity reaches zero. This means that some of the effects of the turbidity maximum are, although implicitly, incorporated in the model. The inverse relationship, for instance, between river discharge and suspended matter concentration for this region as published by Postma (1981) is clearly present.

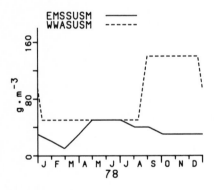

Fig. 4.8. The boundary concentrations of suspended matter in the river Ems (EMSSUSM) and the Westerwoldsche Aa (WWASUSM)

The North Sea boundary condition for silt will be discussed in Section 5.3.4 because this boundary is an integral part of the silt model. The simulated silt concentration is heavily dependent on the actual wind speed. The total organic carbon concentration (TOC) is not constant but follows a seasonal pattern with low values in winter and high values in spring and summer when primary production is high (Fig. 4.9 a).

The oxygen concentrations on the rivers Ems and Westerwoldsche Aa are given in Fig. 4.10. In the river Ems the concentrations follow a more or less sinusoidal curve with high values in winter when biological activity is low and the saturation concentration is high and low values in summer when the biological activity is highest and the saturation concentration is low. The same holds for the river Westerwoldsche Aa. However, there is one big difference between both rivers, this being the decline in oxygen values when the organic waste discharges increase (Fig. 4.9 b).

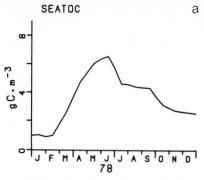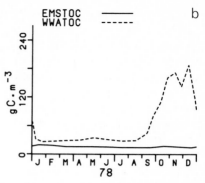

Fig. 4.9. The boundary concentrations of total organic carbon at sea (SEATOC) (a), in the river Ems (EMSTOC) and the Westerwoldsche Aa (WWATOC) (b)

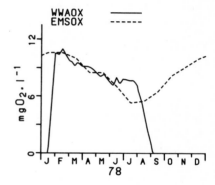

Fig. 4.10. The boundary concentration of oxygen in the river Ems (EMSOX) and the Westerwoldsche Aa (WWAOX)

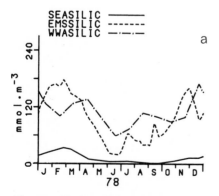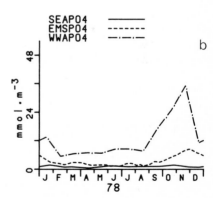

Fig. 4.11. The boundary concentrations of silicate (a) and phosphate (b) at sea (SEA), in the river Ems (EMS) and the Westerwoldsche Aa (WWA)

In the model reactive silicate and reactive phosphate are included as they may act as growth-limiting factors for phytoplankton. The boundary conditions for these nutrients are presented in Fig. 4.11. They generally are high during fall and winter and low during spring and summer.

There are no decisive arguments for choosing phosphorus as a growth-limiting factor above nitrogen, the more so since phosphorus concentrations are high both in the suspended matter and in the sediments. However, phosphorus reached low concentrations at an earlier date than nitrogen and therefore appeared to be more relevant in limiting phytoplankton growth than nitrogen. Including also the potential growth limitation by nitrogen would have led to a complication of the model which seemed unnecessary in view of the lack of evidence on this point for the Ems estuary.

In the model it is assumed that biological constituents such as labile organic carbon (PLOC), pelagic bacteria (PBAC), pelagic flagellates (PFLAG), pelagic diatoms (PDIA), pelagic carnivores (CARN) as well as the microzooplankton (PMIC) behave conservatively in the Ems region between zero salinity and compartment 3. This implies a linear relationship between these state variables and the salt concentration. It also implies a zero concentration of these state variables at 0 salinity (Fig. 4.12) in the Ems. This assumption avoids many complications. Based on these assumptions, the Ems boundary condition for these constituents is calculated from the actual concentrations of these constituents in compartment 3.

A North Sea boundary condition is given for a number of biological pelagic state variables. The values for phytoplankton (PFLAG and PDIA) and meso-zooplankton (PCOP) are derived from Fransz and Gieskes (1984). The boundary condition of the carnivorous zooplankton (CARN) is derived from Van der Veer and Sadée (1984).

For the remaining biological pelagic state variables (PBAC, PMIC) it is assumed that the concentration in the coastal water is equal to the concentration in compartment 5.

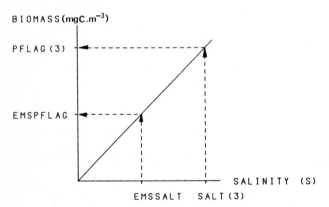

Fig. 4.12. Interpolation of the concentration of a substance [in this case PFLAG(3)] at the boundary between compartment 3 and the river Ems, from the concentration of this substance at 0 S in the Ems and the concentration at the salinity in compartment 3

4.3 Forcing Functions

In most aquatic ecosystems abiotic factors such as temperature and irradiance act
as the key driving forces. In the Ems estuary the tide and the large input of organic
carbon in the Dollart as discussed in the next section, together with nutrient inputs might be considered as additional driving forces. In the model, however, only
temperature and irradiance are used as driving forces.

Both functions, temperature and irradiance, are not independent, but they will
be treated here separately. In the model temperature and irradiance perform different functions, which can be described as rate regulation for biological processes
(temperature) and driving the photosynthesis of autotrophs (irradiance).

The temperature increase in spring is directly caused by the increasing solar
irradiance. However, the large heat capacity of the water delays the heating in
spring of the estuarine water masses and even more those of the North Sea. The
cooling in autumn is delayed in a similar way. The cooling and warming of the
water masses due to irradiance results in a temperature curve which lags behind
the irradiance curve by 2–3 months. The sediments of the tidal flats are subject
to more extreme temperature fluctuations during emersion.

4.3.1 Temperature

Surface water temperatures were derived from long-term measurements of the
RIZA (Rijksinstituut voor de Zuivering van Afvalwater). These temperatures are
used in the pelagic model. Annual temperature curves are given in Fig. 4.13. A
slight difference in water temperature exists between the inner and the outer compartment due to the smaller heat capacity of the inner compartments.

Tidal flat sediment temperatures are calculated independently of the water
temperature as a cosine function of time. The reason for the different concepts
for water and sediment temperatures is obvious: during emersion the sediment
temperature deviates strongly from the water temperature. The temperature in
the anaerobic layer follows that in the aerobic layer with a delay and also has a
smaller amplitude. Different functions are used for the aerobic (surface) and the
anaerobic (deeper) sediment layers. The amplitude, mean and the time lag with
the irradiance curve of the temperatures in both layers were derived from field

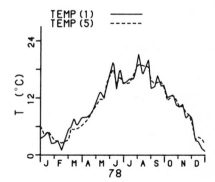

Fig. 4.13. Water temperature in compartment 1
[TEMP(1)] and compartment 5 [TEMP(5)]

measurements (Van Es, 1982 a; Schröder, unpublished). The sediment of the inner compartments (the Dollard) has a larger temperature amplitude than the outer compartments of the estuary. This is due to the longer tidal exposure of these sediments by day to direct heating and cooling by night.

The cosine functions for the sediment temperatures in compartment 1 are:

ACOS   = -COS[π · (TIME-TSA)/180];
ANCOS = -COS[π · (TIME-TSAN)/180];

where TSA = 15 and TSAN = 45 (for the aerobic and the anaerobic layers respectively) are the delays in days after January first. The mean annual temperature in compartment 1 is 11 °C and the half-amplitudes for the temperature in the aerobic (TA) and anaerobic (TAM) layers are 13.5 °C and 11.5 °C respectively:

TA   (1) = 11+13.5 · ACOS;
TAM(1) = 11+11.5 · ANCOS.

Consequently in the aerobic layers the minimum temperature (on day 15) is −2.5 °C and the maximum 24.5 °C, whereas in the anaerobic layer the minimum (on day 45) and maximum temperatures are −0.5 °C and 22.5 °C respectively. In the other compartments both the mean annual temperatures and the amplitude of the temperature curves are lower.

The temperature used in the epibenthos submodel is the water temperature, the same as in the pelagic submodel. Temperature acts directly on biological functions, such as growth rates of phytoplankton and phytobenthos, and food uptake and respiration rates of all heterotrophs (mesozooplankton, microplankton, meiofauna, deposit feeders, carnivores). To make these rates temperature-dependent, the rate constants in the expressions are defined at the mean annual temperature and then corrected for the ambient temperature on a specific day. These functions generate a value of 1 at the mean temperature, and the general form is:

$$A = 2^{[(TEMP - MEANT) · C1]}$$

where C1 is a slope constant, MEANT is the mean annual temperature and TEMP the ambient temperature. Because the metabolic rates do not keep rising exponentially with increasing temperatures for some state variables, a second expression is introduced which flattens the temperature correction at high temperatures:

$$A' = C2 · A / (C2 + A - 1),$$

where C2 is a second constant, representing the theoretical maximum value of the correction factor. For some groups of organisms, especially in the pelagic submodel, the temperature correction factor is a simple exponential function of form A given above. The use of such an expression gives a doubling of the temperature correction when the increase above the mean temperature (in degrees Celsius) is equal to the reciprocal of C1. Since C1 usually is 1/10, this boils down to a q10 of 2.

4.3.2 *Irradiance*

Irradiance data were collected with a Kipp solarimeter in 10-min intervals. These data were integrated to an actual daily irradiance sum $(J \cdot cm^{-2})$. The daily values are used in the model to drive the primary producers.

4.3.2.1 *Tidal Interference with the Light Regime*

The interference between daylight and tidal submersion/emersion of the flats is a peculiar aspect of tidal regions (Fig. 4.14). The coincidence of the daylight period and the submersion of the tidal flats causes a shortening of the effective photoperiod for the phytobenthos and an enlargement of the light-receiving water surface.

For the phytobenthos on the tidal flats it obviously makes a difference whether the flats are dry or covered with turbid water during the day. Also for the phytoplankton it is important whether during the daylight hours the water is spread out over the flats or concentrated in the channels, because in the turbid Ems estuary the euphotic zone is only in the order of 1-m thickness. By the shift of the tidal phase the light conditions change with a periodicity of 2 weeks, superimposed on the annual variation in day length. Even if the tidal influence is aver-

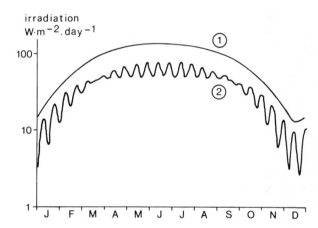

Fig. 4.14. Irradiance at the water surface (1) and on the surface of the tidal flat (2)

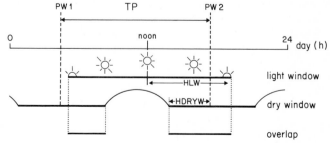

Fig. 4.15. Schematic representation of the interaction of the daylight period and the tidal period and the resulting "overlap", when tidal flats are exposed directly to light

aged out to a large extent due to the high frequency, we take it into account for three reasons:

1. Due to nonlinearities in the dependence of primary production on the light, the production average is not the same as the production at average light.
2. Especially in winter and summer the tidal influence leads to modifications which cannot simply be simulated by parameter changes.
3. The influence is different in the different compartments.

The tidal influence on the benthic primary production is simulated by calculation of the overlaps in hours of the daily "light window" and of the two daily "dry windows" (Fig. 4.15):

OVERLAP	=	MAX[O, MIN(HLW, PW1 + HDRYW) − MAX(-HLW, PW1 − HDRYW)] + MAX[O, MIN(HLW, PW2 + HDRYW) − MAX(-HLW, PW2 − HDRYW)];
TP	=	tidal period (12 h 25 ');
TPS	=	$2 \cdot TP - 24\,h = 50' = 0.833\,h$, daily shift;
PW1	=	MOD[MOD(date · TPS + PWO, 24), TP]; position of the centerpoint of the first dry window in a day, depending on the date;
PW2	=	PW1 + TP;
PWO	=	position of the dry window on date 0;
date	=	number of the day in the year;
HLW	=	SUNQ/2, half light window;
SUNQ	=	daylight in hours, depending on the date;
HDRYW	=	[1 − ptWET(I)] · TP/2, half dry window, depending on the compartment I;
ptWET(I)	=	fraction of the time in which the flats are covered with water.

In the production formula for phytobenthos, OVERLAP must be used instead of SUNQ. For the pelagic system the situation is slightly more complicated. We may approximate the water distribution during the daylight hours in the following way:

1. During OVERLAP all the water is in the channels, partly in the next downward compartment;
2. During SUNQ − OVERLAP a part of the volume is spread out over the flats.

Let DEPPROD be the center of mass of the depth of the phytoplankton production in deep water. This depth depends on the surface illumination and on the turbidity. For the following model calculations we simplify the distribution over depth of the production by the assumption of a light entrance depth DEPFOT = 2 ∗ DEPPROD: above DEPFOT the production is uniform, below DEPFOT it is zero. Then the production is proportional to the relative volume above DEPFOT. The change in the fraction of the illuminated volume due to the shift of the tidal phase can be taken into account by the correction factor CORRF:

CORFF	=	CHVOL · CHDEP + FLVOL · [SUNFL · CHDEP + (1-SUNFL) · FLDEP];
with SUNFL	=	OVERLAP/SUNQ;

CHDEP   = MIN[1, DEPFOT/depch(I)];
FLDEP   = MIN[1, DEPFOT/depfh(I)];
depfh(I)   = water depth over the flats at high water;
depch(I)   = average depth in the channels;
CHVOL   = 1 − FLVOL;
FLVOL   = CORRVOL · PWAFL(I);
PWAFL(I)   = fraction of the volume of compartment I which at high water
               is above the flats;
CORRVOL   = 2/3; this takes into account that the average depth over the flats
               during the wet window is less than the maximal depth
               depfh(I).

The combined effect of daylight and tidal phase on phytoplankton production is contained in

CORR = CORRF · SUNQ/24.

4.4 The Nature and Flux of Organic Matter

The organic matter in the Ems estuary has two sources: local primary production and import from both the North Sea and fresh water sources. In the estuary the organic matter is subjected to two transport mechanisms: part of the organic material is transported diffusively and another part is subjected to particulate transport mechanisms, sedimentation and resuspension. To model the impact of transport processes, it was necessary to divide the organic matter into a dissolved and a particulate part. However, this physically defined division is not suitable to describe the production and consumption of organic matter in the estuary. From a biological point of view it is necessary to know which part can be utilized by heterotrophic organisms: the fauna and the bacteria. In the water phase, 20–50% of the particulate organic matter consists of living material (mainly phytoplankton and bacteria), the rest being nonliving organic matter (Laane, 1982a). In the sediments the contribution of living organic carbon to total organic matter is less than in the water phase (Van Es, 1982a). With the separation into living and nonliving (detrital) organic matter something general is said about the nutritional value: living (fresh) organic matter (phytoplankton, bacteria, etc.) has a high nutritional value for herbivores as compared to detritus (degraded, sometimes even fossil, organic matter). In the pool of detritus there is of course a spectrum of degradability ranging from fresh phytoplankton remains to peat.

Previous simulation studies on marine ecosystems made a distinction between the availability of living carbon and detrital carbon for heterotrophic organisms, assuming a better availability of living carbon compared to the detrital part of organic matter (Kremer and Nixon, 1978; Cuff and Tomczak, 1983). Following these older studies, in the Ems model the detrital organic matter was treated separately from the living carbon, but in the present study classes of detrital organic matter with different degradation times have been introduced.

4.4.1 Partition of Organic Matter

Following the fate of organic matter in the estuarine model requires the division of organic matter into physically and biologically defined categories. In practice, a physical distinction is made between particulate and dissolved organic matter. The fraction smaller than 0.45 μm is usually called dissolved (Cauwet, 1981). In the Ems estuary most of the organic matter (50–90%) in the pelagic system is dissolved (Laane, 1982b), in contrast to the sediments, where most of the organic matter is particulate, since the volume of interstitial water is small. The cut-off of the filters used to harvest particulate and dissolved fractions is not very exact, but most of the living organisms are retained by a filter with a pore size of 0.45 μm; so the filtrate essentially contains only nonliving, extremely small particles and dissolved organic substances. In contrast, the distinction between living and detrital particulate matter is sharp, but technically it is difficult to make a separation. For the Ems estuary, Laane (1982a) calculated the fraction of detrital particulate organic matter by subtracting the estimated biomass of phytoplankton and zooplankton from the total particulate organic carbon. In Fig. 4.16 it is shown that all over the estuary and in all seasons, the contribution of living organic matter to particulate organic matter is relatively low with a maximum of 64% in the outer part of the estuary in summer. For the sediments this ratio is only a few percent (Van Es, 1982b).

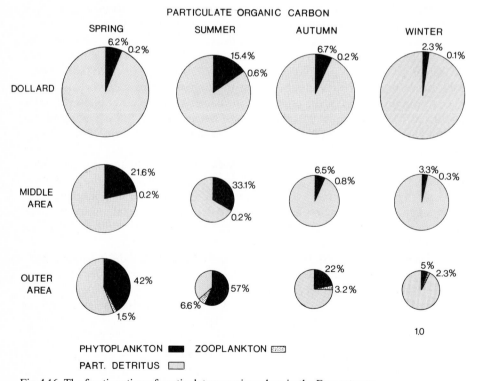

Fig. 4.16. The fractionation of particulate organic carbon in the Ems estuary

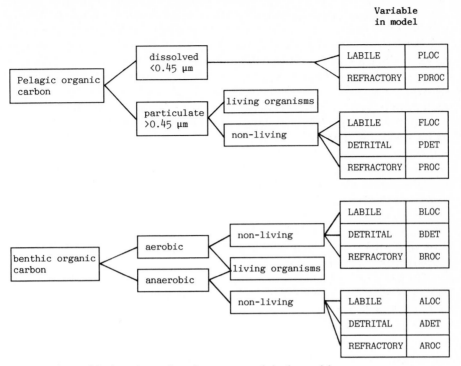

Fig. 4.17. The partitioning of organic carbon compounds in the model

Figure 4.17 shows the subdivision of organic matter in the water phase and in the sediments as used in the model. Pelagic organic matter is divided into dissolved and particulate organic matter, particulate organic matter is divided into a living and a detrital fraction. In the sediments this division is not made as completely. Here only total organic carbon is described, which is divided into a living and a detrital fraction.

The continuous spectrum of detrital organic matter with different decomposition rates is arbitrarily divided into three classes: LOC (labile), DET (detrital) and ROC (refractory). The mineralization takes place in days for LOC, in weeks to months for DET and in (many) years for ROC. The production and consumption of these classes of organic matter will be treated in the sections dealing with the various producers and consumers.

The organic matter at the boundaries of the North Sea, Ems and Westerwoldsche Aa was divided in the same way as described above. However, the physical and biological composition of the allochthonous organic matter is sometimes quite different from that in the estuary and this asked for some modifications.

From the North Sea an enormous amount of particulate organic matter is imported into the estuary. The North Sea phytoplankton entering the estuary seems to survive the changes in salinity (Colijn, 1983). So, the exchange of organisms between the North Sea and the outer estuarine compartment was assumed not to contribute to a transition from living material to detritus by mortality. The total concentration of detrital organic matter at the North Sea boundary was divided into detrital (SEAPDET), refractory particulate (SEAPROC) and refractory dissolved organic matter (SEAPDROC). It is assumed that the North Sea does not import labile organic matter, because its turnover is more rapid than the rate of exchange. The dissolved organic matter present in the North Sea is biologically inert (Van Es and Laane, 1982), so in the model it is called SEAPDROC.

Van Es and Laane (1982) demonstrated that also the dissolved organic matter of the river Ems is biologically inert. So, in the model the Ems introduces only refractory dissolved organic carbon into the estuary (EMSPDROC). An increase in the concentration of dissolved organic carbon has been measured during different cruises in the fresh water tidal compartment of the river Ems (Cadée and Laane, 1983). The source of this dissolved organic matter is not clear, but probably part of the particulate organic matter dissolves due to the increase in salinity (Eisma et al., 1985; Laane et al., 1985). This process is modelled by assuming that 50% of the particulate organic matter from the Ems dissolves to refractory dissolved organic matter.

The partitioning of the organic matter in the water discharged by the Westerwoldsche Aa is more complicated. It is assumed that the Westerwoldsche Aa does not introduce living organic matter to the estuary. During the waste water discharge period a larger part of the dissolved organic matter was mineralized quickly: 70% within 1 day (Laane and Ittekkot, 1983). In the model the dissolved organic matter was divided accordingly into a labile (WWAPLOC) and a refractory part (WWAPDROC). Also part of the particulate organic matter was mineralized quickly (Van Es and Laane, 1982; Laane and Ittekkot, unpublished results).

It is assumed that this labile particulate organic matter settles as floccules and a part of this sedimented material is transformed into detrital organic matter (BDET) during sedimentation and bacterial colonization. In Table 4.3 an overview is given of the partition of organic matter in the Ems model. For most state variables a range value is given. The absolute values are extracted from the published literature (Laane, 1984; Van Es and Laane, 1982).

Table 4.3. Fraction of dissolved organic carbon (DOC) in total organic carbon (TOC) and the partition of dissolved and particulate organic carbon into different subclasses: labile, detrital and refractory for the external organic carbon sources in the EMS estuary

Source	DOC TOC (%)	DOC		POC (nonliving)		
		Labile (%)	Refractory (%)	Labile (%)	Detrital (%)	Refractory (%)
WWA	95	10–75	25–90	10–50	10–40	25–80
Ems	40	–	100	–	8–40	60–92
North Sea	17–84	–	100	–	9–94	6–91

4.4.2 Transport of Organic Matter

The transport of organic matter through the estuary is greatly affected by its occurrence in either the dissolved or particulate state. In the Ems estuary a linear relation between the concentration of dissolved organic carbon and salinity is found, with high concentrations at low salinity and low concentrations at high salinity (Laane, 1980). This inverse linear relation could mean that import of dissolved organic carbon from the rivers dominates the local production and consumption.

Little is known about any contribution of dissolved organic carbon diffusing from the sediments into the water phase. However, if this contribution is high it would not be possible to find a linear relation between salinity and dissolved organic carbon (Laane, 1980).

Particulate organic matter is distributed over the estuary more or less as the dissolved organic carbon: high concentrations at low salinity and low concentrations at high salinity (Laane et al., 1985). In this case the main source is probably not input from the rivers but input from the North Sea (Favejee, 1960; Salomons, 1975; Rudert and Müller, 1981). Particulate (organic) matter is transported into the estuary by bed-load transport and accumulates at low salinity. During transport through the estuary a part of the particulate (organic) matter settles out. So, particulate organic carbon mainly originates from the sea and is transported into the estuary, with the estuary acting as a particle trap. For dissolved organic carbon the opposite was found. It is introduced in the estuary by the rivers and is flushed to the adjacent sea by advective and diffusive transport.

4.4.3 The Nutritional Value of Detrital Organic Matter (Detritus)

The nutritional value of detritus for the fauna has been studied in different ways. In laboratory experiments Williams (1981) showed that the tissue weight of mussels increased slightly when detritus was added to seawater rich in organic matter. He concluded that both aged and unaged detritus had a poor nutritional value for mussels. Many authors have suggested that aquatic organisms may be nourished completely or partly by the bacteria living on detritus (Fenchel and Jørgensen, 1977). Seiderer et al. (1984) found that bacteria play an important role as a nitrogen source for detrivores.

Another way to describe the nutritional value of suspended matter is to study the chemical composition. Laane et al. (1985) expressed the nutritional value of particulate organic matter as the ratio between the sum of the carbohydrates and proteins to total particulate organic carbon. This approach assumes that these constituents are better available than the unidentified part of particulate organic matter. Van Es and Laane (1982) estimated the nutritional value of organic matter for bacteria in another way. In decomposition experiments with samples from the Ems estuary they studied the decrease in the concentration of particulate (living and nonliving together) and dissolved organic matter. The fraction of organic matter that was mineralized after 30 days of incubation was called labile. Figure 4.18 shows that 78–94% of particulate organic matter of the outer part of the estuary was subject to bacterial degradation in summer. The degraded frac-

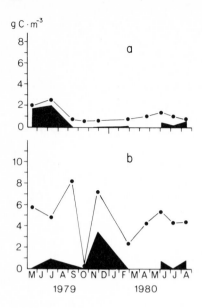

Fig. 4.18. Seasonal variation in total (●) and labile particulate organic carbon (black surface) in compartment 5 (a) and compartment 2 (b)

tion could be mainly ascribed to the phytoplankton bloom. Therefore, this experimentally determined labile fraction is not identical to the labile organic carbon (LOC) which was introduced before, since the contribution made by living organisms still has to be subtracted. The labile fraction of particulate organic matter in the inner part of the estuary was much lower than in the outer part of the estuary. The decomposition experiments confirmed the refractory character of most of the dissolved organic matter. Only ca. 8% was mineralized after 30 days of incubation.

The nutritional value of organic matter in the estuary, determined by chemical criteria (cf. Laane et al., 1985) or by the criterion of bacterial degradability (cf. Van Es and Laane, 1982) is of decisive importance for the growth of heterotrophic organisms. These aspects of the feeding of estuarine bacteria and fauna on dead organic matter have been introduced in the model by formulating five classes of pelagic and six classes of benthic organic carbon (Fig. 4.17). The relative nutritional value of organic matter, as discussed here, is expressed in the ratios of labile, detrital and refractory forms.

Part II: The Blueprint of the Model

5 The Construction of the Transport Submodel

P. Ruardij and J. W. Baretta

5.1 Introduction

In the estuarine environment the transport of material is caused by tidal currents and by fresh water discharge (see Chap. 2). The waters and their dissolved and particulate constituents are continuously mixed and redistributed. The exchange between spatial compartments is caused by two different kinds of transport, closely related to the type of material being transported:
1. Transport of dissolved and suspended material due to diffusion and advection.
2. A tide-induced transport of suspended particulate matter (sediment and biogenic particulates).
The transport submodel calculates the effective exchange of both types of material between the compartments. Some of the transported substances (suspended particulates such as silt) are subject to both transport processes.

The Ems estuary is a well-mixed estuary (Dorrestein and Otto, 1960) and the spatial compartments in the model are boxes of uniform composition, having no horizontal or vertical stratification. The model is tidally averaged, and tidal effects on transport thus have to be modelled implicitly.

5.2 Diffusive Transport

Advection/diffusion models have been described by Zimmerman (1976a, b) for the western Wadden Sea and by Dorrestein and Otto (1960) and Helder and Ruardij (1982) for the Ems estuary. The model presented here is an adapted and simplified version of that by Helder and Ruardij (1982). This simplification is attained by excluding the Ems proper from the system by setting the river-estuary boundary at their confluence and by reducing the number of compartments to five (see Sect. 4.2).

The same set of detailed, long-term salinity data as used by Helder and Ruardij (1982) is used for the calibration of the present model.

The mass flux of a (pelagic) state variable due to advection and diffusion from compartment I to the seaward adjacent compartment $I+1$ is given by:

$$naddif(I,I+1) =$$
$$\frac{Q(I) \cdot 0.5 \cdot [Conc(I) + Conc(I+1)] + RDIFF(I, I+1) \cdot [Conc(I) - Conc(I)+1)]}{V(I)}$$

$$(5.1)$$

naddif: net transport by advection and diffusion, expressed as change of con-
          centration in compartment I $(mg \cdot m^{-3} \cdot d^{-1})$;
RDIFF: exchange coefficient due to diffusion $(m^3 \cdot d^{-1})$;
Q: fresh water discharge rate $(m^3 \cdot d^{-1})$;
Conc: concentration of the transported substance in a compartment
          $(mg \cdot m^{-3})$;
V: tidally averaged water volume in a compartment (m^3).

The first term denotes the advective mass transport which is equal to the fresh
water input. The second term RDIFF(I,I+1) is the effective diffusion constant,
which includes tidal mixing. The extent and the direction of the diffusive transport is dependent on the concentration gradient, expressed as the difference between the concentrations of the transported substance in the two compartments
I and I+1.

Advection and diffusion may work in opposite directions. This is obviously
the case in the transport of salt: diffusion transports it in a landward direction
while the advective transport carries it to the sea. In contrast, nutrients are discharged by the river at higher concentrations than that in the sea and hence both
diffusive and advective transport is directed towards the sea. In this way salinity
and nutrient gradients are formed by the tides and the fresh water discharges.

The model has boundaries with the sea, the Ems and the Westerwoldsche Aa
(Sect. 4.2.1). The discharge of the Westerwoldsche Aa is controlled by sluices,
which precludes diffusive exchange between the Westerwoldsche Aa and the Dollard (i.e. compartment 1). The river Ems enters the system in compartment 3. Because tidal effects extend a long way upriver, it is assumed that between the Ems
and compartment 3 advective as well as diffusive transport does take place. The
same processes also occur between compartment 5 and the sea. These tidal effects
on the external concentrations have been incorporated by taking boundary concentrations, which are interpolations between the concentrations in compartment
5 and the sea and between the concentrations in compartment 3 and in the Ems
at zero salinity. This accommodates the effect of water with a different composition passing across a boundary in the course of a tide.

5.2.1 Determination of the Constants

To establish the exchange coefficients salinity measurements are used because salt
behaves as a conservative variable. An important assumption in computing the
diffusion constants is that the system (the estuary) with respect to salt is in a
steady state. The net exchange rate naddif (I,I+1) for this one-dimensional system then is equal to zero. Such a steady state is reached after a few weeks of constant river discharge (Helder and Ruardij, 1982)

From Eq. (5.1) follows:

$$RDIFF(I) = \frac{Q(I) \cdot [Conc(I) + Conc(I+1)]}{2 \cdot [Conc(I) - Conc(I+1)]}. \qquad (5.2)$$

Using Eq. (5.2) we can compute the exchange coefficients at different flow levels
of the Ems and the Westerwoldsche Aa using long-term salinity observations

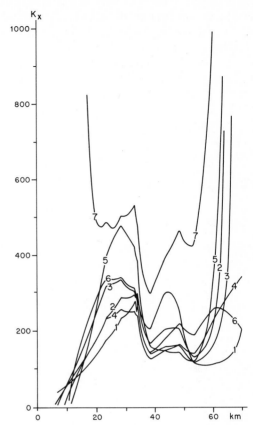

Fig. 5.1. Longitudinal dispersion coefficients Kx ($m^2 \cdot s^{-1}$) for the Ems estuary at increasing discharge rates of the river Ems of 25.6 $m^3 \cdot s^{-1}$ (*1*); 38.5 $m^3 \cdot s^{-1}$ (*2*); 50.0 $m^3 \cdot s^{-1}$ (*3*); 53.3 $m^3 \cdot s^{-1}$ (*4*); 76.2 $m^3 \cdot s^{-1}$ (*5*); 131.2 $m^3 \cdot s^{-1}$ (*6*); 320.0 $m^3 \cdot s^{-1}$ (*7*) (Helder and Ruardij, 1982). River boundary at km = 0

made over the period 1972–1976. From these calculations it became clear that the apparent diffusion increases with high discharges of the rivers (Fig. 5.1). High river discharge itself is not necessarily the cause of the increase of the exchange constants. More likely, a combination of higher average water levels and direct wind-generated turbulence in winter periods might be the cause (Helder and Ruardij, 1982). In the model the fresh water flows are interpolated from a time series of monthly averages. A specific exchange coefficient is computed from the actual flow by interpolation between measured values. In the Dollard compartments (1, 2) only the discharge of the Westerwoldsche Aa is used for that purpose, in the other compartments (3, 4, 5) the sum of both discharges has been used.

5.3 Transport of Particulate Material

The high concentration of suspended matter in the Ems estuary is one of the main features determining the physical and biological conditions in the estuary. Adequate modelling of the silt transport therefore was essential to the whole modelling project. The procedure for silt transport in the Ems model has been designed with the following aims:

1. To describe the tidally averaged movement of particulate organic matter and silt against the gradient and against the fresh water flow from outside (sea) into the estuary.
2. To describe the tidally averaged resuspension and sedimentation of particulate matter.
3. To develop a calibration method for some of the model parameters.
4. To make a budget of the exchange of particulate material between the compartments, the North Sea and the rivers.

This procedure is not intended to fully explain the silt transport, but to describe the seasonal variations in the concentration. The rate of net sedimentation of particulate matter in the model is calibrated with field observations. The model thus is strictly empirical.

5.3.1 Description

Silt in this model is defined as inorganic material of less than 55 µm ESD (Equivalent Spherical Diameter) which either is in suspension or may be suspended by tides or currents. The suspended particulate organic material is closely associated with the silt and this organic/inorganic complex will be defined as mud. So the transport procedure for the particulate material deals with the aggregated constituent mud. The unit of mud is mg dry weight $\cdot$ m^{-3}.

The main difference between mud transport and transport of dissolved material is that in mud transport a large fraction of the material is not in suspension but occurs close to the bottom of channels and is transported along the bottom as fluid mud (Dronkers, 1986). Therefore the total amount of mud [T(I)] is divided into two fractions:
1. Suspended particulate material [M(I)];
2. Mud on the channel bed [C(I)].
I is the index of the compartment.

Tidally averaged, there is a continuous and large exchange between C(I) and M(I). The net transport of mud into the system is closely connected to this continuous redistribution of mud, both horizontally and vertically by tidal effects. Due to the asymmetry of the tidal current velocities more mud goes into suspension during flood than during the ebb tide (Postma, 1954).

The exchange due to resuspension and sedimentation is especially prevalent between the fluid mud layer just above the sediment and the sediment itself, leading to a net transport of mud in the fluid mud layer. In this way mud is transported from the sea into the estuary, despite the fact that the concentration is lower at sea. This phenomenon has been observed in the estuaries of the Thames, the Rhine (Dronkers, 1986) and the Gironde (Eisma, 1986) and is probably associated with the occurrence of a turbidity maximum and a distinct ebb/flood asymmetry which also exists in the Ems estuary.

The fluid mud layer on the channel bed according to our definition does not exist on the tidal flats. The organic part of the fluid mud layer therefore is not available as food for organisms on the tidal flats and the inorganic part does not influence the physical conditions for the benthic organisms in the system, since these are confined to the tidal flats.

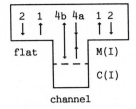 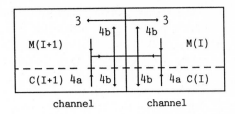

Fig. 5.2. The different processes are presented in schematized compartments. The *numbers* refer to the numbered processes in the text

The suspended part of the mud determines the light extinction coefficient of the water and the organic part of the suspended mud is in principle available as food to organisms. The suspended mud does occur above the tidal flats and sedimentation of mud does take place there.

Mud transport in the model is simulated by four different but interwoven processes (Fig. 5.2):

1. Resuspension;
2. Sedimentation;
3. Diffusive/advective transport;
4. Residual sediment transport.
   This process is a combination of two other processes:
   a) redistribution (transport) of mud;
   b) partitioning of transported mud between suspended mud and mud on the channel bed.

The first two operate within each compartment and on the tidal flats during the time these are submerged. They govern the vertical exchange between the sediment and the suspended material in the water column.

The last two determine the horizontal transport between the compartments and across the seaward boundary.

The advective and diffusive transport only acts on the particulate matter which is in suspension throughout the water column. The advective part transports material towards the sea and the diffusive transport, tending to equalize concentrations everywhere, in this case also transports particulate matter towards the sea because of the lower concentrations there. The residual mud transport, however, because of the tidal asymmetry, is strongly directed inwards. This process affects both the fluid mud layer and the suspended material which thus attain a higher concentration going inwards.

The continuous repartitioning of mud between fluid mud and suspended mud due to sedimentation and resuspension, combined with the replenishing of the fluid mud layer in the channels from more outward compartments, creates a net flow of particulate matter into the system.

All four processes are strongly influenced by wind effects (Postma, 1954; Kamps, 1963; Eysink, 1979). The model assumes resuspension to increase linearly with increasing wind speed (Dücker, 1982), up to a wind speed of $6 \text{ m} \cdot \text{s}^{-1}$, with a consequent reduction in net sedimentation. Especially wave action on the tidal flats is responsible for this phenomenon. At wind speeds over $6 \text{ m} \cdot \text{s}^{-1}$, wave action disturbs the sediment of the tidal flats itself with the consequence that all suspendable material goes into suspension. At the beginning of a period with high wind speeds the concentration of suspended matter in the water increases sharply to a high level but then remains at that level (Slagmolen, 1981) because anything that can be suspended, already is.

Given the scarcity of field observations on suspended matter concentrations during storm events we do not try to model the effects of short-term wind events explicitly, but instead use the monthly mean wind speed to reflect the seasonal differences in suspended particulate matter concentrations [M(I)] due to wind. This approach is strictly empirical; its main merit is that it is fairly successful in reproducing the observed seasonal distributions of suspended particulate matter in the Ems estuary.

In the model wind speed modifies pDOWN which in its turn modifies resuspension, sedimentation and residual silt transport. The dimensionless pDOWN is computed according to the following equation:

$$\text{pDOWN} = \frac{\text{W50}}{\text{W50} + \text{W} - \text{MINWIND}} \cdot \text{pDOWNO} , \tag{5.3}$$

where pDOWNO is a dimensionless parameter (0.8), W50 is the half-value constant of this Monod term (2 m·s^{-1}), W is the monthly mean wind speed, which in the Ems region varies between 3 and 4.5 m·s^{-1} (BOEDE, 1985) and MINWIND is a constant (3 m·s^{-1}). In estuaries where the monthly mean wind speed may be equal or less than MINWIND this equation should not be used. These parameter values result in suspended particulate matter [M(I)] concentrations that are twice as high in periods with strong winds than in calm periods.

5.3.2 Vertical Transport Processes: Sedimentation and Resuspension

Both resuspension and sedimentation on the tidal flats only occur during submersion. In the model these processes occur simultaneously, because the model is tidally averaged. In reality, sedimentation on the tidal flats takes place at low current velocities (around high water slack), whereas resuspension dominates at higher current velocities. The periods with relatively high current velocities on the tidal flats are the periods when the flat is just flooding and when it is just emerging.

The period with low current velocities over the tidal flats is not long enough to let all particulate material from the whole overlying water column settle out. Only the suspended matter from a layer of 0.44 m can do so. The value of this parameter has been calibrated by trial runs with the transport model, for lack of literature data.

The rate of resuspension depends on the concentration of freshly settled suspended matter on the surface of the sediment. This in turn depends on physical processes such as wave action and on the reworking of the sediment by benthic organisms. Bioturbation, describing the vertical transport of sediment organic carbon into deeper layers, is included in the model (Chap. 7). The behaviour of freshly settled inorganic particulates, however, is not explicitly included in the model. Instead, we assume that the concentration of this layer is adequately described as being a function of the total mud concentration (both suspended mud and the fluid mud in the channels) times the resuspension height as a fraction of the average water depth. The equations describing sedimentation and resuspension are:

$$\text{rsed(I)} \cdot \text{M(I)} \cdot \text{rtid} \cdot \text{pDOWN} \cdot \text{pWAFL(I)} \cdot \frac{\text{seddep}}{\text{depfa(I)}} \cdot \text{M(I)}; \tag{5.4}$$

$$\text{resus(I)} \cdot \text{T(I)} = \text{rtid} \cdot (1 - \text{pDOWN}) \cdot \text{pWAFL(I)} \cdot \frac{\text{resdep}}{\text{depfa(I)}} \cdot \text{T(I)}. \tag{5.5}$$

rsed(I):   relative gross sedimentation (d^{-1});
resus(I):   relative gross resuspension (d^{-1});
rtid: number of tides per day (1.95 tides $\cdot d^{-1}$);
seddep: settling water column (0.44 m);
resdep: resuspension height (0.13 m);
pWAFL(I):   fractional volume over the tidal flats (Sect. 4.2.2, Box 4.6);
depfa(I):   average depth above the flat during submersion (m) (Sect. 4.2.2, Box 4.3);
T(I):   total concentration of mud ($mg \cdot m^{-3}$);
M(I):   suspended particulate material ($mg \cdot m^{-3}$).

5.3.3 Horizontal Transport Processes

Advective and Diffusive Transport. Suspended matter is subjected to advective and diffusive transport. The net change in compartment I due to these transport processes between two adjacent compartments is:

$$\text{ad(I)} = \text{naddif(I,I-1)} + \text{naddif(I,I+1)}, \tag{5.6}$$

where I-1 and I+1 are the compartments next to compartment I. Across the boundary between compartment 3 and the Ems only mass transport of the suspended material is assumed to exist (cf. Sect. 5.3.4).

Residual Mud Transport. The tides cause the displacement of a volume of water which is larger than the volume remaining in the estuary at ebb tide. Every tidal cycle this volume is imported and exported from the estuary. Along with this tidal volume, mud is transported against the concentration gradient due to the tidal asymmetry.

The fraction of the total amount of mud in compartment I which is available for residual silt transport to compartment I+1 is estimated as:

$$\text{rex(I, I+1)} = \frac{\text{tidal volume(I, I+1)}}{\text{V(I)}}, \tag{5.7}$$

where V(I) is the volume of compartment I at mid-tide.

In Table 5.1 the tidal volumes across the compartment boundaries are given. Clearly, the tidal volume decreases in a landward direction. The total amount of mud which is redistributed over two adjacent compartments I and I+1,

Table 5.1. Tidal exchange volumes between
adjacent compartments ($10^6 \cdot m^3$)

Compartments	Tidal volumes
Sea ↔ 5	914
5 ↔ 4	549
4 ↔ 3	299
3 ↔ 2	113
2 ↔ 1	15

Am(I,I+1) in $mg \cdot d^{-1}$ is a function of the tidal volume between I and I+1:

$$Am(I,I+1) = \text{tidal volume } (I,I+1) \cdot [T(I) + T(I+1)]. \tag{5.8}$$

Expressed as a mud concentration in compartment I the exchange of mud E(I,I+1) from I to I+1 is:

$$E(I,I+1) = \frac{Am(I,I+1)}{V(I)} \quad (mg \cdot m^{-3} \cdot d^{-1}). \tag{5.9}$$

The residual mud transport is modelled by means of the compartment-dependent parameter redis (m,n). This parameter expresses the asymmetry between compartments in the redistribution of mud by tidal effects. The difference between the value of redis (I,I+1) and redis (I+1,I) results in a residual mud transport in a landward direction. The net redistribution factor or asymmetry factor now becomes:

$$A(I+1,I) = \frac{redis(I+1,I)}{redis(I+1,I) + redis(I,I+1)}. \tag{5.10}$$

Now the net redistribution across the boundary between I+1 and I can be computed for T(I+1) and for T(I):

$$\text{for I+1: netred}(I+1,I) = A(I+1,I) \cdot E(I+1,I) - rex(I+1,I) \cdot T(I+1); \tag{5.11}$$

$$\text{for I: netred}(I,I+1) = A(I,I+1) \cdot E(I,I+1) - rex(I,I+1) \cdot T(I). \tag{5.12}$$

The sum of V(I+1) · netred (I+1,I) and V(I) · netred (I,I+1) is zero, thus satisfying the requirement for the conservation of mass.

Differential Equations for M(I) and C(I). The processes described above result in two differential equations for M(I) and C(I). Advective and diffusive transport, resuspension and sedimentation affect only the suspended mud M(I). Residual transport affects both variables M(I) and C(I).

Let K(I,m) = A(I,m) · E(I,m) (m=I−1 or m=I+1).
For A(I,m) and E(I,m), see Eqs. (5.9) and (5.10).
K is expressed as $mg \cdot m^{-3} \cdot d^{-1}$ and r(I) = rex(I,I+1) + rex(I,I−1).

The following equations are in $mg \cdot m^{-3} \cdot d^{-1}$:

$$\frac{dM(I)}{dt} = ad(I) + resus(I) \cdot T(I) - rsed(I) \cdot M(I) - r(I) \cdot M(I)$$

$$+ (1 - pDOWN) \cdot [K(I, I+1) + K(I, I-1)]; \qquad (5.13)$$

$$\frac{dC(I)}{dt} = -r(I) \cdot C(I) + pDOWN \cdot [K(I, I+1) + K(I, I-1)]. \qquad (5.14)$$

[For compartment 1 there is only $K(1, 2)$].

5.3.4 Boundary Conditions

The input of mud to the river Ems by residual mud transport is equal to the export by diffusion and advection. This assumption is based on the idea that in the river Ems there are no large tidal flats where mud may sediment out and accretion can occur. The particulate matter in the river Ems thus is assumed to be in dynamic equilibrium with the hydrographical conditions.

As stated before in this chapter only advective transport takes place for suspended material between compartment 3 and the Ems. Therefore advective transport of mud is made dependent on the mud concentration in the nontidal reach of the Ems at 0 S. Also there is advective transport from the Westerwoldsche Aa. When modelling mud transport the sea is treated as a sixth compartment. Mud is transported from the sea into the estuary partly as suspended mud, partly as channel mud. Therefore an estimate has to be made of the concentration of channel mud at sea. For that purpose we make use of the mass balance equations (5.13) and (5.14) and assume that at sea there is no advective and diffusive transport, no sedimentation, no resuspension (there are no tidal flats) and that the influence of compartment 5 on the mud concentrations at sea is negligible. This means that redistribution only takes place between M(sea) and C(sea). It also implies that $E(sea) = r(sea) * [C(sea) + M(sea)]$. Now Eqs. (5.13) and (5.14) for the concentrations at sea become:

$$\frac{dM(sea)}{dt} = -r(sea) \cdot M(sea) + (1 - pDOWN) \cdot E(sea) = 0; \qquad (5.15)$$

$$\frac{dC(sea)}{dt} = -r(sea) \cdot C(sea) + pDOWN \cdot E(sea) = 0; \qquad (5.16)$$

with an undetermined $r(sea)$. It is not necessary to know its value, if one assumes equilibrium for C(sea) and M(sea). Then it follows that

$$C(sea) = T(sea) \cdot pDOWN ; \qquad (5.17)$$
$$M(sea) = T(sea) \cdot (1 - pDOWN) . \qquad (5.18)$$

In the model the boundary concentrations at the seaward boundary for those state variables that are subject to the same transport mechanisms as mud are divided by $(1 - pDOWN)$ to establish their concentration near the bottom.

The determination of the boundary concentration of mud itself from field observations has proven to be a difficult task: the BOEDE had only a few measurements from that region, and the data of the RIZA show, just as the BOEDE data would have, a large variability in time and location, to the extent that there is no obvious relation with wind speed. The boundary concentration of total mud [T(sea)] therefore is presumed constant at 42 $g \cdot m^{-3}$. This means that in summer at low average wind speeds the concentration of the suspended fraction is about 14 $g \cdot m^{-3}$, and in winter at high average wind speed about 299 $\cdot m^{-3}$.

5.3.5 Calibration of the Model

The calibration procedure of mud transport in principle is similar to the calibration of the advective and diffusive transport. However, because of the more complicated nature of the mud transport processes the calibration needs some explanation.

To calibrate the parameters the following conditions must be met:
1. There is a steady state situation, where import, sedimentation and resuspension of mud are in balance.
2. Measurements of the average suspended mud concentration during a tidal cycle have been made under these conditions.
3. All other parameters such as wind speed, the water discharge and the associated diffusion constants as well as the sedimentation and resuspension rates are known.

Assuming these conditions to be met, we can compute the associated channel mud concentration as well as the redistribution factors.

Computation of the Channel Mud Concentration. From condition 1 it follows that:

$$\frac{dM(I)}{dt} = 0 \quad \text{and} \quad \frac{dC(I)}{dt} = 0.$$

such that the following equation for C(I) can be derived from Eqs. (5.13) and (5.14):

$$C(I) = \frac{\{[\text{rsed}(I) + r(I)] \cdot M(I) - ad(I) - \text{resus}(I) \cdot M(I)\} \cdot \text{pDOWN}}{r(I) \cdot (1 - \text{pDOWN}) + \text{resus}(I)}. \tag{5.19}$$

Computation of the Redis Parameters. In determining appropriate values for the redistribution parameters redis(I+1,I) the following facts are used:
1. The redis parameters are always used pairwise in calculating the redistribution asymmetry A(I+1,I) and therefore one of each pair of redis(I,I+1) parameters may be given an arbitrary value of one.
2. In compartment 1 there is only redistribution with compartment 2, whereas in the other compartments there are two boundaries across which redistribution takes place.

For compartment 1 Eq. (5.14) reduces to Eq. (5.20). Combining Eqs. (5.10) and (5.20) we arrive at Eq. (5.21). Now A(1,2) can be calculated, and, keeping in mind that redis(2,1) is set to one, redis(1,2) can be determined. Once this value is established, the redistribution between the next two compartments can be calculated and this procedure is repeated for all other compartments. This whole house of cards is built upon the correct determination of the average suspended matter concentration in the various compartments:

$$-r(I) \cdot C(1) + K(1,2) \cdot pDOWN = 0; \qquad\qquad (5.20)$$

$$K(1,2) = A(1,2) \cdot E(1,2) = \frac{rex \cdot C(1)}{pDOWN}. \qquad\qquad (5.21)$$

Actual Calibration. The actual calibration of the mud transport is done for the average conditions of wind and river discharges. The appropriate values for advective and diffusive transport, as determined before, were applied. The sedimentation and the resuspension rate as defined by the parameters seddep and resdep are assigned values that comply with the following requirements:
1. Seddep and resdep cannot exceed the average water depth above the tidal flats.
2. The net annual sedimentation in compartment 2, taking into account the wind conditions and river discharges throughout the year, is equal to the amount of silt contained in the observed annual accretion of 7 mm (RWS, unpubl. data).
3. The net sedimentation during periods of high wind speeds should be equal to zero.

The calibration is done by trial and error, using simulations with different values of seddep and resdep. To every set of seddep and resdep values belongs a different set of redis values. The redis values are calculated in a separate program as described in this paragraph.

The average concentration of M(I) is determined by averaging field measurements from 1972 to 1976. Field observations from later years are used as verification values. The mud concentration varies enormously both in time and in place even within a compartment and during a tidal cycle. This variability only allows a crude estimate of the transport parameters.

Table 5.2. Steady state concentrations of suspended mud M(I) and channel mud C(I) in $g \cdot m^{-3}$, and redis(I) under average conditions of river discharge (for Ems: 144 $m^3 \cdot s^{-1}$ and for WWA: 14 $m^3 \cdot s^{-1}$) and wind (7.1 $m \cdot s^{-1}$) with pDOWN = 0.52

I	M(I)	C(I)	Redis(I)
Sea	20.6	22.3	–
5	21.0	30.2	1.83
4	40.0	60.5	2.62
3	56.0	85.1	1.82
2	100.0	167.5	2.47
1	130.0	258.8	1.69

The calculated C(I) values are higher than the M(I) concentrations. This implies that mud transport along the channel bed is an important factor in the model, as it is in the field (Table 5.2).

5.3.6 Detrital Carbon in the Mud

Though mud is treated as one, highly aggregated, state variable, our main interest is in its particulate organic constituents. These occur both in (M) and (C). Since the behaviour of the organic particulates is thought to be indistinguishable from that of the inorganic particulates, we can calculate the suspended fraction of the particulate carbon:

$$pSUSP(I) = \frac{SILT(I)}{SILT(I) + CSILT(I)}.$$ (5.22)

Thus, the suspended fraction of PDET and PROC is only pSUSP(I).

A consequence of this definition is that C(I) and M(I) cannot be computed by simple addition of the state variables. In Fig. 5.3 the conceptual approach to M and C is illustrated:

M(I) = SILT + the suspended fraction of the particulate carbon state variables PDET and PROC.

C(I) = CSILT + the fraction of PDET and PROC which is not in suspension.

An additional problem when computing the mud concentration is that PROC and PDET are expressed in other units than mud and silt (C, M, SILT and CSILT). Mud is expressed in g dry weight $\cdot$ m^{-3} while PDET and PROC are expressed in mgC $\cdot$ m^{-3}.

All components are brought to the same unit (mg dry weight $\cdot$ m^{-3}) by multiplying the particulate carbon state variables with a conversion factor cCARASD.

After calculation of the concentration of (M) and (C), followed by the computation of the residual mud transport, the transported amounts are decomposed again into their constituents and the various amounts are added to or subtracted from the different state variables after reconversion to mgC, if necessary.

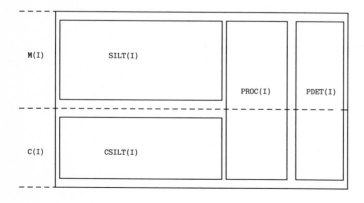

Fig. 5.3. Partitioning in the mud complex

6 The Construction of the Pelagic Submodel

J. W. Baretta, W. Admiraal, F. Colijn, J. F. P. Malschaert and P. Ruardij

6.1 Introduction

In this chapter the construction of the pelagic submodel will be elucidated. All pelagic variables are subject to transport processes; either diffusive transport or siltlike transport. The transport processes not only occur within the system, but also across the system boundaries. This necessitates the definition of boundary conditions. The procedure of incorporating the effects of the boundary conditions into the model links the model system to the outside world, but also enables us to delimit our model system and have it behave as a self-contained unit.

The biological variables belong to the following major groups:

1. The producers, represented by pelagic diatoms and flagellate phytoplankton;
2. The grazers, represented by the microzooplankton (20–200 µm) and the mesozooplankton (200–2000 µm);
3. The planktonic predators, represented by ctenophores;
4. The decomposers, represented by pelagic bacteria.

These variables are the living components of the pelagic. There are also a number of nonliving state variables of biological origin: the detrital state variables. The detritus has been disaggregated into a number of variables, as discussed in Section 4.4.

There are three noncarbon state variables in the pelagic system: the nutrients silicate and phosphate and oxygen dissolved in the water. Through the oxygen dynamics all the respiratory, oxygen-consuming processes express themselves, partially balanced by the oxygen production through primary production, partially by reaeration from the air.

The relationships between this set of state variables and the resulting dynamics represent the significant features of the pelagic system of the Ems estuary. Some of the connections and feedbacks determine to a large extent the performance of the model. Without these feedbacks the system simply cannot be made to behave as observed in the field. An example of a regulating feedback is the relationship between phytoplankton and nutrients. Phytoplankton growth in the model may be limited by low levels of phosphate and/or silicate. When this is the case, phytoplankton has an increased excretion. This in turn leads to an increased bacterial activity. The model recognizes two phytoplankton groups: diatoms and flagellates. Diatoms need both silicate and phosphate, whereas flagellates only need phosphate. This difference in requirements generates spatial and temporal differences between the two groups.

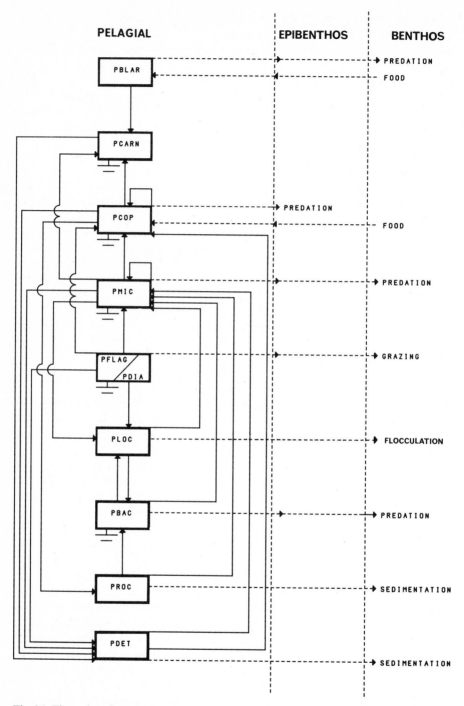

Fig. 6.1. The carbon flows in the pelagic submodel between the state variables

Phytoplankton is the main food source for both the microzooplankton and the mesozooplankton. The mesozooplankton, however, also feeds on the microzooplankton and, in case of low food concentrations, also cannibalistically. These feeding relationships indicate a food web structure. The food web damps the strong fluctuations in zooplankton biomass which would result from straight Lotka-Volterra predator-prey modelling. Moreover, since a shift by mesozooplankton from feeding on phytoplankton to feeding on microzooplankton entails a shift from filter feeding to raptorial feeding, the model also incorporates seasonal succession in the mesozooplankton.

All biological state variables contribute to the detrital pools by excretion, mortality or sloppy feeding (zooplankton). Depending on the degradability of the detrital products, they are recycled immediately by bacteria, slowly degraded or persist for a long time in a dissolved or particulate form. The carbon flows between the state variables are given in Fig. 6.1.

Since particulate detritus is influenced strongly by the transport processes governing the transport of silt, which are directed towards the inner part of the system, allochtonous detritus is continuously added to the detritus formed in the system itself. The sedimentation of detritus, both allochtonous and autochtonous, is the major external carbon source for the benthic system.

6.2 Phytoplankton and Nutrients

During the last decades efforts have been made to model the growth and production of phytoplankton in the oceans, coastal seas and estuaries (Winter et al., 1975; Kremer and Nixon, 1978; Platt et al., 1977; Falkowski, 1980; Platt, 1981). Also detailed growth models of phytoplankton species have been constructed, e.g. *Skeletonema costatum* (Cloern and Cheng, 1981). Basically these models contain formulations for the irradiance-dependent growth of phytoplankton combined with measured (or assumed) respiration and excretion rates. The models also contain terms for the influence of limiting nutrients (Kremer and Nixon, 1978). We have tried to put our knowledge of the primary production of phytoplankton in the Ems estuary (Colijn, 1983) into a model in which also literature data on phytoplankton physiology and ecology are incorporated.

The phytoplankton in the Ems estuary mainly consists of diatoms, in terms of biomass, cell number and number of species (Colijn, 1983). However, only the species distribution in the outer compartments was studied in detail. The species composition is comparable to that of the Wadden Sea and the nearby coastal zone (Cadée and Hegeman, 1974; 1979).

The spring bloom of diatoms is often followed by a dense bloom of *Phaeocystis pouchetii*, a colonial haptophyte, which forms large gelatinous colonies visible to the naked eye. The density reached by this organism can be more than 20 mg chlorophyll-a $\cdot$ m^{-3} or more than $5 \cdot 10^3$ cells per ml. Occasionally large numbers of benthic diatoms are observed in the phytoplankton as a result of wind-induced suspension of sediments (De Jonge, personal communication). In the inner parts of the estuary the influence of the fresh water input by rivers is shown by the presence of typical fresh water species (*Scenedesmus spp., Pediastrum*). The same was observed in the Marsdiep area (Cadée and Hegeman, 1979). The highest biomass of phytoplankton is observed in the outer compartments of the estuary, where light conditions for the growth of phytoplankton are favourable, because turbidity decreases towards the North Sea (Colijn, 1982; 1983).

Although the direct occurrence of nutrient limitation has not been studied we assume that throughout the main part of the year no nutrient limitation occurs, because of a steady input of nutrients from the rivers and a (possible) release of nutrients from the sediments (Rutgers van der Loeff et al., 1981). Only during the spring bloom of diatoms and the subsequent summer bloom of flagellates such as *Phaeocystis pouchetii* a silicate and/or a phosphate limitation may occur in the outer compartments of the estuary. Circumstantial evidence for these limitations is derived from calculations of the amount of nutrients needed for the growth of the phytoplankton (see Colijn, 1983, p.79 sqq.).

6.2.1 State Variables

The phytoplankton (PHYT) in the model is separated into two groups: the diatoms (PDIA) and the flagellates (PFLAG). The total biomass of the diatoms also contains a small fraction of the benthic diatoms (BDIA), suspended in the water column. The amount of BDIA in the water column is a fixed fraction of the total BDIA. In contrast to PDIA, the benthic diatoms do not grow in the water.

The biology of the two groups (PDIA and PFLAG) is modelled in a similar way: they perform the same elementary physiological processes of assimilation, respiration and excretion but have different constants for these processes. As pointed out before, irradiance, temperature and nutrient concentrations act as key driving forces on the physiology of the phytoplankton. The particulate phytoplankton carbon is a direct food source for the copepods (PCOP), for the microplankton (PMIC) and the benthic suspension feeders (BSF). The phytoplankton also produces excretion products (PLOC), detrital particulate carbon (PDET) and as a result of the sedimentation of dead cells benthic detritus (BDET) is formed. As far as the nutrients are concerned, both silicate and phosphate act as a limiting factor on the diatoms, whereas only phosphate acts on the flagellates. This phosphate limitation acts on the diatoms only when it is stronger than the silicate limitation (Law of Liebig). A striking difference between the growth of diatoms and flagellates is that diatoms grow independently of the salinity, whereas the growth of flagellates is limited at low salinities. Although relevant data on the distribution of flagellates in summer are lacking for the inner compartments, the distribution of *Phaeocystis* is closely tied to the higher salinities, and below 10 to 15 S no *Phaeocystis* has been observed. To obtain the observed distribution salinity was used in a Michaelis-Menten function:

$$\text{miSAL} = \text{SALT(I)} / (\text{SALT(I)} + \text{SALTM}),$$

with SALTM being the salinity where miSAL $= 0.5$. This equation is coupled to the assimilation rate of the flagellates, thus reducing assimilation at low salinities.

6.2.2 Irradiance and Photosynthesis

In the estuary a characteristic distribution of suspended sediment (SILT) is observed which forms the direct cause for the steep gradient in the turbidity expressed as EPS values (cf. Colijn, 1982). To calculate the EPS values (the light extinction coefficient m^{-1}) in every compartment, an empirical equation is derived based on the attenuation of seawater, containing a small amount of dissolved humic compounds (Laane, 1981), silt and phytoplankton. The latter term is an implicit feedback mechanism under bloom conditions (self-shading). This empirical formulation, modified after Riley (1957) and Colijn (1982) is:

$$EPS(I) = 0.4 + ZZSILT \cdot SILT(I) + 0.293 \cdot 10^{-3} \cdot PHYT(I),$$

where PHYT contains both groups of phytoplankton and the suspended benthic diatoms. ZZSILT is a specific silt attenuation coefficient.

The attenuation of irradiance defines the depth of the photic layer (DEPFOT), which can be calculated according to the Lambert-Beer law, knowing the daily irradiance at the surface (LIGHT), a daily/hourly minimal irradiance above which maximal photosynthesis occurs in the upper water layer (LIGHTMIN) and the light attenuation (EPS) value:

$$DEPFOT = LOG\left[\frac{LIGHT}{LIGHTMIN \cdot SUNQ}\right] \cdot \frac{1}{EPS(I)},$$

where SUNQ is the number of daylight hours. The assumptions to use such a simple equation to model the photosynthesis in the water are the following. The high turbidity in the estuary results in a thin photic layer, and the high mixing rate in the estuary causes a rapid cycling of phytoplankton cells between the illuminated and the dark layer. So the water column is divided into two parts, an illuminated layer with a phytoplankton population which photosynthesizes at a maximal rate and a dark layer where no photosynthesis but only respiratory processes occur. The argument for this setup is given in Fig. 6.2. which shows a characteristic photosynthesis vs depth profile, and the geometric construction used to cal-

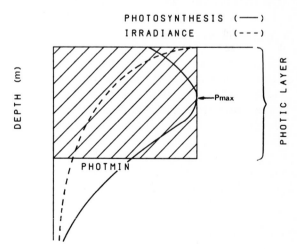

Fig. 6.2. Characteristic photosynthesis vs depth profile

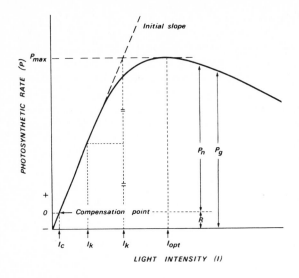

Fig. 6.3. Diagrammatic photosynthesis vs light relationship. (*Pmax*, photosynthetic maximum; *Ic*, light intensity at the compensation point; *R*, respiration; *Pn*, net photosynthesis; *Pg*, gross photosynthesis; *Iopt*, light intensity at Pmax; *Ik*, see text (Parsons et al., 1977)

culate the photosynthesis per m² (the shaded area below DEPFOT equals the shaded area above DEPFOT). In such a way a two-layered system is created with the above characteristics. The concept that we used compares well with those from Talling (1971) and Vollenweider (1965). The assumed DEPFOT in our model is equal to the depth where 0.5 PMAX occurs, at an irradiance of 0.5 I_k (Fig. 6.3). Values for I_k were derived from measurements in an incubator (Colijn, 1983).

The gross photosynthesis in the model, which is defined as the maximal rate in the illuminated layer, is reduced in several ways to obtain an average value for the water column and for the water mass above the tidal flats during submersion (primary production per m²). The correction factors for water volume above the tidal flats and time of submersion are included in the factor CORR (p. 38), and hence also incorporate the additional primary production in the water above the tidal flats (Colijn, 1983).

6.2.3 Nutrients

Both the nutrients phosphate and silicate are used to limit the assimilation rate of the phytoplankton. A Michaelis-Menten equation was derived with Ks values for the uptake or growth of relevant diatom and flagellate species as compiled in the literature (Morris, 1980). Nutrient limitation results in a reduction of the rate of conversion of intermediate photosynthetic products into phytocarbon, which enhances the excretion of these intermediates as PLOC and also the respiration rate. A direct effect of nutrient limitation is the decrease of the assimilation rates.

The nutrient concentrations in the estuary are far beyond the Ks values except in spring during the diatom bloom when silicate concentrations fall below 1 μmol·l^{-1} in the outer compartment (Colijn, 1983). Phosphate shows a similar decrease in concentration during the spring bloom.

Nutrient concentrations in the estuary are determined by input from the rivers and regeneration from sediments, but silicate regeneration from sediments is a relatively slow process (Van Bennekom et al., 1974); regeneration of phosphate, on the other hand, is probably much faster and instantaneous replenishment from phosphate adsorbed on silt also could occur in the water phase.

Low values of nutrients continue to occur up to August when mineralization and replenishment raise the concentrations to much higher values again. In the model we account for the net uptake of nutrients during photosynthesis, but not for any kind of nutrient regeneration. In the case of silicate the uptake is negatively correlated with temperature (Durbin, 1977); the overall fluxes are:

phosphate→flagellates: flKF = MAX (0,ppF – rsF) · PO4CO;
phosphate→diatoms: flKG = MAX (0,ppG – rsG) · PO4CO;
silicate→diatoms: flLG = MAX (0,ppG – rsG) · SILCO ·
$$[1 - 0,033 \cdot \text{TEMP(I)}];$$

in which PO4CO and SILCO are conversion factors for the incorporation of carbon in combination with phosphate and silicate respectively.

6.2.4 Assimilation

The effect of temperature on assimilation is different in both groups of phytoplankton. This is realized by first calculating an uncorrected temperature response (rtemp●) for both groups:

$$\text{rtemp●} = 2^{\text{TEMP(I)}-12} \cdot \text{q10P●},$$

where ● is the code for each of the groups. q10P● has a value of 0.15 for diatoms and of 0.2 for flagellates. This results in a Q_{10} of 2.8 for diatoms and a Q_{10} of 4 for flagellates. Because rtemp● is 1 at 12 °C, this difference in Q_{10} between both groups results in the diatoms performing better at low temperatures and the flagellates performing better at high temperatures (Fig. 6.3 a). These uncorrected temperature responses are then flattened at high temperatures. The corrected temperature response of both groups then is expressed by:

$$\text{RTP●} = \frac{\text{RTMP●} \cdot \text{rtemp●}}{(\text{RTMP●} + \text{rtemp●} - 1)}.$$

The assimilation rates of phytoplankton are dependent on the biomass, the maximal photosynthetic rates, the temperature corrections and the volume and light reduction factor (CORR). In the assimilation rate of the flagellates also the salinity influence (miSAL) is incorporated. Both equations are given below:

asF = PFLAG(I) · PMAXPH · RTPH · miSAL · CORR;
asG = PDIA(I) · PMAXPD · RTPD · CORR.

Data for the maximal photosynthetic rates were obtained from Colijn (1983) after conversion from chlorophyll-a to carbon-based rates. For this purpose a conversion factor of 35 has been used both for diatoms and flagellates. The comparison of the photosynthetic rates is hindered by the different definitions of these rates:

for the model the PMAX rates are potential under 24-h continuous illumination, whereas the rates normally defined as Pmax rates are short-term measurements in an incubator. These measurements at light saturation (Colijn, 1983) have shown that a wide range of Pmax values can be found, but that generally flagellates show somewhat higher values. The use of a constant C to chlorophyll-a ratio for conversion, however, obscures the possible variation in P/B max ratios for different algal groups (Smith et al., 1983). These authors studied the role of photoadaptation during a spring bloom: the increasing irradiance leads to an increase of both Pmax and the initial slope of the P-I curve, which results in a higher value for the irradiance at which saturation occurs. During this study (Smith et al., 1983) the rise in temperature was negligible so that all changes could be ascribed to the increasing irradiance.

6.2.5 Loss Terms

Loss of newly produced carbon occurs in three ways: excretion of dissolved organic carbon going into the PLOC pool, death of cells at limiting nutrient concentrations, resulting in both PLOC and PDET and respiration of assimilated carbon. The formation of excretion products, which occurs under natural conditions at a low rate (Mague et al., 1980) and under stress conditions, in particular nutrient limitation, at an enhanced rate (Lancelot, 1983) is described in the following (for flagellates and diatoms respectively):

$$LOCF = asF \cdot pFDET \cdot (1 - miPO4);$$
$$LOCG = asG \cdot pGDET \cdot (1 - miNU).$$

The LOCG and LOCF are apportioned to both PLOC and PDET according to

$$flFC = caFC \cdot LOCF;$$
$$flFD = LOCF - flFC + asF \cdot pFD \cdot miPO4;$$

and for the diatoms according to:

$$flGC = caCG \cdot LOCG;$$
$$flGD = LOCG - flGC + asG \cdot pGD \cdot miNU.$$

Both terms in the equations containing the assimilation rates are the normal activity excretion rates; all other terms are stress excretion rates. We assume that the most limiting of the two nutrients regulates the assimilation and nutrient stress excretion rates for the diatoms, hence the equation for miNU is as follows:

$$miNU = MIN (miPO4, miSILIC),$$

which results in miNU being assigned the lowest value of the two nutrient limitation values, miPO4 and miSILIC.

Now we can describe the gross primary production for the diatoms and flagellates, which are denoted ppG and ppF respectively:

$$ppG = asG \cdot miNU - flGC - flGD;$$
$$ppF = asF \cdot miPO4 - flFC - flFD.$$

To obtain net photosynthetic rates, respiration has to be included. This process contains three terms: a basal respiration rate, used for maintenance of the cells, which is temperature-dependent and defined at $0°$ C, a second respiration term which is proportional to the assimilation rates and a third respiration term which is related to nutrient stress. The following describes the three types successively:

rest respiration:	rrsF	= PFLAG(I) · RTPH · rs0F;
	rrsG	= PDIA(I) · RTPD · rs0G;
activity respiration:	arsF	= pRSF · asF · miPO4;
	arsG	= pRSG · asG · miNU;
stress respiration:	srsF	= LOCF · pRSFA;
	srsG	= LOCG · pRSGA.

The total respiration for each group is found by summing these terms:

$$rsF = rrsF + arsF + srsF;$$
$$rsG = rrsG + arsG + srsG.$$

Based on the net primary production rates (ppF – rsF and ppG – rsG) the nutrient fluxes given earlier could be calculated.

Three other loss terms not related to the physiological processes of the phytoplankton occur: grazing by microzooplankton (PMIC), mesozooplankton (PCOP) and by benthic suspension feeders (BSF). These fluxes are for diatoms and flagellates respectively:

flGE, flGH and flG7 and flFE, flFH and flF7.

Finally, sedimentation of diatoms on the tidal flats may occur; this results in the formation of BLOC and BDET. The total sedimentation is described by:

$$R = SEDPDQ · PDIA(I) · RTPD · pWAFL(I) · ptWET(I),$$

where SEDPDQ is the sedimentation rate, RTPD is the temperature coefficient and pWAFL is the fraction of the total water volume above the tidal flats. The flux of dissolved carbon to BLOC is given by:

flG8 = R · caGC and to BDET by flG2 = R – flG8.

All fluxes departing from the phytoplankton have now been described. A few remarks, however, have to be made: the model also generates the primary production per m^2 which can be compared directly with measured data (Colijn, 1983), it allows a choice in the way the nutrient limitation works and it calculates the oxygen flux, as a check on the carbon flux. The primary production per m^2 is calculated as:

$$MPPP(I) = (ppF + ppG) · depca(I).$$

This formulation only holds for the channels where the majority of the primary production measurements have been performed (Colijn, 1983). These measurements will be used in the validation of the model.

Under nutrient limitation of two nutrients the problem arises how to deal with this phenomenon: several equations are possible, depending on the assumptions one makes. In our model we used Liebig's law of the most limiting nutrient; this means that only one nutrient limits the assimilation rate.

Other possibilities are to use the products of miPO4 and miSILIC, the sum of both limitations divided by two or the root from the product. However, we do not have any cogent reasons to prefer one equation above the others. Moreover, mathematically the differences to the other equations are quite small (Ebenhöh, personal communication).

To be able to calculate the oxygen flux we introduced PQ and RQ values for photosynthetic and respiratory processes. These variables are named corWAS, corCOp and corCOr (Sect. 6.5.2 and Appendix A-2).

6.3 Zooplankton

The zooplankton part of the pelagic submodel covers a wide size range of organisms, ranging from unicellular microplankton of 20–200 µm in Equivalent Spherical Diameter to cm large jellyfishes. This widely different group of organisms is divided into three state variables ranging from small to large:
1. Microzooplankton, PMIC, containing all heterotrophic organisms in the size class from 20–200 µm;
2. Mesozooplankton, PCOP, consisting of all planktonic herbivores and omnivores with sizes from 200–1000 µm. The major component of this group by numbers and biomass is formed by the calanoid copepods;
3. Carnivorous zooplankton, CARN, consisting of the carnivores over 1000 µm in size.

There is some overlap between microzooplankton and mesozooplankton in the way these state variables have been handled in our model. The naupliar stages of the calanoid copepods of sizes less than 200 µm have been included in the mesozooplankton instead of in the microzooplankton, where they would belong if we had composed our state variables strictly according to size.

This simplified way of defining the state variables is partly born from a lack of knowledge of the groups of organisms we try to model, which forces us to be content with a rough form of input/output analysis and leave it at that.

There is more to this simplicity, however, than sheer ignorance. For one thing, we have to keep in mind that we are dealing with a very heterogeneous biological variable here, encompassing numerous species, some of which (the copepods) have 12 life stages each. This variable constantly changes and adapts to the shift of the environment during the year and to the changes in the biotic variables around it. In trying to model this variable we must take into account the level of aggregation and what this implies. Our contention is that, though the autoecology of the component species and the interrelationships between them is exceedingly complex and impossible to model to any degree of realism, it is precisely the aggregation of all these separate species into one functional one that makes it possible to generalize and model to a fair degree of realism how this functional group performs in the whole system.

Tentative support for this contention may be found in Conover (1978) from his closing remarks in a review of food chain dynamics: "It is not at all clear, to me at least, why some of these simple relationships seem to work so well, particularly in view of the overwhelming complexity which seems to be present in biological systems. Perhaps there really are a few reassuring generalizations to be gleaned from all this detail." ... "It may be that some of us have difficulty seeing the ecosystem for the seaweeds or the copepods or whatever our area of specialty may be."

6.3.1 Microzooplankton

Microzooplankton (PMIC) is defined as heterotrophic planktonic organisms from 20–200 μm in size. It consists for a large part of protozoa (ciliates and zooflagellates) but also pelagic larval stages of meso- and macroplankton and of benthic organisms occur in this size class. In the model these larval stages are not included in the microzooplankton, but in the state variables for mesozooplankton (PCOP) and pelagic benthic larvae (PBLAR) respectively.

Microzooplankton is presumed to form a very important link between the primary producers and larger heterotrophs especially in the temperate regions, where the spring phytoplankton bloom peaks some time before the heterotrophs do (Sorokin, 1981). In systems where allochthonous organic matter is an important complement to local primary production, such as estuaries and coastal waters, they are of prime importance, too (Eriksson et al., 1977). A major role would be to act as very fine filterfeeders, grazing on unattached bacteria. They in turn will be grazed upon by coarse filterfeeders (other micro-planktonic species such as Tintinnidae, copepods, larval stages of benthic animals, benthic filterfeeders) thus in effect making bacteria available to coarse filterfeeders. This is presently known as the "microbial loop". Eriksson et al. (1977) estimated microzooplankton, having 15% of the total biomass of zooplankton in a coastal area off Sweden, to have 65% of the total energy requirement, thus dominating the energy flux through the zooplankton.

The contribution of the unarmored forms of both the zooflagellates and the ciliates to the energy flow through the system is difficult to quantify because these groups are rather averse to the usual fixation procedures. Therefore they have to be studied in intact water samples. Tumantzev (1979), as quoted in Sorokin (1981), found that though loricated ciliates formed 170 of 200 species of ciliates, the loricated forms only made up from 5–15% of the total ciliate biomass.

In view of the fact that tintinnids often are thought to be the major constituent of the marine ciliates possibly because their loricae withstand fixatives well, the results of Tumantsev certainly shift the perspective.

For the Ems estuary, no quantitative data on microzooplankton are available (cf. Admiraal and Venekamp, 1986). The methods neccessary to obtain quantitative data (see Sorokin, 1981, for details) entail immediate counting of fresh samples. Especially Russian scientists have done much work along these lines.

6.3.1.1 Food Sources

The microzooplankton is thought to consist of filterfeeders, feeding on bacteria and phytoplankton, both flagellates and diatoms. In addition they also consume a small fraction of pelagic detritus, labile organic carbon and a very small fraction of pelagic particulate refractory organic carbon (PROC). Because this is a large and diverse group, they also indulge in cannibalism. These food sources are all added together, to determine the total concentration of food available to the microzooplankton:

$$\begin{aligned} \text{FoodE} = {} & \text{pDE} \cdot \text{PLOC(I)} + [\text{pCE} \cdot \text{PDET(I)} + \text{pAE} \cdot \text{PROC(I)}] \cdot \text{pSUSP(I)} \\ & + \text{pGE} \cdot \text{PDIA(I)} + \text{pGE} \cdot \text{SUSBDIA} + \text{pFE} \cdot \text{PFLAG(I)} + \\ & \text{pEE} \cdot \text{PMIC(I)} + \text{pME} \cdot \text{PBAC(I)}. \end{aligned}$$

The p●E parameters are fractions, indicating the availability of the different food sources to the microzooplankton. pSUSP(I) is the fraction of PROC and PDET which is in the water column and thus available to PMIC.

With the next equations the relative rate of uptake from the amount available is calculated:

$$\begin{aligned} \text{UE} \;\; &= \text{rupE} \cdot \text{foodE} / (\text{foodE} + \text{ksmfdE}) \cdot \text{q10E} \cdot \text{PMIC(I)} \\ \text{pupE} &= \text{UE} / \text{foodE}. \end{aligned}$$

The relative uptake pupE from this concentration of food is, through Michaelis-Menten kinetics, dependent on the actual concentration foodE in comparison to a Ks value ksmfdE, which gives the food concentration where half the maximum rate of uptake rupE is realized. The relative uptake is temperature-dependent. It has a Q_{10} of 2.

Sorokin (1981) mentions Q_{10} values for temperatures from 4 °C to 20 °C of 2.3 to 3.5 for ciliates. Since a Q_{10} value for a whole community may be expected to be lower than the Q_{10} for individual species, because of the changes in species composition within the community during the season, a Q_{10} value of 2 is felt to be reasonable. The maximum relative food intake per day rupE, which is set to 2 varies from 0.5 to 8 with a coefficient of assimilation for growth (K_1) of 0.2 to 0.5 for ciliates Sorokin (1981). The assimilation efficiency for bacteria and algae has been estimated to lie between 30 and 60% (Kopylov, 1977 in Sorokin, 1981).

Since we have no information on the assimilation efficiencies of the additional food sources, we have chosen a conservative value of 0.2 for the assimilation efficiency of all the food sources. The microzooplankton has been modelled as a black box and hence the assimilation efficiency implicitly is the K_1, or the assimilation efficiency for growth. That this is the case will be clear from the treatment of respiration and excretion.

6.3.1.2 Respiration/Excretion

In the next equation both the standing stock or rest respiration and activity respiration are calculated and summed into the total respiration rsE:

$$\text{rsE} = \text{rs12E} \cdot \text{q10E} \cdot \text{PMIC(I)} + \text{uE} \cdot (1 - \text{effE}) \cdot (1 - \text{pexE}).$$

The rest respiration rsE is modified by the temperature dependence q10E, which conforms to a Q_{10} of 2. The relative rest respiration rate rs12E is defined to be 0.10 at 12 °C, decreasing to 0.05 at 2 °C and increasing to 0.2 at 20 °C. This high relative rest respiration implies a very high carbon turnover of the whole community: at 2 °C, the turnover time is 20 days and at 20 °C only 5 days. These high rates of turnover are still well within the range given in the literature for generation times. These range from 3 to 48 (Zaika and Pavlovskaya, 1970 in Sorokin, 1981). These values are for ciliates; larger microzooplankton such as rotifers are slower and have daily P/B rates of around 0.2 in summer (Sorokin, 1981).

The activity respiration is taken to be a fraction $(1 - effE)$ of the uptake (uE), multiplied by the fraction of the uptake that is not excreted $(1 - pexE)$. This looks a bit strange, but the reasoning is that the fraction of the uptake which is not assimilated into growth is either respired or excreted. The value of the parameter pexE, the fraction of the uptake excreted, is quite uncertain, but the setting of this parameter is decisive for turning the state variable PMIC either into an organic carbon respiration machine when pexE is low, or into a carbon recycling mechanism through excretion when pexE is high. The total carbon flux through PMIC is not affected by the setting of pexE, since pexE only influences the direction of the carbon flux not assimilated by PMIC. Indirectly, however, it will make quite a difference whether most of the uptake is simply respired to CO_2 or whether PMIC is recycling most of the uptake into labile organic carbon (LOC) and detritus (DET). LOC and DET form the substrate for pelagic bacteria, and pelagic bacteria in turn are the major food source for PMIC. The excretion is given by:

$$exE = uE \cdot (1 - effE) \cdot pexE + [(1 - zOX) \cdot morOXE + morE] \cdot PMIC(I)$$

This equation contains two terms. The first term defines the uptake-dependent excretion. The second term calculates the mortality of the microzooplankton due to oxygen limitation and also the natural mortality. Mortality has been included in the excretion equation because both excretion and mortality are carbon fluxes from the microzooplankton into detritus and LOC.

6.3.2 Mesozooplankton

The state variable PCOP represents some ten different species of calanoid copepods. There are four species of *Acartia*: *A. tonsa*, *A. bifilosa*, *A. clausi* and *A. discaudata*. Of these four, only *A. tonsa* and *A. bifilosa* are true estuarine species, tolerating salinities as low as 5. The other two *Acartia species* occur in densities of less than 1000 m^{-3} in the outer regions of the estuary. Other neritic copepods occurring more or less abundantly in the estuary are *Temora longicornis*, *Centropages hamatus*, *Paracalanus parvus* and *Pseudocalanus minutus*. The most abundant species in the estuary is *Eurytemora affinis*, which occurs abundantly up to salinities of 20 in spring and the whole year round in salinities below 5.

The seasonal distribution and abundance of the calanoid copepods in the Ems estuary is well known (Baretta, 1980, 1981) making the prospect of incorporating a detailed copepod part in the pelagic submodel very tempting. In view of the very limited importance of this group in the total carbon flux through the system, however, the detailed approach was abandoned and the whole group has been aggregated into one state variable: PCOP.

6.3.2.1 Feeding Modes and Food Sources

The modelling of this group incorporates the following assumptions: the whole group is omnivorous, has an opportunistic feeding strategy and is also cannibalistic.

Two different feeding modes have been distinguished in the model: filter feeding and raptorial feeding. The treatment of the feeding of mesozooplankton in the model is complicated because of the presence of these different feeding modes. First, the total amounts of food available for both feeding modes are calculated. Then, the potential gain in energy from either of the feeding modes determines what fraction of the population feeds in either mode. When we have obtained this apportioning of the population as to their feeding mode we calculate the food uptake in relation to the total food concentration, and prevailing temperature. At high densities of mesozooplankton, a density-dependent limitation due to double filtering starts to affect the food uptake.

The particle spectrum which is suitable for filter feeding is considered to consist of all the pelagic diatoms (PDIA) and a fraction of the total amount of flagellate phytoplankton (PFLAG) as well as a small fraction of pelagic detritus (PDET). A major component of the flagellate phytoplankton consists of the colony-forming *Phaeocystis pouchetii*, whose colonies are probably too large for filter feeding (cf. Weisse, 1983). The fraction (0.5) of the biomass of PFLAG suitable for filter feeding is a guess. Detritus consists of particles of a wide range of sizes and biological origin of which only a small fraction (0.01) is thought to be ingested while filter feeding. These food sources are quantified by the following equation:

$$\text{foodHf} = \text{pFH} \cdot \text{PFLAG(I)} + \text{PDIA(I)} + \text{SUSBDIA}$$
$$+ \text{pSUSP(I)} \cdot \text{pCH} \cdot \text{PDET(I)}.$$

This variable contains all edible particles of a size range thought to be most efficiently captured by filtering the water.

The other food variable, food suitable for raptorial feeding, contains the larger particles, most efficiently captured by raptorial feeding. This source is quantified in the next equation:

$$\text{foodHr} = \text{PMIC(I)} + \text{pHH} \cdot \text{PCOP(I)}.$$

About one-third of the standing stock of PCOP is in the size range suitable for cannibalism (nauplii + small copepodids) which is expressed in the term pHH · PCOP(I) (pHH = 0.333).

To determine what fraction of the mesozooplankton will be filter feeding and what fraction feeds raptorially, assuming the population to optimize its food uptake, we need to take into account what fraction of the different food sources can be assimilated. Therefore, the utilizable fractions of both foodHf and foodHr are calculated and summed:

$$\begin{aligned}
\text{utilHf} &= \text{effFH} \cdot \text{pFH} \cdot \text{PFLAG(I)} + \text{effGH} \cdot \text{PDIA(I)} \\
&\quad + \text{eff4H} \cdot \text{SUSBDIA} + \text{effCH} \cdot \text{pCH} \cdot \text{pSUSP(I)} \cdot \text{PDET(I)}; \\
\text{utilHr} &= \text{effEH} \cdot \text{PMIC(I)} + \text{effHH} \cdot \text{pHH} \cdot \text{PCOP(I)}.
\end{aligned}$$

The assumption underlying this approach is that the mesozooplankton can shift between filter feeding and raptorial feeding.

The fraction of the total population filter feeding is determined by the next equation:

pHf = utilHf / (utilHf + utilHr).

To keep the treatment of food uptake, ingestion, sloppy feeding and assimilation in both feeding modes as uniform as possible, raptorial feeding in the model is treated as large-scale filtering, which means that a much larger volume is "filtered" in this feeding mode than in filter feeding. How much larger this volume is, is not clear but combining literature data from Corner (1961), Paffenhöfer (1971) and Corner et al. (1972), we assume it to be from 30–100 times as large as in the filter feeding mode.

6.3.2.2 Food Uptake

The relative food uptake by zooplankton in the model is food-dependent, according to the following equation for filter feeding:

$$FUPT1f = rupH \cdot q10H \cdot MAX \left[0, \frac{foodHf - minfdHf}{ksmfdHf + foodHf - minfdHf} \right].$$

This is a Monod function, incorporating a lower threshold minfdHf below which no uptake occurs. Laboratory experiments, with *A. tonsa* and *E. affinis* showed this lower threshold to lie in the vicinity of 40 mgC·m^{-3} (Malschaert, unpublished data). Though it is by no means certain that such a threshold really exists in the field (Frost, 1972a, 1972b), other workers (Petipa et al., 1971; Frost, 1975; Gamble, 1978) have reported clear-cut threshold concentrations from 25–50 mgC·m^{-3}.

The whole question of the functional response of the copepods to changes in food concentrations and its effect on activity respiration is not very clear, certainly if one tries to model it at the community level. For this model, a generalized response has been incorporated, with a lower food threshold above which there is maximal filtration, resulting in increasing ingestion with food concentration, up to a maximum ingestion rate. When this maximal ingestion rate is reached the filtration rate is reduced at higher food concentrations, while keeping the ingestion rate maximal.

A complication arises because of the availability of different food sources, including the nauplii of their own group. Because large particles like these are seized raptorially, much less energy is expended in obtaining a full ration this way than by filtering small particles through their maxillary filter, especially at low densities of small particles. Conover (1981) suggests that alternating between raptorial and filter feeding at food concentrations near the lower threshold may explain the equivocal experimental results of studies of this part of the functional response curve.

In the model we calculate the relative food uptake in dependence on food concentration and temperature for raptorial feeding:

$$FUPT1r = rupH \cdot q10H \cdot MAX \left[0, \frac{foodHr}{ksmfdHr + foodHr} \right].$$

This equation has modifiers for the maximum rate of uptake (rupH) and for temperature (q10H). The only difference between raptorial and filter feeding here is that there is no lower threshold for raptorial feeding.

The temperature dependence, in the model with a Q_{10} of 2 increases weight-specific ingestion at higher temperatures and depresses it at low temperatures. This effect may not be apparent in field populations (Conover, 1981) because of changes in species composition with temperature.

From the total uptake we can determine the fractional volume filtered, by calculating what fraction of the food available is taken up:

FUPTOTf = FUPT1f · PCOP(I) · pHf;
FUPTOTr = FUPT1r · PCOP(I) · pHr.

When the relative rate of uptake is only dependent on food concentration and temperature, a combination of high biomass of grazers and high temperature might well result in food depletion. This does not happen because of the nature of zooplankton grazing. The filtering process results in a rarefaction of the remaining food particles, because these will be redistributed into the volume of water which has been filtered already. This double filtering effect has been formulated in the model as a correction of the fractional volume filtered to a realized fraction filtered:

$$TfVQ = 1 - e^{-(FUPTOTf/foodHf)};$$
$$TrVQ = 1 - e^{-(FUPTOTr/foodHr)}.$$

As long as the grazing rates are small, the corrected volume will be very close to the original volume, but as the grazing rates become larger these two values will diverge to such an extent that when the negative exponent becomes 1 (its maximum value when all available food is grazed) the corrected volume will be reduced to 0.63.

The same reasoning applies both to TfVQ and TrVQ, since both feeding modes deplete a fractional volume of a certain class of particles.

This mechanism provides a density-dependent feedback between the zooplankton and their food; here, competition for food between the copepods is expressed.

Not all the food, acquired through either filter feeding or raptorial feeding, is ingested. A loss of 10% is estimated to be incurred due to breakup of particles while feeding (sloppy feeding). In Coulter counter grazing experiments usually an increase in small particles is observed, partially due to this fragmentation process (Roman and Rublee, 1981; Paffenhöfer and Knowles, 1978). This loss is assumed to consist entirely of labile organic carbon and thus goes into the state variable LOC.

The assimilation efficiency of the food actually ingested depends on the food source. Generally, animal food is assimilated with a higher efficiency than plant food because its composition already is closer to the composition of the grazer (Conover, 1978). The amounts assimilated are calculated for each food source from the following equation:

u•H = eff•H · fl•H · puw,

where u●H is the amount of carbon assimilated from food source ● by meso-zooplankton (H), eff●H, is the assimilation efficiency for ● (ranging from 0.005 for detritus to 0.85 for mesozooplankton). Fl●H is the gross carbon flux from ● to mesozooplankton and puw is the ingested fraction of the total uptake. The total food assimilated thus is the sum of these u●H expressions.

6.3.2.3 Respiration

The respiration of the zooplankton is separated into two components: rest respiration and activity respiration, to accommodate the effect of widely varying food concentrations on the activity respiration. The filtering rate is inversely dependent on the food concentration. The energy expenditure to obtain the daily ration by filter feeding thus decreases at higher food concentrations. This energy expenditure is treated as forming the activity respiration. It is calculated by multiplying the fractional filtered volume RFVf with the energy loss (mgC per m^3 filtered) qrsHf and qrsHr as given in the two following equations:

$$arsHf = qrsHf \cdot RFVf.$$

The values of qrsHf are species-specific and vary from 29 $mgC \cdot m^{-3}$ for *A. tonsa* to 80 $mgC \cdot m^{-3}$ for *T. longicornis* (data recalculated from Jørgensen, 1966). Until we disaggregate the zooplankton model we will have to make do with a single value of qrsHf, which is set at 40 $mgC \cdot m^{-3}$. This value is only applicable to the filter-feeding mode. Obviously activity respiration is not confined to the filter-feeders. As will be clear from the next equation the activity respiration while feeding raptorially has also been related to the volume cleared RFVr:

$$arsHr = qrsHr \cdot RFVr.$$

Here however qrsHr, which has been set to 20 $mgC \cdot m^{-3}$, reflects the energy expenditure of swimming, which is taken to be much less than the energy needed for filtering. Though there are literature data on the energy expended on filtering (Jørgensen, 1966), data on activity respiration for raptorial feeding in omnivores are lacking.

The activity respiration is calculated from the fractional volume filtered, before the correction for double filtering is made. At high densities of zooplankton biomass when the difference between the corrected and the uncorrected volume becomes noticeable, this implies a density-dependent increase in respiratory loss.

The rest respiration of the zooplankton is temperature-dependent. Here, again it is debatable whether the Q_{10} for the community would tend to be flattened ($Q_{10} < 2$) due to successional adaptation over the season or that the tendency of the copepod community to consist of smaller individuals at high temperatures (MacLaren, 1963) would indicate a Q_{10} higher than 2. This is because the weight-specific respiration rate is inversely related to size (Corner, 1972). We have used a Q_{10} of 2 for the effect of temperature on the rest respiration.

The equation for rest respiration rate in the model is:

$$prrsH = rs12H \cdot q10H,$$

where rs12H, the respiratory constant is 0.0125 and q10H reflects the change in respiration rate due to temperature.

The rest respiration then is:

rrsH = prrsH · PCOP(I).

6.3.2.4 Excretion/Egestion

Faecal pellets are the most important source of organic material excreted by the mesozooplankton. Another potentially important source is the soluble products of metabolism. The principal soluble product appears to be ammonia (Parsons et al., 1977), excreted at seasonally varying rates ranging from 2% of body nitrogen in winter to 10% in spring (Butler et al., 1970). Since the N-cycle is not modelled, the excretion of organic carbon compounds by mesozooplankton other than in faecal pellets is ignored.

The calculation of total faecal pellet production necessitates calculating the contributions from all food sources.

ex●H = (1 − eff●H) · fl●H · puw,

where ex●H is the amount of faecal pellets from a food source ●. The term 1 − eff●H is the fraction which is not assimilated but passed as faecal material.

Summing the contributions we calculate the total amount. This amount is then apportioned over pelagic detritus (PDET) (0.5) and pelagic refractory organic carbon (PROC) (0.5).

6.3.2.5 Mortality

Even in the absence of external mortality mesozooplankton clearly will not live forever. Therefore, a temperature-dependent mortality is included in the model. The temperature dependence has the standard Q_{10} of 2 and the relative mortality rate at 12° is 2%. This results in a population with a turnover time of 100 days at 2 °C, 50 days at 12 °C and 25 days at 22 °C (cf. Allan et al., 1976).

The only mortality due to abiotic factors in the model is mortality due to low oxygen saturation. This mortality is zero at 100% oxygen saturation and increases to 50% at 10% oxygen saturation.

Since oxygen saturation in the estuary usually is over 90%, the daily mortality is less than 2%.

6.3.3 Carnivorous Zooplankton

The pelagic carnivores (CARN) are the only state variable that have a semblance to a real taxonomic grouping. They consist of the ctenophores which invade the Ems estuary in large numbers in spring and sometimes show other (lesser) abundance peaks in summer and/or autumn. The possible importance of this group was indicated by the work of Greve at the Biologische Anstalt Helgoland, where he showed the relative abundances of *Pleurobrachia pileus* and *Calanus helgolandicus* to be decisive in determining whether the system would be *Calanus-* or

Pleurobrachia-dominated. Moreover, it was thought that these carnivores were the major predators of the mesozooplankton (Greve, 1971; 1972). The cteno-phores "filter" the water through their extended tentacular nets searching for prey. The following sections deal with those facets of the carnivore model that deviate from the description given in Section 4.1.

This group has been modelled in a rather desultory fashion, everyone being more interested in other variables of the system. Because of the salinity limitation on the feeding activity, the group cannot grow at low salinities but only die. Diffusive transport processes continuously transport the planktonic carnivores across the seaward boundary into the system and the distribution over the compartments reflects the transport processes as much as it does the internal dynamics of this group.

6.3.3.1 Food and Feeding

The carnivores prey on a number of different food sources. These are quantified such as to allow feeding on these food sources to be opportunistic, in other words to concentrate on the most abundant food source:

$$fdI = PBLAR(I)^{xfdI} + PCOP(I)^{xfdI} + [0.5 \cdot PMIC(I)]^{xfdI};$$
$$foodI = fdI^{1/xfdI};$$

where foodI is the amount of food available to the carnivores and xfdI ($= 2$) is an exponent.

The arguments for this approach have been given in Section 4.1.5.

6.3.3.2 Uptake

The relative rate of uptake of the carnivores is dependent on their potential uptake rate predQ, which has a value of 10 in the model and is modified by a temperature dependency ztI and a salinity dependence zsI. These regulating factors are calculated according to:

$$zsI = MAX \{0, [SALT(I) - SALCAR] / [SALCAR + SALT(I)] \cdot 3\};$$
$$ztI = MAX \{0, TEMP(I) / TEMCAR \};$$

where SALCAR, with a value of 12, is the salinity below which only *Pleurobra-chia* were found that had lost their combs (Van der Veer, personal communication). ZsI ranges from 0 at salinities below 12 to 1.4 at a salinity of 32.

TEMCAR, which has a value of 6 °C, is the temperature above which the uptake is stimulated by temperature. Because of their fishing behaviour (drifting along with trailing tentacles) a biomass of 1 mgC at 2% carbon content relative to wet weight (this would mean 50 mg wet weight) can only sweep a fractional volume cVOLI m^{-3}. Therefore, the predation rate can be calculated as:

$$pVOLI = cVOLI \cdot CARN(I) \cdot ztI \cdot zsI.$$

From this fractional volume cleared by the carnivores the uptake can be calculated:

$$uI = pVOLI \cdot foodI \cdot ksmfdI / (ksmfdI + foodI),$$

where pVOLI is the fractional volume cleared from prey, uI is the daily uptake by carnivores and ksmfdI, with a value of 30 mgC·m^{-3}, is the density of prey where the carnivores realize half their potential uptake.

6.3.3.3 Excretion

Gross growth is calculated from the uptake uI, assuming that the uptake is assimilated into growth with an efficiency effI of 10% (Reeve et al., 1978):

$$GCARN = effI \cdot uI.$$

The remaining 90% of the uptake is excreted. Half of the excretion is labile organic carbon (PLOC), the other half is detritus (PDET). The implication of this treatment is, apart from the fact that it is a primitive way of modelling things, that excretion is coupled with feeding activity.

Implicitly, the excretion also contains the carbon that is lost to activity respiration:

$$exI = (1 - effI) \cdot uI.$$

6.3.3.4 Respiration

Respiration in this group is only present in the form of rest respiration. First, in the next equation the prevailing relative rest respiration rate is calculated:

$$RESCAR = rs0I \cdot e^{0.069*TEMP(I)},$$

where rs0I is the relative respiration rate at 0 °C (1%) and the other right-hand term is the influence of temperature on this rate. This formulation is equivalent to a Q_{10} of 2. Then this relative rate is applied to the standing stock of carnivores to obtain the rest respiration:

$$rsI = RESCAR \cdot CARN(I).$$

6.3.3.5 Mortality

The carnivores have a constant natural mortality, CARMOR, of 3% per day. Superimposed on this constant mortality there is a mortality which is oxygen-dependent (CARMOX). At 100% oxygen saturation (OXMIC=1) it is 0, but at 90% it adds already 2% to the constant mortality. At 50% oxygen saturation it results in an extra daily mortality of 12%:

$$flIC = CARN(I) \cdot [CARMOR + CARMOX \cdot (1 - OXMIC)].$$

6.4 Pelagic Bacteria

The modelling of the pelagic bacteria is based on the widely accepted assumption that the growth of bacterial populations is governed by the supply of organic materials suitable as carbon and energy sources. Hence, our main problem is defining what fraction of the various stocks of organic carbon in the estuary is subject to bacterial degradation. A subdivision of the organic carbon, based on biological availability, has been presented in Section 4.4, the bacterial utilization of these hypothetical fractions is described in Section 6.4.2.

The modelling of the losses in bacterial populations such as sedimentation and mortality are still fraught with problems and some of these processes have been modelled by trial and error, thereby reducing our possibilities for independent validation of the submodel. However, data on the growth rate of bacteria were obtained with the thymidine method after the model was constructed (Admiraal et al., 1985) and these can serve very well for validation purposes (Chap. 10).

6.4.1 Biomass

The state variable pelagic bacteria (PBAC) is defined as the biomass of the aerobic heterotrophic bacteria. Aerobic heterotrophs are by far the most important group of bacteria in the Ems estuary, despite considerable numbers of nitrifying (Helder, 1983) and sulphide-oxidizing bacteria (Schröder, unpublished). In the Dollard (compartments 1 and 2) the viable count of the latter group of bacteria sometimes exceeds those of the aerobic heterotrophic bacteria (Schröder and Van Es, 1980); the (re)suspension of sediments mixed with benthic bacteria may be the primary cause of the presence of sulphur bacteria in the water. A considerable part of the sulphide-oxidizing bacteria obtains energy both from the oxidation of reduced sulphur compounds and from organic substrates (Kuenen, 1975). Therefore these mixotrophs (capable of heterotrophic and autotrophic growth) and chemolithotrophic heterotrophs are considered as aerobic heterotrophic bacteria. Nitrifying bacteria oxidize large amounts of ammonium and nitrite to nitrate with a low yield of biomass (Fenchel and Blackburn, 1979; Helder and De Vries, 1983). Because of their small biomass production these nitrifying bacteria were neglected in the model.

The biomass of the total bacterial population in the water was determined by staining the bacteria with acridine orange and counting the cells by epifluorescence microscopy (cf. Hobbie et al., 1977). Because the mean biovolume of the cells can vary by more than an order of magnitude in the Ems estuary (Schröder and Kop, unpublished data) individual cells were classified in 13 size classes and the mean biovolume was calculated. The total biomass can be computed using a conversion factor from biovolume to biomass in the range of $82 \cdot 10^{-15}$ to $121 \cdot 10^{-15}$ gCμm^{-3} (Van Es and Meyer-Reil, 1982). Here a conversion factor of $100 \cdot 10^{-15}$ gCμm^{-3} is used.

6.4.2 Uptake

Pelagic bacteria assimilate dissolved and particulate organic matter present in the water. This material is differentiated into four classes: dissolved labile (LOC), particulate detrital (DET), dissolved refractory (DROC) and particulate refractory (ROC). These categories of organic carbon have been defined according to their degradation rate in the estuary (Sect. 4.4). The labile organic carbon (LOC), including small molecular substrates, is assumed to be degraded generally on the day of production. Indeed, small substrates such as amino acids and glucose show a turnover rate of maximally $0.7\,d^{-1}$ in the Ems estuary (cf. Admiraal et al., 1985). It is often assumed that the excretion by algae and the following consumption by bacteria has a time lag of a few hours only (Iturriaga and Hoppe, 1977). Therefore, the labile organic carbon (LOC) present and the newly produced LOC is assumed to be completely available to bacteria on the day of production according to the equation:

$$upDM = flFD + flGD + flED + PLOC(I) \cdot pDM,$$

where upDM is the total available labile carbon, and flFD, flGD and flED the daily excretion of the planktonic flagellates, diatoms and microzooplankton respectively. The factor pDM sets the availability of the PLOC(I) and is given a value of 1.

For the classes of organic compounds, other than LOC, only a fraction (pCM, pBM or pAM) is available to the bacteria each day.

For pelagic detritus:

$$upCM = pCM \cdot PDET(I) \cdot pSUSP(I), \text{ where } pCM = 0.01.$$

For dissolved refractory compounds:

$$upBM = pBM \cdot PDROC(I), \text{ where } pBM = 0.0005.$$

For particulate refractory compounds:

$$upAM = pAM \cdot PROC(I) \cdot pSUSP(I) \text{ where } pAM = 0.0005.$$

From the particulate classes (A and C) only the fraction pSUSP is suspended in the water column (Chap. 5).

The total amount of organic carbon available for uptake by the bacteria is:

$$upM = upAM + upBM + upCM + upDM.$$

The bacterial populations are not necessarily capable of mineralizing the total amount of available organic matter on the same day. The maximum rate of uptake (upMmax) depends on the bacterial biomass [PBAC(I)], the temperature, the oxygen concentration and a specific maximum uptake rate (rupM) according to the equation:

$$upMmax = rupM \cdot q10M \cdot zOXM \cdot PBAC(I),$$

where q10M is the temperature dependency and zOXM expresses the negative impact of low oxygen saturation of the water.

The maximum rate of the daily uptake (rupM) was estimated at 6.3 (at 12 °C and 100% oxygen saturation). This maximum value seems reasonable as pelagic bacteria can have high in situ growth rates and can double their biomass per day (Van Es and Meyer-Reil, 1982; Azam and Fuhrman, 1984; Hagström and Larsson, 1984). In the Ems estuary 0.5–1 doublings per day were measured at a temperature of 12 °C (Admiraal et al., 1985) and it is imaginable that the daily substrate uptake necessary to sustain this growth rate is slightly less than five times the standing stock of bacteria.

Temperature is assumed to affect the potential uptake rate of bacteria rather than the actual assimilation rate of available substrates. The temperature dependency is given by:

$$q10M = 2^{[TEMP(I)-12]/HTEMPM},$$

where the constant HTEMPM has a value of 10. The resulting temperature correction thus corresponds to a Q_{10} of 2. This Q_{10} is lower than normally observed in natural populations of bacteria incubated at different temperatures (Q_{10} between 2.2 and 4) (Meyer-Reil, 1977; Gocke, 1977). The use of a lower Q_{10} seems reasonable, however, because of the seasonal succession of mesophilic and psychrophilic populations such as Sieburth (1967) found in Narragansett Bay. For the benthic bacteria in the Ems estuary no such seasonal succession was demonstrated (Schröder and Van Es, 1980), but also an acclimatization of bacterial populations to the seasonal temperature regime would result in low Q_{10} values.

The potential uptake of organic substrates by bacteria (upMmax) is also dependent on the concentration of oxygen. During waste water discharges the oxygen saturation (OXMIC) can drop dramatically in compartment 1. The aerobic uptake then is limited, but this limitation is effective only at very low oxygen saturation values as has been expressed in the correction factor zOXM in the formula:

$$zOXM = OXMIC / (OXMIX + ksmoxM),$$

where OXMIC is the oxygen saturation value and the constant ksmoxM was given a low value of 0.01.

Finally, the actual uptake (uM) of the bacterial population is obtained by comparison of the maximum uptake rate (upMmax) and the available substrate (upM):

$$uM = MIN(upMmax, upM).$$

Thus, as long as the uptake capacity of the populations is not exceeded by the supply of the substrate, the latter controls the population growth. When the available substrate exceeds the potential uptake capacity of the bacteria, the four carbon sources are assimilated proportionally to their abundance, as modified by the relative availabilities.

6.4.3 Loss Terms

In microbial ecology large uncertainties and controversies exist with regards to in situ bacterial respiration, assimilation efficiency, mortality and excretion. Therefore, we have estimated the value of the various constants in the model and compared the resulting cell quota with published data.

In our model the equation for respiration (rsM) consists of two terms, rest respiration and activity respiration:

$$rsM = rs12M \cdot PBAC(I) \cdot q10M +$$
$$[1 - effM \cdot OXMIC - effMa \cdot (1 - OXMIC)] \cdot uM.$$

Here rs12M is the relative rest respiration at 12 °C, estimated at 0.2 d^{-1} and corrected for the prevailing temperature with q10M.

The second term of the equation for respiration describes the fraction of the substrate uptake (uM) that is respired. This fraction is expressed in the more usual complement: the assimilation efficiency, composed in this case of an efficiency (effM = 0.3) under aerobic conditions and a (lower) efficiency (effMa = 0.2) under anaerobic conditions. The latter efficiency represents the lower growth efficiency of facultative anaerobes, which might be present at very low oxygen saturation; also adaptation in bacterial metabolism caused by the varying oxygen concentration may result in a lower efficiency.

The equation for respiration (rsM) assumes that bacterial popula- tions growing at a rate of 0.5 d^{-1} in oxygen-saturated water convert their substrate into biomass with an efficiency of ca. 25%. This cell yield is in the range of the 7–37% as measured by Newell et al. (1981, 1983) and Linley et al. (1983) in long-term incubations of natural organic substrates and bacteria.

In support, Lancelot and Billen (1984) compared the direct assimilation of ^{14}C-labelled substrates with cellular synthesis, measured with the ^{3}H-methylthymidine method and indicated a growth efficiency of ca. 30%. Application of this procedure to the Ems estuary showed an efficiency of ca. 50% (Admiraal et al., 1985). These and even higher assimilation efficiencies have been measured in short-term in situ experiments following the uptake and respiration of ^{14}C-labelled substrates (cf. Hoppe, 1978; Williams, 1984; Sepers, 1979; Billen, 1984). However, our conservative estimate of assimilation efficiency (effM = 0.3) is based on two considerations. Firstly, the widely used incubation time of a few hours might lead to an underestimate of respiratory losses. Secondly, the published values of assimilation efficiency may be biased by the use of easily assimilable compounds such as glucose and amino acids.

Excretion of PBAC is not explicitly formulated, but is included in the mortality flMD, which leads to new labile organic carbon (PLOC):

$$flMD = morM \cdot q10M \cdot PBAC(I).$$

The relative death rate morM is set at 0.2 at 12 °C and is temperature-dependent. In the Scheldt estuary and in Belgian coastal waters Servais et al. (1985) observed a daily mortality of pelagic bacteria of 0.2 at 12 °C, which was temperature-dependent and had a Q_{10} between 1.5 and 2.3.

Sedimentation of particulates is important in the Ems estuary and supplies the benthic system with organic material produced or discharged in the pelagic (cf. Hargrave, 1984). Also, pelagic bacteria attached to suspended silt, organic particles or macroscopic aggregates can sediment from the water to the benthos. Authors working in different marine environments often have divergent ideas about the percentage of attached bacteria and the in situ activity of this fraction (Wiebe, 1984; Pomeroy, 1984; Azam and Fuhrman, 1984; Zimmerman, 1977; Kirchman and Mitchell, 1982; Kirchman et al., 1984). Due to these controversies and to our limited knowledge about attached bacteria in the Ems estuary we did not simulate them as a separate class or fraction. The net sedimentation of PBAC

(M3) is arbitrarily calculated according to:

$$M3 = sedM \cdot [1 + PBAC(I) / dsedM] \cdot PBAC(I) \cdot pWAFL(I) \\ \cdot ptWET(I),$$

where sedM is the relative daily sedimentation of PBAC, which is estimated at 0.22. A correction is made for the fraction of the total water volume above the tidal flats (pWAFL) and the fraction of the time there is water above the tidal flats (ptWET).

During waste water discharges we have sometimes observed flocculation and precipitation of organic waste material in compartment 1. This phenomenon is incorporated in the sedimentation equation by means of an enhancement of sedM. This speculative increase of sedimentation is density dependent and causes a doubling of sedM if PBAC equals dsedM. The constant dsedM is set on 250 mgC·m^{-3}, a bacterial biomass reached only during waste discharges in compartment 1 and occasionally in compartment 2 (Van Es, 1984; Schröder and Kop, unpublished).

The deposited bacteria are apportioned to benthic aerobic bacteria (BBAC) and benthic labile organic carbon (BLOC), implying an estimated mortality (pM8s) of 0.9. Hence, the flux to benthic labile organic carbon is:

$$flM8 = M3 \cdot pM8s,$$

and the flux to benthic bacteria is given by:

$$flM3 = M3 - flM8.$$

The daily production of the bacterial population is given by the substrate uptake minus respiration:

$$prodM = uM - rsM.$$

During 1983 we measured the bacterial production, using the ^{3}H-methylthymidine method of Fuhrman and Azam (1980, 1982). In Chapter 10 we discuss these data as they serve as independent validation data, having been obtained after the model was constructed.

6.5 Oxygen in the Water

Obviously, oxygen is a vital parameter in determining the suitability of an aquatic environment for aerobic organisms. In most marine environments the oxygen content of the water is relatively constant. But in estuarine waters the oxygen concentration may fluctuate greatly because of high rates of photosynthesis and mineralization, strong tidal currents and, especially in the Ems estuary, waste water discharges. The solubility of oxygen in water is low (some 10 mg·l^{-1}) and hence the oxygen input processes, i.e. photo- synthesis, reaeration and input of oxygen-rich water by transport must continuously balance the oxygen-consuming processes. Oxygen under-saturation affects the uptake, respiration and behaviour of many organisms. The aim of modelling the oxygen dynamics was twofold:
1. To trace the consequences of oxygen depletion on the biota; and

2. To obtain an independent check on the total carbon flux through the system by means of the total oxygen consumption and production. This is possible because all biological processes, even the anaerobic ones, directly or indirectly produce or consume oxygen.

6.5.1 Physical and Chemical Aspects

The state variable oxygen (OX) is expressed as g $O_2 \cdot m^{-3}$. Oxygen is a dissolved substance and thus is subject to diffusive and advective transport. The exchange of oxygen between air and water is also modified by diffusion and turbulence in the water column (Sect. 5.2). Net oxygen transport from the air to the water occurs when the oxygen concentration in the water is below the saturation value. The saturation concentration (OSAT) is dependent on temperature and salinity according to the next equation (Anonymus, 1964):

$$OSAT = [475 - 2.65 \cdot SALT(I)] / [33.5 + TEMP(I)].$$

The transport between air and water, called aeration, can be positive (reaeration) or negative (deaeration). The aeration rate of a water body, expressed per unit volume in the model, is dependent on the difference between the saturation concentration and the actual oxygen concentration, the oxygen exchange coefficient (REACON in $m \cdot d^{-1}$), the average depth of the water column [depth(I)] and on the temperature (zteOX):

$$AERATION = [OSAT - OX(I)] \cdot REACON \cdot zteOX / depth(I).$$

The temperature correction as given by Anonymus (1964) and O'Kane, (1980) is:

$$zteOX = e^{0.023*TEMP(I)}.$$

Wilson and MacLeod (1974) reviewed the numerous attempts to calculate the oxygen exchange coefficient from physical parameters such as current velocity, temperature, depth of the water column, slope of the water surface and a factor giving the energy dissipation during the passage of the water across the transect under consideration. Unfortunately, under natural conditions these equations yield only an order of magnitude of the coefficient (Wilson and MacLeod, 1974). A further complication in estuaries is that current velocity, mean depth and slope of the water surface change continuously with the tide. Moreover, very few data are available on current velocities above the tidal flats of the Ems estuary. For these reasons we used empirically determined coefficients on oxygen exchange (Anonymus, 1964). In a separate model the waste water discharge into the Dollard was modelled (Van Es and Ruardij, 1982). The authors tested the whole range of REACON values published for the Thames estuary (Anonymus, 1964) and found that a REACON value of 0.96 $m \cdot d^{-1}$ gave the closest match between model predictions and field observations on oxygen concentrations. This value is used in the present study for all compartments of the estuary.

6.5.2 Sources and Sinks of Oxygen

In respiration, pelagic organisms consume oxygen and organic carbon. The primary producers transform inorganic carbon into organic matter and produce oxygen at the same time.

In the model all pelagic processes are expressed in $mgC \cdot m^{-3} \cdot d^{-1}$. To calculate the corresponding rates of production and consumption of oxygen, conversion factors are determined for both processes.

For primary production (photosynthesis) it is assumed that carbo- hydrates are produced first. In this case the conversion factor from carbon to oxygen (cor-COp) is 0.00375 ($g\,O_2 \cdot mg\,C^{-1}$).

For respiration the assumption is that the cell material used for respiration is of a more reduced form (lipids, proteins) than the carbohydrates. Consequently, a smaller ratio (corCOr) has to be used within the range of 0.0026 (Gocke and Hoppe, 1977) to 0.0033 (Van Es, 1977). In this model the value 0.0033 is used. This factor corCOr is used for all respiratory processes, except for bacterial respiration in situations of low oxygen saturation. During periods of waste water discharge the oxygen saturation in compartment 1 is appreciably lowered and presumably the dissolved and particulate organic matter in the discharged water is in a more reduced state than phytoplankton organic material, due to anaerobic fermentation processes. Moreover, the anaerobic waste water contains some sulphide. The oxidation of the more reduced organic substrates and of this sulphide increases oxygen consumption in the water relative to the amount of carbon consumed. This is incorporated in the model by supposing that the major part of this material is decomposed by bacteria and hence by multiplying the bacterial respiration rsM with a factor corWAS. This factor is dimensionless and corrects for the differences in the composition of the substrate under low oxygen conditions:

$$corWAS = maxCO \cdot ksCO\,/\,(ksCO + OXMIC),$$

where maxCO (1.25) is the maximum value of the correction factor corWAS, OX-MIC is the actual relative oxygen saturation ($0 \leq OXMIC \leq 1$) and ksCO is 4.

The value of ksCO is set such that corWAS at full oxygen saturation has a value of 1, increasing to 1.25 at an oxygen saturation of 0. After computing corWAS all respiration processes (O2) and production processes are added together and then converted from mgC to gO_2 per m^3:

$$
\begin{aligned}
O2 &= rsE + rsF + rsG + rsH + rsI + rsM \cdot corWAS; \\
consOX &= O2 \cdot corCOr; \\
prodOX &= (asG + asF) \cdot corCOp.
\end{aligned}
$$

The consumption and production of oxygen is also computed in the two other submodels. These oxygen flows are computed in an analogous way and partly (benthic) or fully (epibenthic) added to the pelagic oxygen source variable [SOX(I)] (Sect. 7.5).

6.5.2.1 Biological Oxygen Consumption

The oxygen concentration and the oxygen consumption (BOC in $mg \cdot m^{-3} \cdot d^{-1}$) was measured in the water of the estuary during 19 surveys in 1975, 1976 and

1977. BOC was determined by incubating samples for 1 (BOC1) or 2 days (BOC2) in the dark at the in situ temperature according to Van Es (1977). The extended data set of BOC2 measurements led to the introduction in the model of a variable (BOC2) which was computed according to:

$$BOC2(I) = (rrsF + rrsG + rsE + rsH + rsM \cdot corWAS) \cdot corCOr/cBOC2,$$

where BOC2(I) is the biological oxygen demand in water incubated for 2 days and cBOC2 = 0.6. This relation was derived by comparing the results of 1- and 2-day incubations (Van Es and Ruardij, 1982). BOC2(I) as well as OX are used in analyzing the model results (Chap. 10).

7 The Construction of the Benthic Submodel

W. Admiraal, M. A. van Arkel, J. W. Baretta, F. Colijn, W. Ebenhöh,
V. N. de Jonge, A. Kop, P. Ruardij and H. G. J. Schröder†

7.1 Introduction

Studies of the marine benthos are usually hampered by special problems of sampling and harvesting of organisms or by difficulties in measuring chemical parameters. Analogously the simulation of the estuarine benthos offers a few tantalizing problems not encountered in simulating the plankton.

Firstly, the sediment shows pronounced vertical gradients. The illuminated top layer is only a few mm thick and the primary producers, mainly diatoms, are unlike the estuarine phytoplankton capable of positioning themselves in the light gradient by vertical migration (Sect. 7.2). Microbial processes in deeper sediment layers are characterized by the reduction of sulphate instead of oxygen consumption (Sect. 7.5). Consequently, the reader will be confronted in this chapter with state variables and processes specified separately for aerobic and anaerobic conditions.

Secondly, the metabolism in the deeper parts of the sediment is completely dependent on the transfer of organic material from the surface; the nature of this transport, by molecular diffusion, by bioturbation or by storm-induced mixing is not well understood. The downward flux of oxygen, the upward flux of free sulphide and sediment accretion pose further problems that are treated in Section 7.6. This section also describes the simulation of the sulphide horizon, which varies continuously, equilibrating variations in the supply of organic matter to the sediment and the consequent changes in mineralization.

Thirdly, the food relations in the benthos are not simply a repetition of those in the plankton. The food particles in the benthos are dispersed among enormous numbers of inedible particles and dissolved organic nutrients may easily absorb on the particle surface. The feeding of deposit-feeding macrofauna (Sect. 7.4) and meiofauna (Sect. 7.3) was carefully considered in this respect. The degradation of detritus by microorganisms in the bottom extends over long periods of time so that the kinetics of the microbial activity have to be defined with some precision.

Figure 7.1 depicts the carbon flow in the benthic submodel, also showing the numerous interactions with the pelagic. The benthic diatoms occupy a key position due to their autotrophic production of organic carbon. The simulated benthic system is only 30 cm thick so that an open boundary exists with deeper sediment layers. The sediment accretion in the Ems estuary (see Chap. 2) therefore results in a steady loss to burial of refractive organic carbon and pyrite, the inert end product of sulphate reduction.

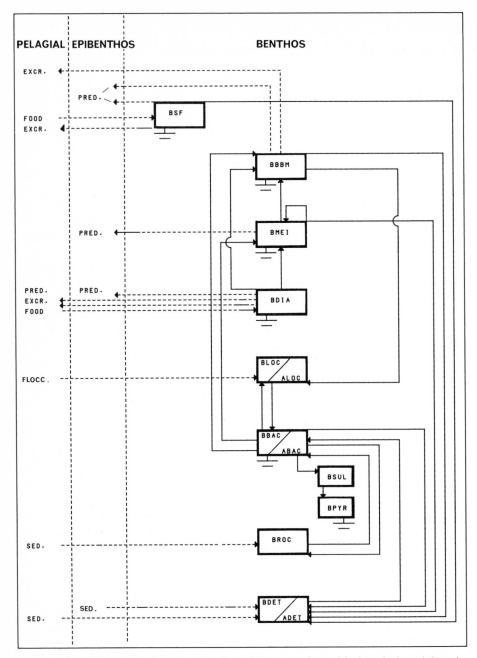

Fig. 7.1. Carbon flow in the benthic submodel and its connections with the pelagic and the epibenthos. Anaerobic state variables are lumped with their aerobic counterparts (e.g. ABAC/BBAC). Transport between anaerobic and aerobic layers is not shown

7.2 Benthic Primary Producers

Benthic microalgae have not been subjected to the same intensive modelling effort as their planktonic relatives. The integration and calculation of phytoplankton production and the use of models for light- and nutrient-limited growth (cf. Sect. 6.2) are well-known aspects of phytoplankton research that have not yet found their application for the microphytobenthos.

The definition of benthic primary production is subject to discussion and its measurement is fraught with technical problems. Especially for the benthic microphytes the upward and downward fluxes of CO_2 and O_2 pose severe problems (for technical details see Colijn and De Jonge, 1984; Ludden et al., 1985). Therefore, this chapter describes the construction of the submodel microphytobenthos supported only by one previous model analyzing the flow of carbon and oxygen in microphytobenthos (Ludden et al., 1985).

The part of the model describing the microphytobenthos must represent a variety of microphyte species. Merely the cell volume of the various species extends over at least four orders of magnitude (from ca. 10 to ca. 100,000 μm^3). Also the migratory and "mechanical" properties of the dominating diatoms vary widely. In addition to diatoms we found in compartments 1 and 5 summer blooms of cyanophytes. The diversity of microphytic organisms in the estuary and the succession of species through the seasons hampers the choice of universally applicable parameters for the growth rate, vertical migration, etc. of estuarine microphytobenthos. The present model refers mainly to assemblages of motile, intermediately sized diatoms; this is the dominant component of the microphytobenthos in the Ems estuary.

7.2.1 Vertical Distribution of Biomass

Benthic microalgae in estuarine sediments are not restricted to the upper illuminated layer and this phenomenon is of critical importance in simulating their photosynthesis. The layer of sediment, reducing the irradiance and consequently the photosynthesis of the algae to insignificant levels, is usually less than 3 mm thick (Colijn, 1982). Living diatoms were found in deeper layers of the sediment even when these are anaerobic (for references see Admiraal, 1984). The growth of microalgal populations in the thin illuminated top layer of the sediment is counteracted by physical disturbance of the sediment, caused by e.g. tidal currents and waves or by the activity of macrobenthic organisms. Tidal currents are also responsible for the suspension of benthic diatoms as tychoplankton (De Jonge, 1985; De Jonge and van den Bergs, 1987). Diatom species show various strategies to cope with these conditions; some species (belonging to the so-called epipsammon) stick firmly to sand grains thereby avoiding suspension, but they are unable to migrate to the sediment surface once they are buried in deeper sediment layers. Other species (belonging to the so-called epipelon) adhere more loosely to the sediment, and protect themselves by vertical migration up to and away from the photic layer of the sediment. Figure 7.2 shows that the large and intermediately sized cells of the epipelon are abundant on the silty sediments of the compartments 1 and 4 and are accompanied in compartment 5 by large numbers of small

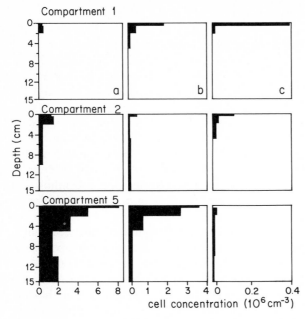

Fig. 7.2. Examples of distribution patterns of benthic diatoms in sediments as observed by epifluorescence microscopy. Populations are divided into three categories: a small non-motile cells adhering to particles (mostly < 10 μm large); b 10–50 μm large motile cells; c > 50 μm large motile cells. Compartements 1 and 2: Oct. 197; compartment 5: April 1978

cells adhering to sand grains. In the sandy sediments of compartment 5 microphytes were dispersed deeply into the sediment in contrast to the superficial occurrence of microphytes on the anoxic silty sediments of compartment 1. The vertical distribution of microphytobenthos in the sediment shows large variations, probably due to the interaction of production of the microphytes, consumption by herbivores and mixing of sediments. These interactions are not fully understood and hence are difficult to incorporate into a dynamic benthic model with only two layers. Therefore we have assumed a standard distribution pattern of microphytes for each of the compartments 1 to 5. The vertical distribution of biomass at depth d is described by the equation:

$$f(d) = f(0) \frac{CBO^2}{CBO^2 + d^2},$$

where CBO is a constant, depending on the compartment, and f(0) is the biomass at the surface (Fig. 7.3). The values of CBO for the various compartments have been chosen in order to obtain a reasonable fit between simulated and observed patterns of vertical distribution (Figs. 7.2 and 7.3). They vary between 0.2 cm in compartment 1 and 1 cm in the outer compartments. The biomass per m² between the surface and a depth d (XDIA) is determined by an integral over the function for distribution with the result:

$$XDIA = BDIA(I) \cdot arctan(d/CBO) \cdot 2/\pi,$$

where BDIA is the total biomass in the top 30 cm of the sediment. This integration is executed for various values of d, representing the different layers and results in the following biomass values:

FBDIA: biomass in illuminated top layer (d = LAL);

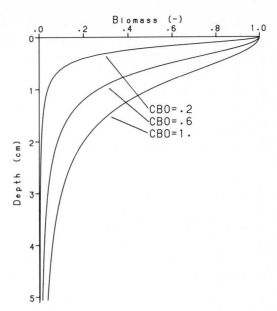

Fig. 7.3. Vertical distribution of
diatom biomass in sediments as
calculated in the model for three
values of the parameter CBO

CDIA: biomass in top 0.5 cm layer (d = 0.5), used for comparison with field ob-
        servation on a similar layer;
UDIA: biomass in the "aerobic" layer of the sediment (d = BAL);
LDIA: biomass in the "anaerobic" layer of the sediment (LDIA = BDIA-
        UDIA).

The actual depth distribution of diatoms varies over the seasons. In winter, when
the biomass values are low, sediment reworking by macrofauna is at a low level
and the light penetration is minimal, the diatoms tend to a more superficial distribution. This effect has been introduced by an increase of CBO at low values
of BDIA:

$$CBO = CBOF(I) \cdot \frac{BDIA(I)^2}{BDIA(I)^2 + KCBO^2} .$$

Without such a dynamic CBO the ratio of illuminated biomass (FBDIA) to the
total biomass (BDIA) would attain too low values in winter, and a severe decima-
tion of benthic microphytes in winter, not observed in the estuary, would be the
consequence. On the contrary, minor winter blooms of diatoms were occassion-
ally seen on the tidal flats.

7.2.2 Effects of Light

The response of microphytobenthos to light is simulated essentially in the same
way as for phytoplankton. The benthic as well as the planktonic populations have
each been divided into a photic and an aphotic part: the photic part carries out
light-saturated photosynthesis and the aphotic part does not show photosynthesis. The light-saturated photosynthesis of microphytobenthos has been modelled

in a slightly different way from that of the phytoplankton. The estuarine phytoplankton is taken to be completely mixed over the illuminated and dark layers of the water; hence, the light absorption by the cells is assumed to proceed linearly over the light period. However, a number of microphytobenthos species can change their position in the light gradient. Because they do this only very slowly and over a short distance, part of the population receives excess light and another part is not illuminated at all. In this case the photosynthesis does not proceed linearly with time, as has also been shown in phytoplankton incubated for 5–10 h in the light (Lancelot, 1984). For phytobenthos, short light periods of a few hours were sufficient for a high specific growth rate and longer illumination periods were no longer effective (Admiraal and Peletier, 1980b; Admiraal et al., 1982). The minimum daily irradiance at which the microphytobenthos shows at least 50% of their light-saturated growth (Photmin) is derived from culture experiments (Admiraal, 1977b; Admiraal et al., 1984).

The depth (LALL) at which the irradiance Photmin is reached can be calculated from the incident radiation and the light extinction parameter Kd given for the sediments of the compartments 1 to 5 (Colijn, 1982). We assume that vertical migrations of benthic diatoms enable them to reach the illuminated top layer. We have modelled this phenomenon by adding a migration distance (LAD) to the photic layer. It seems logical that LAD is proportional to the length of the light period and proportional to the fraction of motile species (cf. Fig. 7.2) in the populations. In very sandy sediments the migration path can be up to a few mm (Round, 1979); it is probably less in silty sediments. On the basis of these considerations LAD is estimated at about $0.01 \text{ cm} \cdot (\text{h illumination})^{-1}$.

7.2.3 Specific Growth Rate and Primary Production

Relative growth rates of the microphytes in the Ems estuary, such as cell division rates and photosynthetic rates have been measured independently from production. These two sets of measurements have played a different role in modelling the microphytobenthos. The data on specific production and loss rates were derived from experiments on cultures and natural populations and were used to estimate the constants used in the model. Some observations on the in situ production of the microphytes (together with observations on the seasonal fluctuation in the biomass) were used in calibrating the model and its constants, but most field observations were used to validate the model (see Chap. 11).

The maximum rate of photosynthesis of illuminated benthic diatoms expressed per day and per unit biomass (RPP = 1 d^{-1}) is derived from a variety of measurements on cell division rates (in situ and in cultures; Admiraal and Peletier, 1980b; Admiraal et al., 1982, 1984; Admiraal, 1977a) and on photosynthetic rates of microphytobenthos (Admiraal et al., 1982; Admiraal, 1977b). The specific rate of productivity has been determined for a temperature of 12 °C and its value is modified for other temperatures with the factor zt4.

The potential photosynthesis of the benthic microphytes (PPR) is calculated according to the equation:

PPR = FBDIA · RPP · zt4,

where FBDIA is the biomass of the illuminated microphytobenthos.

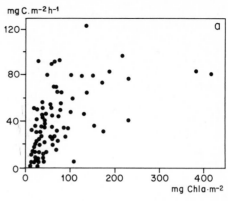

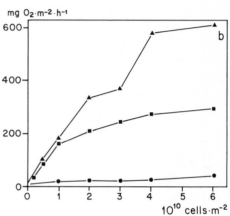

Fig. 7.4. Relation between photosynthetic rate and population density in natural (a) and cultured (b) populations of benthic diatoms. $6 \cdot 10^{10}$ cells equal about 150 mg chlorophyll-a (a). The effects of various concentrations of bicarbonate tested in b (Δ: 250 mg $NAHCO_3 \cdot 1^{-1}$; ■: 50 mg $NAHCO_3 \cdot 1^{-1}$; ●: 0 mg $NAHCO_3 \cdot 1^{-1}$) indicate carbon-limited photosynthesis in dense diatom mats. a Colijn and De Jonge (1984); b Admiraal et al. (1982)

It was found that in nutrient-rich mud flats such as in the Dollard the diffusion of the substrate for photosynthesis, bicarbonate is the rate-limiting step that dominates over similar effects of N, P and Si (Admiraal et al., 1982). Experiments on photosynthesis and multiplication in cultures and natural populations (Fig. 7.4) pointed to a similar "self-inhibition" of photosynthesis in natural and artificial populations. In the present model we introduced the self-inhibition of photosynthesis by a reduction factor, employing two constants, PPH and PPM. PPH is the maximum photosynthetic rate that an algal mat can attain. This rate is determined by the diffusion rate of inorganic carbon into and oxygen out of algal layers (Ludden et al., 1985). Several conditions such as the flow rate of the water film on top of the microphytobenthos affect the maximum rate of diffusion of these substances, so that the value of PPH can be somewhat variable. Field and laboratory observations show that ca. 120 $mgC \cdot m^{-2} \cdot h^{-1}$ or 1500 $mgC \cdot m^{-2} \cdot d^{-1}$ is a reasonable estimate.

The actual photosynthetic rate (PP) is calculated from the potential photosynthetic rate (PPR) using the formula:

$$PP = \frac{PPR + PPH + PPM}{2} - SQR \left\{ \left[\frac{(PPR + PPH + PPM)}{2} \right]^2 - PPH \cdot PPR \right\}$$

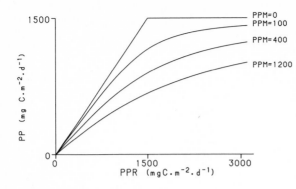

Fig. 7.5. Relation between the photosynthetic rate *PP* and the potential photosynthetic rate *PPR* for different values of the slope parameter *PPM*. In the model the value of *PPM* varies between 800 (under fully aerated conditions) and 1200 (under oxygen depletion)

Figure 7.4 shows how the value of the slope parameter PPM influences the primary production. A large value of PPM results in only a small part of the potential photosynthesis (PPR) being realized. PPM is made dependent on the oxygen saturation in the water (zox4) to mimic the inhibition of benthic photosynthesis by high concentrations of ammonia (Admiraal et al., 1984):

$$PPM = PPM0 + (1 - zox4) \cdot PPMox,$$

PPM0:   the slope parameter at 100% oxygen saturation;
PPMox: the maximum addition to PPM reached under oxygen depletion.

Under anoxic conditions the parameter PPM reaches a maximum of 1200 $mgC \cdot m^{-2} \cdot d^{-1}$; consequently, the photosynthetic rate (PP) increases only slowly with increasing potential rates (Fig. 7.5).

The primary production as it is measured in the field (MBPP) in the upper 0.5 cm of the sediment encompasses the actual photosynthesis minus the activity related respiration and minus the rest respiration in that layer (see definition of CDIA) according to the equation:

$$MBPP(I) = PP - ars4 \cdot PP - brs4 \cdot \frac{PPR - PP}{PPR} \cdot PP - rrs4 \cdot zt4 \cdot CDIA(I).$$

Activity respiration will be treated later, together with the other loss factors.

7.2.4 Excretion and Respiration

The loss factors excretion and respiration consist of three terms: one related to photosynthesis (activity losses), one related to the degree of self-limitation in populations (stress loss) and another fraction proportional to the biomass of the population (rest excretion or rest respiration).

The fractions of the photosynthesis that are excreted (aex4 = 0.10) and are respired (ars4 = 0.10) were chosen similarly to those of the phytoplankton on the basis of unpublished results.

Any divergence between actual (PP) and potential photosynthesis (PPR) indicates stress. The model assumes that such a stress leads to larger losses of photosynthetic products according to the equation:

$$stress\ respiration = brs4 \cdot PP \cdot \frac{PPR - PP}{PPR},$$

whereby brs4 defines arbitrarily the highest fraction (0.2) of the actual photosynthesis that can be respired. The stress excretion (bex4 = 0.2) is formulated similarly to the stress respiration.

The biomass related terms of excretion (rex4 = 0.02 including mortality) and respiration (rrs4 = 0.01) were chosen in view of the large stocks of photosynthetically inactive cells in the sediment. The low value of rest respiration (0.01) is inferred from dark incubations of diatoms, showing 50% survival after 20–80 days (Admiraal and Peletier, 1979).

Several processes contribute in the biomass-related excretion: the formation of slime trails or mucilage pads by mobile cells and the mortality in populations. The true excretion of organic matter by the large stocks of buried diatoms is probably low as was indicated by the survival of long dark periods. Furthermore, it is shown that benthic diatoms are effective scavengers of small organic molecules (Admiraal, 1984) and this may counterbalance their excretion. An average mortality in populations is extremely difficult to assess. Observations and experimental analysis of the seasonal succession patterns in natural populations indicated that mortality occurs in restricted periods of time (Admiraal et al., 1984). The triggering conditions may be e.g. desiccation, high temperatures, high sulphide concentrations, frost or fungal infections (Admiraal, 1984). We estimated an average mortality rate rex4 = 0.02.

The excretion and mortality of benthic diatoms is assumed to lead for 70% to slowly decaying detritus (P42 = 0.7), the rest of the products constitute labile organic carbon. This subdivision is supported by culture experiments with ^{14}C-labelled diatoms that excreted roughly 70% macromolecules (MW > 1800) and 30% as small molecules (MW < 1800) as measured by gel filtration (unpublished results).

7.2.5 Suspension

Significant numbers of benthic diatoms are usually found in estuarine plankton. De Jonge (1985) estimated for the Ems estuary that the numbers of suspended benthic diatoms ranged from 16–50% of the numbers present in the benthos. These percentages depended strongly on the degree of exposure of the mud flats to wave action and on the occurrence of storms. For the present model we assumed that 10% of the benthic microphytes are temporarily suspended. The somewhat higher percentages of 16–50% of benthic diatoms observed in the plankton were assumed to be metabolically identical to phytoplankton diatoms. Indeed, suspended benthic microphytes were able to carry out photosynthesis (unpublished results), but wind-induced suspension of benthic microphytobenthos also increases the silt content of the water, thereby reducing the photosynthesis of the suspended algae. As a simplification we assumed that the 10% of suspended benthic microphytes were metabolically inactive.

Suspended microphytobenthos was subject to planktonic grazing and filter feeding by macrobenthos. Filter-feeding macrobenthosis was assumed to prefer suspended microphytobenthos over phytoplankton, since the suspended benthic algae are concentrated in the water layer covering the tidal flats inhabited by the filterfeeders.

7.3 Meiofauna

Meiofauna is one of the organismal groups in the sea, which has often been treated as a "black box". Lee et al. (1975) were probably the first who tried to analyze the feeding relationships mathematically in the meio- and microbenthos. For the present purpose we treat the meiofauna (all Metazoa passing a sieve with a mesh size of 1 mm) as one compound state variable. As in many other areas the well developed meiofauna in the Ems estuary is dominated by nematodes, in numbers of species, individuals and in biomass (Bouwman, 1981). Second to the nematodes are the harpacticoid copepods and the oligochaetes (cf. Bouwman, 1983a). The latter group was occasionally abundant on the brackish mud flats in the inner estuary (Bouwman, et al., 1984). The meiofauna species of the Ems estuary have been characterized with respect to their feeding biology in a series of investigations (Romeyn and Bouwman, 1983; Romeyn et al., 1983; Admiraal et al., 1983; Bouwman, Hemmes, Jansen and De Jonge, unpublished; Bouwman 1983b). This work has shown that the three different feeding modes in the nematode population, bacteriovorous, herbivorous and carnivorous, occur in very variable proportions, both temporally and spatially. The relative abundance of the different types of food seems to be the decisive factor determining this proportion. The implicit adaptation of the meiofaunal species composition to its food has been included in the model (see Sect. 4.1).

7.3.1 Food Uptake

Available Food. Diatoms and bacteria are probably the main food source for nematodes (Romeyn and Bouwman, 1983; Montagua et al., 1983; Montagua, 1984), although truly carnivorous species (cf. Heip et al., 1978) also exist. Some species are also capable of absorbing ^{14}C-labelled organic substrates, such as glucose (cf. Lopez et al., 1979), but this type of feeding as well as the direct utilization of (bacteria-free) detritus were neglected in the model. Harpacticoid copepods also exploit a wide variety of food sources (Hicks and Coull, 1983). The oligochaete species in the Ems estuary could be divided into diatom consumers and detritus- or bacteria-feeding species (Bouwman et al., 1984). In summary, nematodes, harpacticoid copepods and oligochaetes all operate at similar trophic levels and thus may be modelled in the same way.

Though the concentration of potential food particles is nearly always very high, the problem for the selectively feeding meiofauna is to discriminate between edible and inedible particles and to select attractive microniches with high concentrations (Joint et al., 1982). The abundance of diatoms in the sediment of the Ems estuary led to the blooming of specialized herbivorous nematodes (Bouwman et al., 1984), indicating that the model also should allow for successional adaptation to the food spectrum.

The food uptake of the meiofauna is simulated in three steps: first the food concentrations are calculated, then the feeding rate is determined from the food concentration and finally the actual uptake rates are calculated.

The total food concentration is calculated from the concentrations of bacteria and diatoms in the upper layer (which is BAL cm thick), the concentration of meiofauna (BMEI) and the bacteria and diatoms in the lower (anoxic) layer as given by the following formula:

for aerobic bacteria: PW35   = [BBAC/BAL]^{POWM};
for aerobic diatoms: PW45   = [UDIA/BAL]^{POWM};
for meiofauna (cannibalism):   PW55   = [BMEI/BAL]^{POWM};
for anaerobic bacteria: PW35A = [ABAC/TAL-BAL(I)]^{POWM};
for anaerobic diatoms: PW45A = [LDIA/TAL-BAL(I)]^{POWM}.

The total food concentration in the benthos thus is:

$$PW5S = PW35 + PW45 + PW55 + PW35A + PW45A.$$

The division by BAL implies that the amount of available food (PW5S) is given in $mgC \cdot m^{-2} \cdot cm^{-1}$. This is a dimension, which tries to account for those situations where the upper layer (BAL) is smaller than 1 cm, enabling the meiobenthos to perceive more food in the small remaining layer.

Regulation of Food Uptake. There is considerable uncertainty on the regulation of in situ feeding and growth of meiofauna, since only recently field studies on this topic have been started (cf. Montagua, 1984; Cammen, unpublished). Experimental studies indicate suboptimal feeding at low food concentrations. Schiemer (1982) reports a threshold of $0.03\ mgC \cdot ml^{-1}$ of *E. coli* cells in agar for the growth of a nonmarine nematode species. Tenfold higher concentrations of bacteria were not yet saturating to another bacteriovorous nematode species from fresh water (Schiemer et al., 1980). The brackish water nematode *Eudiplogaster pararmatus* required more than $0.1\ mgC\ diatoms \cdot ml^{-1}$ to grow rapidly in agar cultures, whereas $0.05\ mgC\ diatoms \cdot ml^{-1}$ hardly sustained growth (Romeyn et al., 1983). Considerably lower concentrations of diatoms saturated the growth of harpacticoid copepods from the Ems estuary (Ustach, 1982; Bouwman, Hemmes, Jansen and De Jonge, unpublished).

These experimental studies prompted us to include food limitation in the model in addition to food saturation. Food limitation and food saturation levels have been included by two correction factors, respectively zFOOD5L and zFOOD5. A standard food concentration STANFO5 ($550\ mgC \cdot m^{-2} \cdot cm^{-1}$) is defined where 1 mgC meiofauna consumes upMEI (0.072) $mgC \cdot d^{-1}$. STANFO5 is defined as such because of experimental data (Romeyn et al., 1983) which could be translated into these dimensions, whereas we do not have data relating food concentrations in the sediment to the amount of sediment "cleared" by meiobenthos. Figure 7.5 shows that the resulting uptake rates at high food concentrations amount to ca. 0.2 mg per mg biomass $\cdot d^{-1}$ (at a temperature of 12 °C). Bause (1982) noticed that the intrinsic rate of growth of meiobenthic metazoans is low compared to their planktonic relatives. The rates compiled by Bause (1982) range from 0.08 to $0.3\ d^{-1}$, generally at a temperature of 20 °C. Our estimated maximum rate of food intake ($0.2\ d^{-1}$) agrees with the data given by Bause (1982), if we take into account both the lower temperature and the loss factors. The potential uptake rate (PupMEI) is given by:

$$PupMEI = FOOD5\ /\ STANFO5 \cdot upMEI.$$

Now saturation of feeding is given by:

$$zFOOD5 = 2 \cdot F5 / (PupMEI + F5),$$

where F5 ($0.075 \; mgC \cdot m^{-2} \cdot d^{-1}$) is a constant, determining at which level of Pup-MEI saturation will take place.

Food limitation is assumed to be determined by the energetic requirements in scavenging the sediment for edible particles. Therefore the function for food limitation (zFOOD5L) is calculated by comparison of the uptake rate (minus activity-dependent losses) with the maintenance energy. These equations again include the factor zFOOD5 and PupMEI, discussed earlier:

$$FOOD5X = zFOOD5 \cdot [PupMEI \cdot (1 - pex5 - prs5) - asWORK5],$$

where pex5 (0.2), prs5 (0.3) and asWORK5 (0.02) express the fraction of the uptake that is lost in excretion, respiration and energy loss through locomotion respectively.

The distribution of ingested food over the excretion and respiration as set by the parameters pex5, prs5 and asWORK5 conforms to data obtained by Marchant and Nicholas (1974) and Schiemer (1982) of grazing by fresh water nematodes on bacteria. Apart from activity related loss terms, there are also respiratory and excretory losses due to the standard metabolism:

$$rsex5r = rrs5 + rex5,$$

where rrs5 (0.008) and rex5 (0.005) express the daily fraction of the biomass that is respired and excreted respectively.

Finally, now that the potential gain from feeding activity as well as the potential energy losses from this feeding activity are known, the food concentration (FOOD5X), below which no net gain is to be expected from this energy expenditure, is calculated. Depending on the exponent POW5 (2), the function zFOOD5L not only defines the lower threshold where feeding commences, but also the extent of the transition zone to where it becomes maximal. High values of the exponent POW5 make it an on/off switch function, by reducing the transition zone.

It is somewhat illogical to include the standing stock metabolic losses in this function, since these losses are not dependent on the feeding activity. The effect is that meiobenthos in the model only start to feed at food concentrations which enable them to recoup all their metabolic losses:

$$zFOOD5L = FOOD5X^{POW5} / (FOOD5X^{POW5} + rsex5r^{POW5}).$$

Under conditions of low oxygen levels in the sediment, food uptake by the meiobenthos is decreased, and excretion increased by the oxygen stress function zox5.

The influence of temperature on food uptake is incorporated through the temperature correction function zt5.

Calculation of Feeding Rates. The actual food uptake of meiofauna (W5) is calculated by the formula:

$$W5 = BMEI \cdot PupMEI \cdot zFOOD5 \cdot zt5 \cdot zox5 \cdot ZFOOD5L,$$

where BMEI and PupMEI are respectively the biomass and the potential food uptake of the meiofauna. The factor PupMEI and the correction factors zFOOD5, zt5, zox5 and zFOOD5L control the actual regulation of the food uptake.

The uptake from the five food sources is given as a fraction of the total food uptake (W5):

for aerobic bacteria:	W35	$= PW35/PW5S \cdot W5;$
for aerobic diatoms:	W45	$= PW45/PW5S \cdot W5;$
for meiofauna (cannibalism):	W55	$= PW55/PW5S \cdot W5;$
for anaerobic bacteria:	W35A	$= PW35A/PW5S \cdot W5;$
for anaerobic diatoms:	W45A	$= PW45A/PW5S \cdot W5.$

This approach causes the meiofauna in the model to feed opportunistically, feeding predominantly on abundant food sources. Hence, the anaerobic bacteria and diatoms rank very low on the diet of the meiofauna. Another consequence is that the meiofauna is subject to a density dependent feedback through cannibalism if its population density approaches that of its main food sources: the bacteria and the diatoms.

7.3.2 Loss Factors

Excretion and respiration, as related to the biomass and food uptake have been treated in the preceding section because of their impact on food uptake. These factors will be surveyed only briefly.

The excretion (including mortality) encompasses both rest excretion and any mortality due to oxygen depletion (zox5) and is temperature-dependent (zt5):

$$W58r = BMEI(I) \cdot [rex5 + rex50 \cdot (1 - zox5)] \cdot zt5.$$

Excretion mostly leads to labile organic carbon (LOC), except for the activity excretion associated with the consumption of diatoms; a part p42 (0.7) here is liberated as detritus (DET). This detail was introduced for reasons of completeness, thus:

$$W52 = W45 \cdot pex5 \cdot p42.$$

Hence, the total excretion of LOC is written as:

$$W58 = W58r + W5 \cdot pex5 - W52,$$

where W5 is the total food consumption.

Total respiratory losses in the meiobenthos have three components:
1. Standard metabolic loss (W50r), depending on biomass and temperature (zt5);
2. Losses due to feeding activity, in the model coupled to uptake (W5);
3. Respiratory loss, due to the specific dynamic action of food (FACTOR5·asWORK5) (Valiela, 1984).

The whole set of losses is then accounted for in the next equation:

$$W50 = W50r + W5 \cdot prs5 + FACTOR5 \cdot asWORK5,$$
with $W50r = BMEI \cdot rrs5 \cdot zt5.$

The other two components also contain the same temperature dependency as the standard respiration.

7.4 Macrobenthos

The macrobenthic fauna consists of larger animals living in the sediment. They are traditionally studied as one group because they have important biological characteristics in common which distinguish them from other organismal groups. Most macrobenthos have a generation time of at least 1 year, reproduce only once a year and have a life span of several years. The macrobenthos live predominantly sedentary, in burrows that may penetrate both the aerobic and the anaerobic layer. Still the animals are quite independent of the conditions in the sediment. They may oxygenate their burrows by circulating water through them (worms) or extend their siphon above the sediment (molluscs). The animals may influence the local conditions in the sediment by the presence of their burrows and by the bioturbation which is a result of their (feeding) activities.

Finally, most species have pelagic larvae. These pelagic larvae act as a mechanism for dispersal.

The total macrobenthic fauna of the Ems estuary consists of about 40 species and is comparable to the macrofauna of the Wadden Sea (Van Arkel and Mulder, 1979; Michaelis, 1981; Beukema, 1976). Yet the biomass is generally lower than in corresponding areas in the Wadden Sea. The largest number of species is found in compartments 4 and 5. Most species that prefer higher salinities and/or the accompanying environmental conditions do not penetrate further into the estuary than compartments 3 and 4. In the Dollard only ten species occur, which are very tolerant to extreme conditions. In compartment 1 almost no macrobenthos are found (Table 7.1). There are very few macrobenthos living in the sediment of the

Table 7.1. Composition of the macrobenthic fauna of the compartments. Only the species that contribute substantially to the biomass are given (mgC^{-2})

Compartment		5	3/4	2	1
Deposit feeder	Arenicola marina	500	490	40	0
	Heteromastus filiformis	150	190	160	0
	Scoloplos armiger	230	10	0	0
	Hydrobia ulvae	70	150	30	0
	Scrobicularia plana	20	0	70	0
	Macoma balthica	1040	910	720	50
Suspension feeder	Mya arenaria	990	3950	560	0
	Cerastoderma edule	460	90	0	0
	Mytilus edulis	70	0	0	0
Epibenthic carnivore/omnivore	Nereis diversicolor	350	160	720	750
	Corophium volutator	30	310	200	90

channels, except for some mussel banks in shallow parts of compartment 5. The macrobenthos of the channels are unimportant in comparison to that of the flats. Therefore the model only concerns itself with the fauna on the flats.

Different feeding types can be distinguished in the group of macrobenthic animals (see Table 7.1). Deposit feeders ingest sediment including all kinds of small organisms and organic matter. Suspension feeders live in the sediment but feed on suspended matter in the pelagic. These groups are represented as two different state variables in the model.

By modelling the fauna on a time scale of days it is possible to describe the changes of the biomass through the seasons. These changes are mainly determined by growth. The effects on the fauna, due to changes in composition and recruitment, are in these long-living and slowly reproducing animals usually changes on a time scale of years. Modelling of such changes necessitates detailed population models. The present ecosystem model primarily concerns changes in the fauna caused by growth-related processes.

7.4.1 Biology

7.4.1.1 Metabolic Rates

In the model only two state variables represent all the different macrobenthos species. A set of general physiological characteristics is needed, as detailed in Section 4.1, to describe the behaviour of deposit feeders and suspension feeders in the model. It is difficult to generalize the physiology of these state variables, especially their behaviour under field conditions, because obviously physiological investigations have been carried out with single species and under specific conditions (Newell, 1979). There may be large differences between the results of the same experiments with different species. Even with animals of the same species there may be considerable differences in size and age. For these reasons it is difficult to choose parameter values that are representative for "the" deposit feeders or "the" suspension feeders.

To overcome these difficulties another approach has been chosen. The basic metabolic characteristics are deduced from a general budget of gain and expenditure of energy over a period of 1 year (Table 7.2). This budget is based on field data on production and biomass from the estuary (Van Arkel, unpublished data) combined with literature data. Both production and biomass values comprise all species and year classes, so they are a reliable generalization of the whole deposit-feeding or suspension-feeding fauna. Moreover, biomass and production are the outcome of several processes that are modified by environmental conditions. These conditions may differ locally and over time. In biomass and production these variations have one integrated outcome.

For the suspension feeders the P/B ratio is lower than for the deposit feeders because of the preponderance of older animals in the total biomass of the suspension feeders. Older animals are generally less productive. They are less efficient too, therefore the net growth efficiency is set at 10%, whereas 15% is assumed for deposit feeders. These estimates are rather moderate (cf. De Wilde and Beukema, 1984; Gray, 1981).

Table 7.2. Annual energy budget of 1 000 mgC of macrobenthos

	A. Deposit feeders	B. Suspension feeders
P/B ratio	0.8	0.5
Net growth efficiency	15%	10%
	$mgCy^{-1}$	$mgCy^{-1}$
Somatic production	800	500
Regenerating parts	200	100
Gonad output	200	200
Total production	1 200	800
Assimilation	8 000	8 000
Total metabolism	6 800	7 200
Standard metabolism	1 100	1 100
All activity processes	5 700	6 100
Foraging	3 200	4 100
Specific dynamic action	2 500	2 000

In production measurements only somatic production is measured. Production of regenerating parts and gonad output is estimated after De Vlas (1979); Beukema and De Vlas (1979); Beukema (1981); Chambers and Milne (1975); De Wilde and Berghuis (1978); Hughes (1970a); Hughes (1970b); De Wilde and Beukema (1984). The production of regenerating parts is less important in suspension feeders than in deposit feeders, but the relative gonad output is the same in both groups.

The assumed net growth efficiencies for both groups lead to 8000 mgC assimilated per year for each 1000 mgC of biomass. All the energy taken up and not used for production (e.g. in deposit feeders: 8000–1200 = 6800 mgC·y^{-1}) is used in standard metabolism and activity metabolism. By setting the relative energy demand for standard metabolism at 3% d^{-1}, resulting in 1100 mgC·y^{-1}, the ratio of activity metabolism to standard metabolism exceeds 4. The value of 3% d^{-1} for standard metabolism seems rather low (cf. Newell, 1979), but higher values would decrease the ratio of standard to activity metabolism to values below 4, which was deemed improbable. The activity-related metabolic processes have two components: the energy used in foraging for food and the energy required to digest the food: the specific dynamic action (Warren, 1971). The first component is modelled as being dependent on food availability and the second component is directly related to the amount of food consumed.

The problem is that no data are available to make a proper division between these components for a general deposit feeder, so the following reasoning has been used:
1. From Table 7.2 we can calculate the energy used in activity-related processes. Active processes = assimilation – production – standard metabolism = 5700 mgC·y^{-1}; 5700/8000 = 0.71. The activity-related processes take 71% of the assimilated food.

2. The maintenance ration can be calculated to be 4500 mgC·y^{-1}. At this ration the animals do not grow; they use all the assimilated food for standard metabolism, gonad output and activity metabolism (4500 = 1100 + 200 + 3200; where 3200 = 0.71 x 4500).

3. At maintenance ration much of the energy for activity metabolism will probably be used for foraging, because living at a maintenance ration implicitly means a short food supply. Under normal conditions, for which the budget is made, the activity metabolism is higher mainly due to a higher digestion and only slightly to a higher foraging. Therefore 3200 is used as an estimate of the foraging energy used under normal conditions. Relative to the biomass it is 3200/1000/360 = 0.9%·d^{-1}.

4. The specific dynamic action now can be calculated as 2500 mgC (activity metabolism – foraging = specific dynamic action; 2500 / 8000 = 0.31). This means that 31% of the assimilated food is used for digesting it.

The estimates for the standard metabolic rate and the values derived under steps 3 and 4 are used as parameters in the model.

For the suspension feeders the partitioning of the energy consumed between standard metabolism, foraging and specific dynamic action is calculated in the same way as for the deposit feeders.

7.4.1.2 Food Uptake

The food uptake of deposit feeders and suspension feeders differs from the food uptake of the other functional groups in the model. They do not feed on separate food items but the deposit feeders ingest a certain amount of sediment and suspension feeders clear a volume of water of all kinds of particles and small organisms. So they cannot select between different food sources but they eat the different food items proportionally to their abundance. To model the food uptake of deposit feeders and suspension feeders estimates have to be made of their feeding rates. A general feeding rate cannot be deduced from the annual energy budget because then an intrinsic link between energy uptake and expenditure would be built into the model. The general feeding rates have to be estimated independently. Based on data of Cadée (1976, 1979) on two deposit feeders of quite different size, a weight-specific feeding rate has been calculated. This allometric function is $FR = 0.00186 \cdot W^{-0.25}$ (FR = feeding rate in cm·m^{-2}·y^{-1}·mgC^{-1} W = individual weight in mgC). The exponent of -0.25 is commonly found for weight-specific physiological rates (Platt and Silvert, 1981). This equation has been generalized for all deposit-feeding species. By means of this equation the food taken by the deposit-feeding species in each compartment has been calculated according to the mean individual weights for each species as observed in the field. The total amount of food eaten by all deposit feeders in one compartment was compared to the biomass of the deposit feeders. From these comparisons one simple biomass-dependent feeding rate was derived for the general deposit feeders of the estuary: $0.038 \cdot 10^{-3}$cm·(mgC·m^{-2})$^{-1}$·d^{-1}. In the model the feeding rate is dependent on temperature like the other activity related processes.

The food uptake is not independent of the food supply. The relation between the feeding rate and the food supply is modelled as an optimum curve. Also the assimilation efficiency decreases at increasing food uptake thus forming another food-dependent feedback mechanism. Together these mechanisms limit the growth at high food levels and minimize possible energy losses at low food levels.

A general weight-specific filtration rate was derived from the weight-specific filtration rates of different species of suspension feeders, as given by Newell (1979): $FR = 0.756 \cdot 10^{-3} W^{-0.32}$ (FR is feeding rate in $m^3 \cdot mgC^{-1} \cdot d^{-1}$; W is the individual weight of an animal in mgC). The median weight of the suspension feeders differs considerably between the compartments. Therefore the filtration rate of the generalized suspension feeder will differ too between compartments. The suspension feeders of compartments 3 and 4 are used as a standard with a median individual weight of 72 mgC. The rate values refer to this standard weight. The suspension feeders are subjected to the same kind of food-related feedback mechanisms in the model as discussed for the deposit feeders.

7.4.1.3 Influence of Temperature

We have adopted the concept of Newell (1979) that the activity related metabolic processes are modified by temperature in another way than the standard metabolic rate. Many animals living in the intertidal zone are able to acclimate especially their standard metabolism to seasonal changes in temperature. Thus standard metabolism stays constant over a wide temperature range which enables them to conserve energy at higher temperatures. Only at very high temperatures often a steep increase of the standard metabolism occurs indicating a heat stress.

The energy required for active processes, as can be measured by oxygen consumption, generally increases with temperature. This relation is often described by an exponential function with an increase over a temperature interval of 10 °C of a factor Q_{10}. Common Q_{10} values range between 1.2 and 3.5 with a mean of approximately 2 (Newell, 1979; Cadée, 1976). From the energy budget it follows that the standard activity of the animals is defined at the mean temperature in the estuary (ca. 11 °C). A Q_{10} of 2 would mean that the animals are still half as active at 0 °C. This is not realistic because observations at low temperatures indicate that activity almost stops (Kinne, 1970). Therefore a temperature dependence of the activity related processes was formulated that results in a Q_{10} of 2 at higher temperatures but is nearly 0 at 0 °C.

The result of the interaction between both temperature dependencies is that the animals have a maximum scope for growth at intermediate temperatures. At those temperatures the standard metabolic rate is still low, whereas the rates of the activity-related processes already increase.

7.4.1.4 Other Regulating Factors

Deposit feeders and suspension feeders react to changes in the oxygen saturation of the water. At reduced oxygen levels they often show regulation of their uptake by compensating mechanisms, but they only can do this to a certain level. After the data of Mangum and Van Winkel (1973) the general deposit feeder is inhibited at O_2-saturation values below 60%.

Natural mortality is difficult to estimate because of a lack of quantitative data. Most mortality is probably due to predators but sometimes the remains of dead animals are found in the sediment which indicates the existence of natural mortality. The mortality rate is set at $0.0008 \cdot d^{-1}$, which implies that the animals have a life span of 3 to 4 years.

To compensate for the differences in median weight of the suspension feeders in the compartments a conversion factor CIND(I) is used. It consists of $Wx^{-0.32}$ / $Wstandard^{-0.32}$. Wx is the median weight of the suspension feeders of a certain compartment and Wstandard the median weight of the animals in compartments 3 and 4. The exponent used is the same as that for the weight-dependent filtration rate. It is close to the theoretical value for aquatic animals (Platt and Silvert, 1981).

The feeding activity of deposit feeders is independent of the tide. They are supposed to be constantly active. Contrary to the deposit feeders the suspension feeders only can feed during immersion of the tidal flats.

The regulation of the food uptake of suspension feeders at high food levels is modelled as a function of the particle concentration. At high particle concentrations the feeding rate decreases. Phytoplankton and silt provide the largest quantities of particles. The number of phytoplankton cells present is calculated from the phytoplankton biomass, using $1.2 \cdot 10^{-7}$ mgC as the weight of a single cell (personal communication W. Admiraal). Not all silt particles influence suspension feeders. Part of them are too small to be retained. Although it is known that the particle composition of silt is not constant and may differ from compartment to compartment too, a mean value (see PARSIL) was chosen to convert silt weight into the concentration of particles that are relevant for the feeding of suspension feeders. As a simplification for the model the relative contribution of silt and phytoplankton to the particle concentration have not been taken into account and regulation by means of pseudofaeces production has been neglected.

A second regulation of the feeding concerns double filtering. Water taken in by an animal is cleared of particles and then returned to the water mass present above the flats with which it mixes. Part of this water may be used again, thus diluting the food. There is little exchange of water between the channels and the water covering the flats at high tide. So the suspension feeders only feed on the latter volume. It depends on the ratio between this volume and the volume of water cleared by the suspension feeders how strong the dilution effect is.

7.4.2 Model Description: Deposit Feeders

To calculate the actual value of the regulation factors, first the potential food uptake (FOOD6) of a standard animal (biomass = 1 mg, standard temperature and oxygen conditions) is calculated at the prevailing concentrations of the different food sources:

$$FOOD6 = [BMEI(I) + UDIA + BBAC(I)]/BAL(I) \cdot BSL + [LDIA + ABAC(I)]/LBBBM \cdot BTL.$$

The feeding rate is divided into two parts: one for feeding above the sulphide horizon (BSL) and one below this layer (BTL). This division is based on the share in the total biomass of the deposit feeders of species like *Macoma balthica* that feed exclusively in the upper layer. The potential food uptake (FOOD6) is calculated as the product of the feeding rates and the actual food concentrations. The concentration of LDIA and ABAC is not homogeneous over the total anaerobic layer; it is higher in the upper part. As deposit feeders mainly use the upper part of the anaerobic layer to feed, the higher concentration of food items there is accounted for by using a smaller depth (LBBBM) than the depth of the anaerobic layer in calculating the concentrations:

$$ZFOOD6 = 2 \cdot F6 / (FOOD6 + F6) \cdot zox6.$$

ZFOOD6 is the function in which the potential food uptake is compared with the value F6, which is the daily food uptake of a standard animal living under standard conditions with an assimilation efficiency of 70%.

If the potential food uptake exceeds the standard food uptake, indicating a rich food supply, the function ZFOOD6 decreases below 1.

The function is combined with the regulation factor for O_2 saturation (zox6). ZFOOD6 forms the right side of the optimum curve regulating the feeding rate. The function describing the left side depends on an estimate of the net gain or loss of energy (biomass) after feeding and metabolic processes. To make this estimate the assimilation efficiency is needed, so this is calculated first:

$$pex6A = pex6 \cdot [1 + FOOD6 / (FOOD6 + F6)],$$

where pex6 (0.25) is the excreted fraction of the uptake. Pex6A is the fraction of the food that is not assimilated, but excreted as faeces. The assimilation efficiency thus is 1-pex6A. Pex6A is a function dependent on a comparison of the potential food uptake and the standard food uptake F6. As the food supply increases the unassimilated fraction increases too. At the maximum it may be doubled:

$$FOOD6X = (1 - pex6A) \cdot zFOOD6 \cdot FOOD6 \cdot (1\text{-}prs6)$$
$$\cdot zFOOD6 \cdot asWORK6.$$

FOOD6X is a preliminary energy balance. The gain and loss factors connected with food uptake of a standard animal are calculated. The potential gain from food uptake is corrected for the actual concentration of the food sources and the actual assimilation efficiency minus the fraction of the food needed for digesting it (1- prs6) and for foraging (asWORK6·zFOOD6). FOOD6X can be negative at low food concentrations because the losses exceed the gains. In those cases no feeding occurs: the animal minimizes its loss:

$$REST6 = (rrs6 + rex6) \cdot zt6c / zt6a.$$

REST6 is the temperature corrected sum of the standard metabolic rate and natural mortality:

$$zFOOD6L = FOOD6X^{POW6} / (REST6^{POW6} + FOOD6X^{POW6}).$$

FOOD6X and REST6 are compared in function zFOOD6L (Fig. 7.6 a). The function zFOOD6L is almost 1 when an energy gain is expected (FOOD6X > REST6) and it is very small when feeding is insufficient to cover both activity and standard metabolism (FOOD6X < REST6). Then feeding is stopped and so is the expendi-

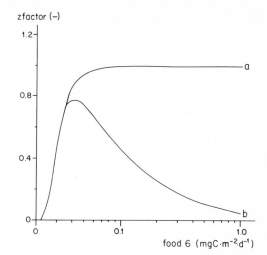

Fig. 7.6. Curve *a* is the dependence of restriction factor zFOOD6L on the potential food uptake (FOOD6). Curve *b* is the combined effect of zFOOD6L and zFOOD6, regulating the food uptake

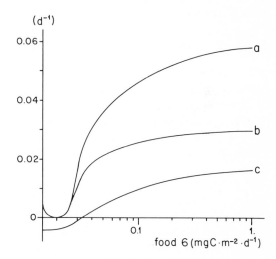

Fig. 7.7 The dependence between food uptake (a), assimilation (b) and growth (c) at different levels of potential food uptake

ture of energy needed for feeding, thus restricting the loss of energy to the standard metabolism. The steepness of the curve around FOOD6X = REST6 is determined by the value of POW6. Here POW6 = 8 is used, which gives a steep increase. Therefore zFOOD6L acts as a switch function.

The functions zFOOD6 and zFOOD6L together form the optimum function regulating the feeding rate (Fig. 7.6 b).

Figure 7.7 shows the combined effect of the three food-related feedback mechanisms at different levels of (potential) food uptake. The area between a and b represents the fractions of the food excreted as faeces. The area between b and c represents the part of the food used for metabolic processes. The area under curve c is the production. At very low food levels feeding stops but standard metabolism continues, resulting in a negative production.

The fraction of the assimilated food that is used for metabolism decreases at increasing food levels. It consists of standard metabolism, specific dynamic action, which is a constant fraction of the assimilated food, and energy for foraging. The last term decreases with an increasing food supply, which brings the total metabolic fraction close to a constant value.

After calculating the value of the food-dependent regulation factors, according to the food concentrations, the actual feeding and metabolic expenditures of the deposit feeders can be calculated:

DIFF6 = BBBM(I) · zFOOD6 · zt6 · zFOOD6L;
FACTOR 6 = DIFF6 · zt6a / zt6.

The biomass of the deposit feeders is combined with the regulation factors into a "virtual" biomass: FACTOR6. DIFF6 is an intermediate value with another temperature correction factor, used in transport calculations (Sect. 7.5.2). By means of the virtual biomass all the values of biomass-dependent rates and activities under the specific conditions on that day, can be calculated directly. For example the amount of sediment eaten from above and below the sulphide horizon is respectively:

BSLX = BSL · FACTOR6;
BTLX = BTL · FACTOR6.

Possible ingestion of a certain parcel of sediment more than once is not corrected for as the daily amount ingested is always much smaller than the total amount of sediment available.

The actual carbon fluxes from the different food sources to the deposit feeders are calculated from the feeding rate and the concentration of these food sources. As explained before the concentration of bacteria and diatoms in the anaerobic layer is calculated over a smaller depth (LBBBM).

The actual values of rest respiration (W60r), natural mortality (W68r) and faeces (FAEC) are calculated from the relevant parameter values. Mortality includes a second term for mortality due to low oxygen concentrations. This term is regulated by (1-zox6). This function is one at anoxic conditions and decreases at increasing O_2-saturation values. Parts of the faeces that stem from the uptake of diatoms are refractory (W62A) and therefore go to benthic detritus (BDET). The remainder of the faeces, including remains from natural mortality, are attributed to BLOC (W68A):

W60A = W6 · prs6 · (1-pex6A) + asWORK6 · FACTOR6,

where W60A is the activity respiration. It consists of a term for foraging activity (asWORK·FACTOR6) and one for specific dynamic action. The expression is comparable to that of FOOD6X, but here the real biomass and food uptake are used. Finally, the release of gonad output is given by SPAWN6:

SPAWN6 = BBBM(I) · RSPAWN · zSPAWN.

The function zSPAWN distributes the amount (RSPAWN), which is defined on an annual basis, into daily portions over two periods of spawning, one in spring and a smaller one in autumn. The biomass released is attributed to the state variable PBLAR, the benthic larvae in the pelagic. The spawning implies a transition of biomass from the benthic to the pelagic. Therefore the biomass is converted from $mgC \cdot m^{-2}$ on the tidal flats, to $mgC \cdot m^{-3}$ in the water volume covering the flats at high tide by $1/depfa(I)$.

In the pelagic the larvae are subject to transport processes, by which they are dispersed over the estuary and the adjacent North Sea. Settlement and recruitment to the population are not included in the model because their contribution to the biomass is very small.

7.4.3 Model Description: Suspension Feeders

The modelling of the suspension feeders follows the same setup as the modelling of the deposit feeders. First, the food-dependent regulation factors are determined. Then the potential uptake is calculated from which two more regulation factors are deduced. Together with these two the real uptake can be determined. Finally, the physiological activities (respiration, growth, excretion) and the fluxes from the different food sources are calculated.

The food uptake of the suspension feeders is regulated by an optimum function. In fact it is the combination of two functions each describing the decrease at one of the sides of the optimum. The right side is described in function zPAR:

$$zPAR = 2 \cdot OCP / (2 \cdot OCP + PART).$$

In suspension feeders there is a clear relation between feeding activity and suspended particulates (Winter, 1978). zPAR decreases at increasing particle concentrations. The particle concentration (PART) consists of food particles (phytoplankton) as well as suspended sediment particles. Thus the adverse influence of high silt concentrations on suspension feeders is included in the mechanism limiting the food uptake. The coefficients in zPAR are chosen such that $zPAR = 0.5$ when the particle concentration is twice the optimum concentration (OCP) for the feeding activity of suspension feeders (Winter, 1978).

The left side of the optimum curve is described in function zFOOD7L. It is based on a preliminary energy balance (FILB) of a standard animal (1 mgC biomass; activity defined under mean temperature conditions) under the prevailing food conditions, so this balance is calculated first:

$$FILB = (1\text{-}pex7) \cdot RRFIL \cdot ptWET(I) \cdot FOOD7 \cdot zPAR \cdot zox7 \cdot (1\text{-}prs7) - FACT7 \cdot zPAR \cdot zox7,$$

where FILB consists of the potential food uptake (RRFIL·FOOD7). FOOD7 is the sum of all the food sources for suspension feeders. The potential uptake is corrected for the oxygen saturation (zox7), for high food and silt concentration and for the time available for feeding (ptWET). FILB is corrected for two loss factors:
1. The fraction of the food that is not assimilated but excreted as faeces (pex7);
2. The fraction needed for the specific dynamic action (prs7).
Thus the first term of FILB is the potential uptake. From this the second term (FACT7·zPAR·zox7) which is the foraging/filtering energy is subtracted. It is a fraction of the biomass corrected for the prevailing conditions. The result of this

preliminary balance is compared with the energy needed for standard metabolism $(rrs7 \cdot zt7c/rt7a)$ in the function zFOOD7L:

$$zFOOD7L = FILB^8 / [(rrs7 \cdot zt7c / zt7a)^8 + FILB^8],$$

where zFOOD7L is a switch function comparable to the function zFOOD6L used for deposit feeders. It has a very small value when $FILB \leq$ energy needed for the standard metabolism (potential negative production). It approximates one when $FILB >$ standard metabolism.

Now the actual food uptake can be calculated. In FACTOR7 (virtual biomass) the biomass of the suspension feeders is combined with regulation factors for temperature, oxygen and a factor for the differences in the median individual size of the suspension feeders in different compartments:

$$FIL = FACTOR7 \cdot RRFIL \cdot zFOOD7L \cdot ptWET(I) \cdot zPAR,$$

where FIL is the volume of water that the suspension feeders in 1 m² of tidal flat filter in 1 day. It is the product of the (virtual) biomass and the relative filtration rate supplemented with the optimum function for food uptake and the available time for feeding. The dimension of FIL is a volume per unit surface area: $m^3 \cdot m^{-2} \cdot d^{-1} = m \cdot d^{-1}$. It also can be regarded as the thickness of the layer of water that will be filtered in 1 day.

The food uptake is subject to two more regulation mechanisms. First, the effect of double filtering is considered:

$$COR0 = FIL / depfa(I),$$

where COR0 gives the ratio of the filtered volume to the water volume above the tidal flat during the period of immersion. If this ratio increases, the chance of filtering water more than once increases too, thus reducing the amount of water effectively filtered. This is represented by the function COR2:

$$COR2 = \frac{1 - e^{-CORO}}{CORO}.$$

The effectively filtered volume is given by $COR2 \cdot FIL$:

$$FILA = FIL \cdot FOOD7 \cdot COR2,$$

where FILA is the amount of food contained in the effectively filtered volume of water.

The second regulation of the feeding is a maximum value for the daily food uptake. In reality there are several feedback mechanisms operative at high food levels (decreasing assimilation efficiency, pseudofaeces production) but in the model this is simplified into an upper limit for the food uptake:

$$
\begin{aligned}
MAXBSF &= pup7 \cdot zt7a \cdot BSF(I); \\
zFOOD7 &= MIN[1, 2 \cdot MAXBSF / (FILA + MAXBSF)]; \\
FFIL &= FILA \cdot zFOOD7;
\end{aligned}
$$

where MAXBSF is the maximum daily uptake. It is a fraction of the biomass with a seasonal adaptation through the temperature factor zt7a. The filtered amount of food is compared with this maximum in zFOOD7. Finally, FFIL gives the real food uptake.

After introductory calculations of the regulation factors and the food uptake, the calculation of the metabolism of the suspension feeders is simple and straightforward. W62 are the faeces. It is a constant fraction of the food, because the assimilation efficiency is kept constant. Variation in the assimilation efficiency is already accounted for in the food uptake limitation MAXBSF. The faeces are attributed to benthic detritus (BDET):

$$W70a = prs7 \cdot (1. - pex7) \cdot FFIL + FACT7 \cdot FACTOR7 \cdot zFOOD7L \cdot zFOOD7;$$
$$W70r = rrs7 \cdot BSF(I) \cdot zt7c \cdot CIND(I).$$

The activity respiration (W70a) consists of a fraction depending on the food uptake representing the specific dynamic action and on feeding activity which depends on the (virtual) biomass. This second term is corrected for the food uptake regulation factors (zFOOD7L·zFOOD7) as these are not included in the virtual biomass.

The rest respiration (W70r) depends on the real biomass corrected for the temperature dependence of standard metabolism and individual weight.

Spawning is modelled in the same way as for the deposit feeders.

PR7 is the production of BSF. It is the net outcome of all gains and expenditures and it may be positive as well as negative or zero.

W72S is mortality due to oxygen deficiency. The definition is the same as that for the deposit feeders. It is subtracted from the production in the final mass balance but it is not used in the production calculation to make PR7 comparable to production figures based on field measurements.

The suspension feeders are assumed to be mainly active during the period of submersion of the tidal flats. Therefore the oxygen needed for their activities is added to the oxygen demand in the water (CONSOX). Standard metabolism, on the other hand, is a constant process which goes on regardless of the tide. During low tide the suspension feeders are supposed to obtain their oxygen from the air and only for the duration of the immersion period their oxygen consumption is added to CONSOX:

$$pFIL = FFIL / depfa(I) / FOOD7 \cdot pWAFL(I).$$

To determine what fraction of the available food is eaten by the suspension feeders, pFIL is calculated from:
1. FFIL/depfa(I), the amount of carbon filtered from 1 m^3 of water;
2. FOOD7, the food available (although the depth of the water over the flats may be less than 1 m, the quantities are expressed per m^3 to keep the dimensions comparable);
3. pWAFL(I), converting the value for 1 m^3 of water over the flats into the value valid for the average 1 m^3 of water in the pelagic. pFIL is used to calculate the loss terms for all the food sources, except for the suspended benthic diatoms. Resuspension and deposition of benthic diatoms mainly takes place in the water above the flats during high tide. So for this food source the grazing is calculated only for the water mass above the flats without applying the mixing term.

7.5 The Spatial Structure of the Sediment and Vertical Transport

The processes that govern aerobic decomposition of organic matter in the sediment are quite different from those ruling anaerobic mineralization. To model these processes it was necessary to divide the sediment into an aerobic and an anaerobic zone.

A consequence of such a layered benthic system is the occurrence of vertical transport processes which transfer organic matter, both dissolved and particulate, from one layer to the other.

Three different kinds of transport processes contribute to the vertical flux of material in the sediment.

1. Active transport: physical mixing and bioturbation. The activity, both in feeding and in burrowing, of benthic organisms moves organic matter from the surface to deeper (anoxic) layers and the other way around. This continuous reworking of the sediment extends to a depth of 30 cm. Mixing by wave action and storm events also contribute to the vertical transport of matter in the sediment.
2. Diffusive transport: The oxygen demand in the sediment determines the size of the diffusive flux into the sediment as well as the penetration depth of the oxygen and hence the location of the transition zone between aerobic and anaerobic sediment. The presence and the activity of larger benthic organisms, inhabiting well-oxygenated burrows or tubes, influences the diffusion of oxygen into the sediment.
3. Transition and burial: A shift in the location of the aerobic/anaerobic interface causes a transition of material and organisms from the aerobic layer into the anaerobic layer or vice versa. The location of this interface is indicated by the presence of free sulphide.

   A second form of transition is the accretion of sand and silt by sedimentation. This accretion not only transfers material across the aerobic/anaerobic interface, but also across the boundary of the biologically active zone at 30 cm depth. In passing this last boundary material leaves the system and is effectively buried.

7.5.1 Oxygen Demand and Sulphide Production

The respiratory processes of the benthic system cause an oxygen demand that has to be met by oxygen diffusing downward from the overlying water or, when the tidal flats are emerged, directly from the air. An additional oxygen demand is caused by the oxidation of sulphide formed by sulphate-reducing bacteria. This sulphide production is a most prominent feature of anaerobic bacteria in marine sediments (cf. Sect. 7.6).

The sulphide partly remains dissolved as free sulphide in the anaerobic pore water. The dissolved free sulphide is dependent on pH according to:

$$H_2S \leftrightarrow (H^+ + HS^-) \leftrightarrow (2H^+ + S^{2-}).$$

In the Ems estuary (pH normally between 7 and 8) most of this sulphide is present in estuarine pore water in the form of HS^-. It is diffusible and it can be oxidized

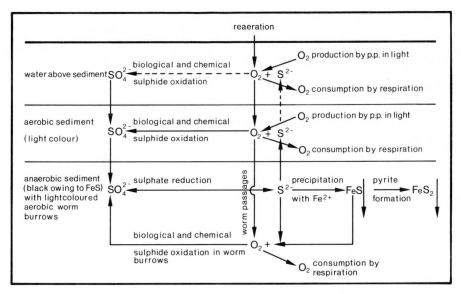

Fig. 7.8. Schematic representation of the sulphur cycle in the sediment showing processes in which sulphate and sulphide are involved. Sulphate is transported via the interstitial water to the anaerobic sediment, where the sulphate is consumed by bacterial sulphate reduction and is converted to free sulphide. This free sulphide dissolved in the interstitial water can diffuse to the surface, can be oxidized or can be precipitated to iron sulphide (FeS). Iron sulphide can be oxidized or can form pyrite. Oxygen in the sediment originates from reaeration and from primary production. It is transported downwards via the interstitial water and is consumed there by the respiration of organisms and by chemical oxidation (BOEDE, 1985)

back to sulphate after diffusion into an aerobic environment. Oxidation can occur either chemically or biologically (by sulphide-oxidizing bacteria, see Sect. 7.6) consuming mainly molecular oxygen.

The remainder of the sulphide is precipitated, at least temporarily. Because the iron content is usually high in marine sediments, the most important precipitation product is the insoluble ferrous sulphide (FeS) (Berner, 1964; Goldhaber and Kaplan, 1974):

$$Fe^{2+} + HS^- \rightarrow FeS\downarrow \text{ (black)} + H^+.$$

This compound which is still oxidizable can accumulate in the sediment in appreciable amounts. In the Ems estuary the amount of FeS present in the intertidal sediments is equal to the total sulphide production of 100–400 days (Schröder, unpublished data) and hence forms a large buffer against oxidation of the anaerobic layer (Fig. 7.8). Thus oxidizable sulphides characterize the anaerobic layer, whereas above the sulphide horizon oxygen is thought to be present at least during part of the tidal cycle (see p. 143).

7.5.1.1 The Aerobic/Anaerobic Interface

Measurements with microelectrodes have shown that in coastal sediments oxygen generally penetrates no more than a few millimeters into the sediment (Revsbeck et al., 1980).

However, in worm tubes oxygen penetrates much deeper (Jørgensen and Revsbeck, 1985) (Fig. 7.9). The penetration of oxygen can vary continuously and penetration is dependent on the pumping rhythm of the worm. In these tubes oxygen can reach depths of up to 30 cm in the sediment (Fig. 7.10).

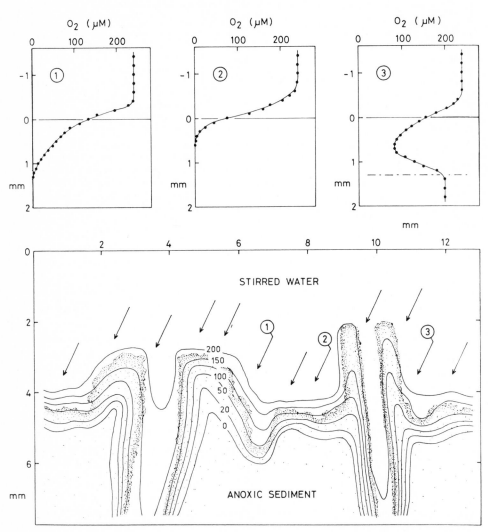

Fig. 7.9. Vertical section through a sediment surface (Limfjorden) including two polychaete tubes. *Numbers* on the oxygen isopleths indicate $\mu mol\ O_2 \cdot l^{-1}$. Oxygen from the aerated seawater penetrates through the diffusive boundary layer into a thin, oxic sediment zone which follows the surface topography (Jørgensen and Revsbeck, 1985)

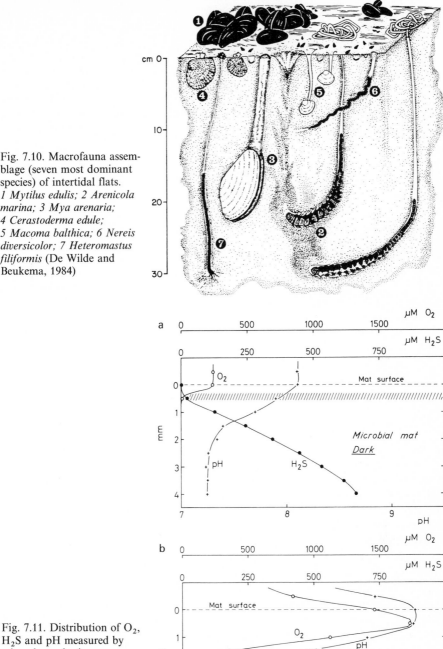

Fig. 7.10. Macrofauna assemblage (seven most dominant species) of intertidal flats. *1 Mytilus edulis; 2 Arenicola marina; 3 Mya arenaria; 4 Cerastoderma edule; 5 Macoma balthica; 6 Nereis diversicolor; 7 Heteromastus filiformis* (De Wilde and Beukema, 1984)

Fig. 7.11. Distribution of O_2, H_2S and pH measured by microelectrodes in a cyanobacterial mat sediment at night (a) and during the day (b). The *hatched areas* indicate the zones of dynamic co-existence of O_2 and H_2S (Jørgensen, 1983)

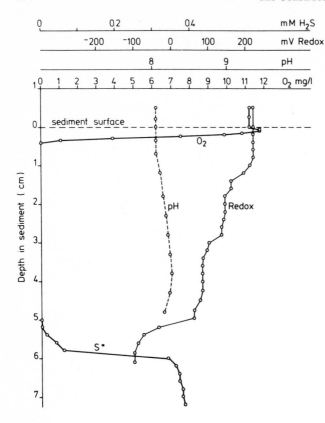

Fig. 7.12. Distribution of oxygen and hydrogen sulphide, and the redox potential and pH gradients in the sediment of sampling station Hals on December 17, 1982 (Lindeboom and Merks, 1983)

On the other hand, free sulphide (HS⁻) and FeS are generally found only in the deeper layers and in intertidal sediments. It may occur at depths varying from a few millimeters in fine-grained sediments to a few centimeters in coarse grained sediments (Lindeboom et al., 1984; Schröder, unpublished data). Especially in sediments with a high input of easily degradable organic carbon, such as sediments covered by a thick algal mat, the sulphide production by sulphate-reducing bacteria will be large. Consequently, the free sulphide can reach upwards to a depth of about 0.5 or 2 mm, whereas oxygen reaches zero concentration at approximately the same depth. In such environments the exact depth of the small zone of dynamic coexistence of O_2 and H_2S can change diurnally depending on the light conditions and consequently on the oxygen production in the algal mat (Fig. 7.11). Generally, however, the interface of oxygen and free sulphide in intertidal sediments is much more heterogeneous due to the activity of macro- and meiofauna and to tidal pumping. Here, the two compounds even may not coexist at all; often a zone of a few centimeters is present without any detectable amounts of O_2 and H_2S (Fig. 7.12) (Jørgensen, 1983; Lindeboom and Merks, 1983).

Due to this large temporal and spatial heterogeneity it is difficult to pinpoint the depth of the O_2–H_2S interface at a certain horizon and to distinguish a clearly aerobic and anaerobic layer in the sediment.

In the benthic submodel we divide the total biologically active sediment layer along the sulphide horizon into two layers, which simply are assumed to be homogeneous. The sulphide horizon is at, or somewhat above, the upper boundary of the visually dark sediment, which is discoloured by FeS and/or pyrite.

The layer below the sulphide horizon is defined as the anaerobic zone. The light coloured top layer of the sediment above the sulphide horizon comprises not only the true oxic zone (of a few millimeters thickness) but additionally the adjacent layer down to the sulphide horizon. This layer may contain some free sulphide, yet this is negligible compared to the oxidizable amount of sulphide present in the deeper anaerobic layer. Nevertheless it is better to avoid the term "aerobic zone" to indicate this layer.

On the other hand, the thickness alone of the oxic zone as measured by microelectrode is not very meaningful in an averaged model. It has to be enlarged to represent the occurrence of well-oxygenated worm tubes in the anaerobic zone and to allow for the effects of short-term fluctuations in oxygen penetration. In such a way the spatially and temporally heterogeneous oxygen penetration in the sediment is smoothed into a horizontal plane at a defined depth (see p. 143 and Box 7.1).

7.5.2 Nondiffusive Transport Processes

In the model there is a nondiffusive transport of pore water, sediment particles and smaller organisms across the sulphide horizon, caused by macrobenthic activity, by storm-induced mixing of the sediment and by sediment accretion due to sedimentation. Accretion only transfers sediment in the direction of the anaerobic layer. Macrobenthic activity and storm-induced mixing on the contrary cause the exchange of sediment in both directions across the interface. Hence these processes tend to equilibrate the concentrations of organic matter on both sides of the interface (BDET/ADET and BLOC/ALOC). The refractory organic matter (BROC) is supposed to be evenly distributed over the benthic profile. This assumption is justified because of the long turnover time of this class (see Sect. 4.4.3).

The storm-induced mixing of the sediment varies seasonally. In reality the variation is both wind strength and wind direction dependent (see Sect. 2.1). However, in the model the dependence of both processes on the season is simplified by modulating the mean mixing rate (RML in $cm \cdot d^{-1}$) seasonally with a simple cosinus function:

$$zMIX = 1 + AZMIX \cdot COS\left[\pi / 180 \cdot (TIME + 30)\right];$$
$$zSED = 2 - AZMIX;$$

where AZMIX is an amplitude constant ($=0.4$).

In this way it is unnecessary to define the wind dependencies exactly. This simplification is important as even the mean rates themselves are inaccurate. As it is modelled now, zMIX reaches a maximum amplitude of 1.4 on day 330 (zSED then is 0.6) and a minimum amplitude of 0.6 on day 150 (zSED = 1.4). Thus storm mixing (zMIX) is high in autumn/winter and low in spring/summer, corresponding to the average wind strengths. Sediment accretion, on the other hand, decreases as storm mixing increases and hence the accretion amplitude (zSED) is inversely proportional to zMIX.

The daily sediment mixing rate (SMLX) is then calculated from:

$$SMLX = SML \cdot zMIX \cdot zSM \text{ with } zSM = BAL0 / [BAL0 + BAL(I)].$$

The mean mixing rate SML $(0.01 \text{ cm} \cdot d^{-1} = 3.6 \text{ cm} \cdot y^{-1})$ is arbitrarily estimated.

In the model the sediment mixing process is only important if it effects a transfer of sediment across the sulphide horizon. Therefore sediment mixing as a transfer process is insignificant when this horizon is deep below the surface. This is taken into account by the factor zSM, calculated from the actual BAL value and a reference position BAL0 (3 cm).

The macrobenthos contributes to sediment transport across the interface mainly through deposit feeders. Filter feeders only displace sediment by the extension and retraction of their siphons. Sometimes they also move horizontally through the sediment over a short distance. However, these activities of filter feeders are assumed to be negligible with respect to the activities of deposit feeders. Deposit feeders actively burrow through the sediment and eat it on the go (see Sect. 7.4). Meiobenthic animals are too small in size to have a large contribution to sediment displacement. Moreover, they are assumed to live mainly in the aerobic zone and therefore they do not transfer sediment across the interface. Epibenthic animals, living in the sediment, are also mainly restricted to the surface zone of the sediment (e.g. *Corophium*) or they live in deeper burrows (e.g. *Nereis*). But *Nereis* merely creep in and out of these tubes and use them as hideouts (Fig. 7.12).

In Section 7.4 the sediment-eating activity of the deposit feeders is calculated. The amount of sediment eaten in the upper layer (BSLX) is computed as well as the amount eaten below the sulphide horizon (BTLX). It is assumed that only the deposit feeders feeding in the anaerobic layer contribute to a transfer of sediment across the aerobic/anaerobic interface. Hence only the feeding rate on anaerobic sediment (BTLX in $\text{cm} \cdot d^{-1}$) is used in the vertical transport equations. However for sediment transport not only the sediment passing the mouth of deposit feeders is important, but also the sediment displacement due to creeping, burrowing and to the collapse of a tube or funnel. The simplest way to incorporate these other sediment displacements is achieved by amplification of the sediment eating rate BTLX with a constant (BIOAM). This constant may be compartment-dependent as the species composition of the deposit feeders is different in each compartment. Especially species such as *Arenicola marina* and *Heteromastus filiformis* are supposed to contribute significantly to this transport process, in contrast to the species *Macoma balthica*, which eats mostly surface deposits. The deposit feeder term (BIOTUR in $\text{cm} \cdot d^{-1}$) of the sediment transfer across the interface is then:

$$BIOTUR = BTL \cdot BIOAM(I) \cdot DIFF6,$$

where the amplifier BIOAM has values between 1 and 10.

The net transfer of organic matter between the state variables BDET and ADET and between BLOC and ALOC is calculated according to:

T2 = BDET(I) / BAL(I) · (BIOTUR + SMLX + SDLX) – ADET(I) · pDISEXP;

T8 = T8 + BLOC(I) / BAL(I) · (BIOTUR + SMLX + SDLX) – ALOC(I) · pDISEXP.

Here pDISEXP is that fraction of the anaerobic layer in which the organic matter is thought to be concentrated. T2 and T8 are the net transfer rates of organic carbon (in $mgC \cdot m^{-2} \cdot d^{-1}$) for benthic detritus and labile organic carbon respectively. If a transfer rate is positive the transfer is in the direction of the anaerobic layer. The transfer rate T8 is added to the diffusive transfer of labile organic carbon (see Sect. 7.6). SDLX is the transition of material due to sediment accretion. The transition process is detailed in Section 7.5.4.

Analogously to the organic carbon transfer, benthic bacteria (BBAC and ABAC) and sulphide (BSUL) are transported across the sulphide horizon:

T3   = BBAC(I) / BAL(I) · (BIOTUR + SMLX + SDLX);

T3A = ABAC(I) · pDISEXP;

TSUL = BSUL(I) / [TAL -BAL(I)] · (BIOTUR + SMLX).

Here no net transfer in the form of a difference occurs as bacteria transported across the interface presumably die and the sulphide transported is oxidized within a time step.

The transfer rates T3 and T3A are subtracted from BBAC and ABAC. These rates are added partly (pM32) to BDET and ADET and partly (1-pM32) to BLOC and ALOC respectively. Here the fraction pM32 (0.2) represents the bacterial cell walls as a fraction of the total bacterial biomass (see Sect. 7.6). The transfer rate TSUL is subtracted from BSUL and added to the benthic oxygen demand by sulphide oxidation (BOXS).

7.5.3 Diffusive Transport Processes

In the benthic submodel the position of the sulphide horizon (BAL) is a state variable. It is recalculated every time step from the oxygen concentration at depth x = 0 cm, from the total benthic oxygen consumption, including sulphide oxidation and from a seasonally varying apparent diffusion constant. The basic idea is simple: oxygen penetrates deeper if the diffusion constant is larger and if the oxygen demand is lower. In Box 7.1 this qualitative statement is transformed into an equation for BAL with the help of some simplifying assumptions.

One of the main problems in making a diffusion model for oxygen and free sulphide in intertidal sediments is the spatially heterogeneous distribution of abiotic environmental conditions, of biota and of the activities of organisms. Heterogeneity not only exists for oxygen and sulphide compounds (as stated above) but also for the distribution of salinity, pH, nutrients, organic matter, clay content, benthic diatoms, aerobic and anaerobic bacteria, meiofauna, etc. These heterogeneous distributions in intertidal flat sediments originate to a large extent from local differences in sedimentation/erosion (hydrography, morphology) that result in a spatially heterogeneous sediment composition. In addition, especially the activity of the macrobenthos (deposit feeders as well as filter feeders) can intensify patchiness. The more so as the spatial distribution of these organisms is often contagious and this distribution in its turn is more or less dependent on the sediment composition. To model the diffusion process in the sediment therefore requires the averaging of concentrations, gradients, biomasses, activities, etc. and also the simplification of environmental conditions.

7.5.3.1 Oxygen at the Sediment Surface

For the computation of the location of the sulphide horizon (BAL) the oxygen concentration at the sediment surface (OXO) has to be known (see Box 7.1). This oxygen concentration OXO changes during the day, depending on the tide and the light conditions. The fraction of a day the tidal flats are emerged in daylight is:

$$pOVERLAP = OVERLAP / 24 \quad \text{(for OVERLAP see p. 38)}.$$

Submersion occurs during the fraction of the day ptWET and consequently the flats are emerged in the dark during the fraction of the day:

$$[1 - ptWET(I) - pOVERLAP].$$

During submersion the oxygen concentration at the sediment surface is equal to the concentration actually present in the pelagic (OX), whereas during emersion of the flat, the oxygen concentration at the sediment surface is equal to the saturation value of oxygen in water (COX in mg $O_2 \cdot dm^{-3}$) at the prevailing water temperature and salinity:

$$COX = [475 - 2.65 \cdot SALT(I)] / [33.5 + TEMP(I)]$$

(see Sect. 6.5 for description). Implicitly, diffusion from the air into the pore water in the top layer of the emerged sediment is modelled by assuming a thin layer of oxygen-saturated water above the sediment surface. Reaeration of this thin layer then occurs instantaneously.

When the tidal flats emerge in daylight the situation is more complicated. Due to primary production by the phytobenthos, oxygen is then produced in the top layer of the sediment. This oxygen production may result in a supersaturated oxygen concentration in the surface pore water (Fig. 7.14). This oxygen not only diffuses to deeper benthic layers, but it also diffuses from the supersaturated top layer of pore water into the air. The diffusion from the water into the air in the model is defined as a reaeration process, using the same relative reaeration rate (REACON in $m \cdot d^{-1}$) and the same temperature correction factor as in the transport submodel.

The total oxygen production by phytobenthos is PP (expressed in $mgC \cdot m^{-2} \cdot d^{-1}$) (see Sect. 7.2). The amount of oxygen available per day for diffusion and reaeration to the air is PPox (in $mgC \cdot m^{-2} \cdot d^{-1}$), which is the total oxygen production minus the activity respiration by phytobenthos. This respiration is subtracted because it occurs in the same period and layer as the production of oxygen:

$$PPox = PP \cdot corCOr / corCOp - W40a.$$

When the sulphide horizon [BAL(I)] is close to the surface of the sediment it comes near the photic layer of the sediment (LAL); then a large fraction of the oxygen produced is immediately consumed in the sediment, reducing the amount diffusing into the overlying water layer:

$$PPox' = PPox \cdot [1 - 0.5 \, LAL / BAL(I)].$$

In the transport model the amount of oxygen reaerated per time step and per m^2 across the air/water interface was calculated as REACON times the concentration difference of the actual oxygen concentration relative to the saturation concentration divided by the water depth in meters.

Analogously to this equation we calculate the amount of oxygen produced in a day during the time pOVERLAP in a 1-cm-thick water layer:

$$PPox' \cdot corCOr / pOVERLAP = REACON \cdot (COXP - COX) / 0.01,$$

where corCOr is the conversion factor [in $mgO_2 \cdot m^{-3} \cdot (mgC \cdot m^{-3})^{-1}$] (see Sect. 6.4).

By rearranging this formula the oxygen concentration COXP in this layer during pOVERLAP can be calculated:

$$COXP = COX + PPox' \cdot corCOr / (pOVERLAP \cdot REACON / 0.01).$$

The oxygen concentration OXO present in a thin (1-cm-thick) layer above the sediment surface then is:

$$OXO = \{ptWET(I) \cdot OX(I) + [1 - ptWET(I) - pOVERLAP] \cdot COX\} \\ / corCor \cdot 0.01 + pOVERLAP \cdot COXP,$$

where OXO is calculated as a weighted mean concentration from the relevant oxygen concentrations during three periods: the submersion period of the flats (ptWET), the emerged + dark period (1 − ptWET - pOVERLAP) and the emerged + light period (pOVERLAP).

7.5.3.2 Oxygen Diffusion

In principle the molecular diffusion coefficient of oxygen in aqueous solutions can be measured, but the apparent diffusion coefficient will be larger and will vary in space and time in intertidal sediments due to bioturbation, tidal pumping and wave-induced disturbances in the pore water. In the model we have to average the diffusion coefficient over a day and over a large area. Therefore we simply enlarge the molecular diffusion coefficient (DIFFox) in order to obtain the apparent diffusion coefficient (Dox) according to equation:

$$Dox = DIFFox \cdot f(Temp) \cdot [1 + f(biotic factors) + f(abiotic factors)],$$

in which f(Temp)   = temperature-dependent correction factor of DIF-Fox;
f(biotic factors)   = bioturbation-induced enlargement of DIFFox;
f(abiotic factors)   = tidal pumping and storm-induced enlargement of DIFFox.

The diffusion coefficient Dox is used in Box 7.1 for the derivation of a formula for the position of the sulphide horizon:

$$BAL = OXO \cdot Dox / [BOC / 2 + Jsul(BAL)].$$

Here BOC is the total oxygen consumption in the aerobic layer and Jsul is the upward sulphide flux.

Generally BOC and Jsul will depend on BAL, hence this equation determines BAL only implicitly. With an appropriately chosen nonlinear dependence of Jsul and BOC on BAL, multiple solutions are possible and hysteresis effects in the seasonal shifts of BAL can be modelled.

7.5.3.3 Sulphide Diffusion

In the anaerobic layer sulphide is produced by sulphate-reducing bacteria (see Sect. 7.6). This free sulphide is assumed to be in equilibrium with FeS. In the model free sulphide and FeS together are considered as one state variable BSUL (expressed as carbon equivalents in $mgC \cdot m^{-2}$).

There is only one place in the model where it matters that BSUL is a compound state variable: FeS does not "diffuse" in the same way as free sulphide (dissolved in the pore water) (Fig. 7.13).

This has been solved by letting only the free sulphide part (FLSO) of BSUL diffuse with a diffusion constant Dsul. Dsul is an apparent diffusion constant in the same way as Dox (p. 139). It depends on temperature and on turbation (see below).

As stated in Box 7.1, the downward oxygen flux at depth BAL is equal to the upward flux of sulphide at depth BAL:

$$Jox(BAL) = Jsul(BAL).$$

Furthermore, in Box 7.1 a formula for Jsul(BAL) is derived under the assumption that the diffusion coefficient and the sulphide production are independent of depth within the anaerobic layer. In reality the sulphide production is concentrated a few cm below BAL (Fig. 7.13; Schröder, unpublished). Yet, because the effective diffusion constant Dsul is the product of several not very well substantiated constants, we have left this equation as it is.

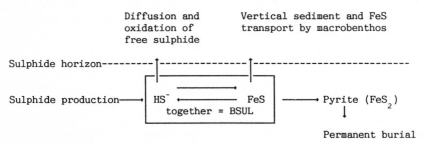

Fig. 7.13. Modes of vertical transport of HS$^-$ and FeS (see text)

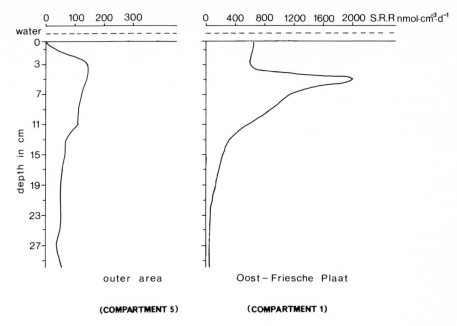

Fig. 7.14. Sulphate reduction rates (SRR in nmol·cm^{-3}·d^{-1}) versus depth in the sediment. Sampled at the station in the outer area on July 17, 1979 and on the Oost-Friesche Plaat on September 12, 1979. Note the difference in scale between the two stations (*water* = overlying water) (BOEDE, 1985)

7.5.3.4 Bioturbation and Diffusion

In previous sections we have already argued that the apparent molecular diffusion coefficients of oxygen and free sulphide should be enhanced to incorporate the effect of hydrodynamic phenomena and biological activity (cf. Billen, 1982). The hydrodynamic phenomena include pore water drainage by tidal pumping and storm-induced disturbances in the pore water. The burrowing and feeding activities of benthic invertebrates results in extensive reworking of the sediment and irrigation of pore water, but the mere existence of burrows or tubes already contributes to the enhancement of molecular diffusion into the sediment. In the model the apparent diffusion coefficient for dissolved oxygen (DIFFX in cm^2·d^{-1}) then is the temperature corrected (zt0) molecular diffusion coefficient (DIFF $= 1.73$ cm^2·d^{-1} $= 2·10^{-5}$ cm^2·s^{-1}) times the enhancement factor AMIX:

   DIFFX = DIFF · zt0 · AMIX.

The factor AMIX is the sum of three terms:

   AMIX = 1 + APHYS + ABIO,

where the 1 corresponds to the molecular diffusion, APHYS to the enhancement by physical factors and ABIO to the enhancement by biological activity. The latter can be subdivided further into their respective causative agents:

APHYS = storm-induced mixing + tidal pumping;
ABIO = sum of contributions by deposit feeders, suspension feeders, meiobenthos and epibenthos.

The various functional groups contribute to the enhancement of diffusion by bioturbation to a varying degree, depending on their biomass, their activity and their depth of occurrence. The contribution of the deposit feeders to ABIO is the most important. They not only transport organic matter across the sulphide horizon by their feeding activity, they also indirectly change diffusion simply by the presence of their burrows. An exact and explicit form of all these contributions to AMIX cannot be given yet; lack of knowledge forces us to resort to qualified estimates. In the model the enhancement factor AMIX can reach maximal values of up to ten when biomass and activity of the macrobenthos is high.

The diffusion coefficient for free sulphide (DIFFY) is calculated in almost the same way. However, here some biological activities and storm mixing are omitted as these are thought to be less important to the enhancement of the sulphide diffusion than to the oxygen diffusion because of the larger depth. In our model the molecular diffusion coefficients for oxygen and sulphide (DIFF) are assumed to be approximately equal and the used value ($20 \cdot 10^{-6}$ cm$^2 \cdot$s^{-1}) corresponds roughly with values published for oxygen ($16 \cdot 10^{-6}$ at 11 °C: Broecker and Peng, 1974) and for sulphide ($1.2 \cdot 10^{-6}$: Ostlund and Alexander, 1963 and $28 \cdot 10^{-6}$: Kleiber and Blackburn, 1978).

Box 7.1

Under special circumstances the one-dimensional diffusion equation has simple solutions: if the production or consumption rate and the diffusion constant are independent of time and of the spatial variable x, the stationary concentration c(x) of the diffusing substance depends on x in a parabolic way. The parameters of the parabola c(x) are determined by boundary conditions: concentrations or fluxes at both ends of the considered spatial interval. Let us assume a simple model for the biologically active sediment layer of a thickness TAL. There shall be a well-defined sulphide horizon at depth BAL with a constant oxygen consumption rate above and a constant sulphide production rate below BAL. Further, the diffusion coefficients Dox and Dsul are taken to be independent of depth x. The oxygen concentration ox(x) and the free sulphide concentration sul(x) both approach 0 at the sulphide horizon BAL. There the sulphide oxidation takes place. Hence the sulphide flux upward and the oxygen flux downward cancel there (both oxygen and sulphide are measured in units of carbon, determined by the stoichiometric coefficients in CO_2 and SO_4^{2-}). The fluxes are

Jsul(x) = Dsul · sul'(x) (upward) and
Jox (x) = Dox · ox'(x) (downward),

where the prime (') denotes the spatial derivatives. Thus, with the boundary conditions

ox(0)   = OXO   (surface value, see p. 138)
ox(BAL) = sul(BAL) = 0

Jox(BAL) = Jsul(BAL)   (pp. 138–140)
sul'(TAL) = 0

the full functional forms of ox(x) and sul(x) and the position of the sulphide horizon
BAL can be calculated from the sulphide flux Jsul(BAL) upward and the oxygen consumption BOC above BAL:

$$BAL = \frac{OXO \cdot Dox}{BOC/2 + Jsul(BAL)}.$$

The two parabolas for ox(x) and sul(x), together with the boundary conditions, are
shown in Fig. 7.15.

More general, we may want to decompose the oxygen consumption into its
different constituents (index i):

$$BOC = \sum_i r(i) \cdot BOC\ i.$$

Each of the r(i) is the thickness of the layer below the surface relative to BAL, in which
the oxygen consumption process i is active. This is useful when taking into account the
contributions of the benthic primary producers such as e.g. the respiration in CBO (see
Sect. 7.2), and the oxygen production (as negative consumption) (see p. 138) in LAL:

BOC = W50 + W60 + W70r + W30 + TSUL
        + W40r · CBO/BAL-0.5 · PPox · LAL/BAL.

In the stationary situation the flux Jsul(BAL) must be equal to the sulphide production
minus the sulphide sinks in the anaerobic layer. Due to the assumed parabolic shape
of sul(x) it is also possible to connect the total sulphide per square meter (BSUL) with
the flux:

$$Jsul(BAL) = 3 \cdot Dsul \cdot (FLSO \cdot BSUL) / (TAL - BAL)^2.$$

Note that FLSO · BSUL is the freely diffusible part of the total sulphide (see p. 140).

7.5.3.5 Calculation of the Position of the Sulphide Horizon (BAL)

The time-averaged description of the oxygen and sulphide diffusion with a slow
moving (quasi-stationary) sulphide horizon (Box 7.1; Fig. 7.15) seems to be in
agreement with some observations (Fig. 7.11), but disagrees with other observations (Fig. 7.12). The results of Lindeboom and Sandee (1983) show the existence
of a gap between the horizon of detectable levels of oxygen and the black sulphide
horizon. When the location of the black discolouration in the sediment (FeS) is
used as indicative also of the presence of free sulphides, this gap cannot be accommodated in our approach. If this gap is present permanently, this implies the absence of concentration gradients of oxygen and sulphide and hence the absence
of diffusive transport of these substances. Since the position of the sulphide horizon is stable (in the daily average) over quite long periods (weeks), while there
is continuous sulphate reduction, there must be sulphide oxidation on this horizon, otherwise the horizon would rise. In the absence of concentration gradients,

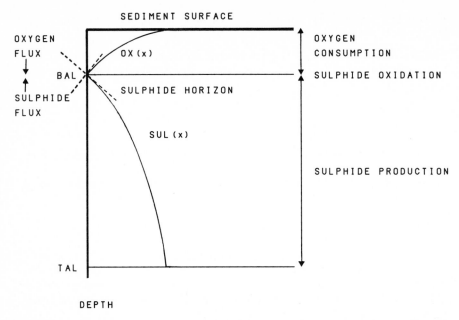

Fig. 7.15. The stationary concentrations of oxygen [ox(x)] and sulphide [sul(x)] in the model as functions of the depth x (cm). At depth *BAL* the oxygen flux downward and the sulphide flux upward cancel by sulphide oxidation. Depth *TAL* is the lower boundary of the biologically active sediment layer

what transport mechanism(s) could explain the flux of oxygen to the sulphide horizon? There may be other oxidizing substances as the gradual decrease with depth of the redox potential indicates (Fig. 7.12). They could act as carriers, as a kind of hidden oxygen. On the other hand, the discrepancy between measurements with microelectrodes of the penetration depth of oxygen and the location of the sulphide horizon do not necessarily indicate a time-averaged discrepancy between the two because the microelectrode measurements are necessarily spot measurements, representing the situation of the moment. Keeping in mind the existence of potential transport mechanisms such as tidal pumping, we postulate that for our purpose of simulating system-wide exchanges of material between the benthic layers our approach, as given in Box 7.1, suffices.

With the formulas from Box 7.1 it is possible to calculate BAL at every time step, assuming equilibrium between the opposing fluxes. Yet the shifts of BAL itself demonstrate that there is no static equilibrium. For a correction of the equilibrium formulas to handle the nonequilibrium situation we have to take into account that the upward flux of free sulphide Jsul(BAL) not only depends on the actual BAL but also on its history. This is due to the buffering action of the large amount of FeS during changes of the sulphide horizon. We model this buffering action by introducing empirically Jsul(BAL) · (1 – DL / DAMP) instead of Jsul-(BAL) in the formula for BAL. Here DL is the change of BAL on the actual day and DAMP is a parameter, describing the buffering. DAMP has the meaning of a maximally allowed change of BAL per day. With that DL is implicitly given by:

$$BAL + DL = \frac{OXO \cdot Dox}{BOC/2 + Jsul(BAL) \cdot (1 + DL/DAMP)}.$$

Here BAL is the position of the sulphide horizon on the day before and BAL + DL the new value. The approximate solution of this equation is:

$$DL = \frac{OXO \cdot Dox - BAL \cdot [BOC/2 + Jsul(BAL)]}{BOC/2 + Jsul(BAL) \cdot (1 + BAL/DAMP)}.$$

With this equation the daily shift of BAL is calculated in the model.

7.5.4 The Transition of Organic Matter Between the Benthic Layers

7.5.4.1 Transitions due to Changes in the Aerobic Layer

When the calculated value of DL is positive the location of the sulphide horizon (BAL) moves downward, if negative, upward. A change of BAL in the model only affects detritus (ADET and BDET), benthic bacteria (ABAC and BBAC) and sulphide oxidation. Macrobenthos is not influenced at all and meiobenthos is thought to migrate vertically thus escaping potential asphyxiation.

When BAL moves downward there is a transition of material from anaerobic conditions into aerobic conditions. Obviously, when BAL moves upward the reverse is true.

7.5.4.2 Transitions due to Sedimentation

The steady accretion of sediment accomplishes a passive shift in the location of the sulphide horizon and thus causes a transition of material from aerobic into anaerobic conditions.

Sedimentation and thus accretion of the sediment varies seasonally. This variation is mainly caused by seasonally varying average wind speeds. In the model this wind effect has been included by the following equations:

$$zSED = 1 - AzMIX \cdot \cos[\pi / 180 \cdot (TIME + 30)].$$

The daily accretion rate is then calculated from:

$$SDLX = SDL \cdot zSED,$$

where the mean accretion rate SDL ($0.0022 \text{ cm} \cdot \text{d}^{-1} = 0.8 \text{ cm} \cdot \text{y}^{-1}$) is derived from Reenders and Van der Meulen (1972), who have calculated this value for the tidal basin of the Dollard. In the western Wadden Sea a nearly similar value of $0.7 \text{ cm} \cdot \text{y}^{-1}$ was found by Van der Goes et al. (1979). The treatment of SDLX in the model is identical to the handling of storm mixing (Sect. 7.5.2).

The sediment accretion due to sedimentation not only transfers material across the sulphide horizon but also over the boundary of the biologically active zone TAL ($= 30$ cm). By this process materials such as benthic refractive organic carbon (BROC), pyrite (BPYR) and sulphide (BSUL) (all in $mgC \cdot m^{-2}$) are buried and leave the system.

7.5.4.3 Transition Effects on the Pyrite Equilibrium

The precipitated FeS ultimately can be transformed into pyrite (FeS_2) by addition of sulphur ($S°$) (Goldhaber and Kaplan, 1974; Krouse and McCready, 1979; Jørgensen, 1983) according to:

$$FeS + S° \rightarrow FeS_2.$$

This is a very slow process and FeS and $S°$ are therefore able to coexist in sediments. The exact mechanism of pyrite formation is, however, still not quite clear (Jørgensen, 1983). Pyrite is the stable form of iron sulphide under natural reducing conditions. Hence it is resistant to oxidation at neutral pH and it is the main sink of sulphide in marine sediments. In contrast to FeS, pyrite dissolves only in oxidizing acids (Jørgensen, 1983).

Pyrite may reach high concentrations in sediments with active sulphate reducers. In compartment 1, pyrite constitutes about 2% of the wet weight of the sediment and it exceeds the amount of sulphate by more than an order of magnitude (Schröder, unpublished data). Thus the sediment acts as a sulphur trap (see Fig. 7.8) by extracting and concentrating the diffusible sulphate sulphur in the form of nondiffusible pyrite sulphur which eventually is buried.

In the model, pyrite formation is described by a simple time constant FLSP (d^{-1}) according to:

$$Ppyr = FLSP \cdot BSUL,$$

where Ppyr = pyrite formation rate (in $mgC \cdot m^{-2} \cdot d^{-1}$). Pyrite (BPYR in $mgC \cdot m^{-2}$) leaves the system only due to the accretion of the sediment. This burying of pyrite (SEDPYX) is expressed in $mgC \cdot m^{-2} \cdot d^{-1}$. Assuming a pyrite equilibrium, the pyrite formation rate Ppyr has to be equal to the pyrite burying rate (SEDPYX). This equilibrium implies:

$$\frac{BSUL}{BPYR} = \frac{SDLX \cdot FLSP}{TAL}$$

in each compartment. Since the only compartment-dependent variable on the right-hand side of this equation is the sedimentation rate SDLX, the ratios in the compartments should vary with the sedimentation rates. In Chapter 11 we will return to this question.

7.6 Benthic Bacteria

The interaction between vertical transport processes in the sediment and the dynamics of the benthic bacteria determines the horizon of the oxygen boundary and hence the division between aerobic and anaerobic mineralization. The vertical transport of organic matter and oxygen into deeper sediment layers and the upward transport of e.g. hydrogen sulphide thus plays a key role for the benthic bacteria. The treatment of these transport processes is given in Section 7.5. Here we describe the modelling of the carbon flow through aerobic bacteria in the oxygenated surface layer and through anaerobic bacteria in the deeper anoxic sediment layer.

The modelling of the benthic bacteria has to accommodate incompletely explored processes of aerobic and anaerobic metabolism. The carbon flows between organic matter and bacteria in the benthos, as incorporated in this model, are given in Fig. 7.1. The benthic bacteria have been modelled in essentially the same way as the pelagic bacteria (see Sect. 6.4), but the mineralization of refractory organic matter by benthic bacteria involves sequential degradation by various metabolic groups over a longer time span than usually occurs in the pelagic. Therefore we introduced a number of regulations in the benthic submodel, such as excretion and cometabolism of refractory organic matter that were omitted from the pelagic.

The degradation of organic matter in marine sediments has been presented in static models by Billen (1982a, b), using three classes of organic matter with different decomposition rates; the same concept has been used in the present model. Billen (1982b) calculated the degradation rates of the three classes of organic matter, assuming first-order kinetics. In the present model, however, we model the degradation dynamically as a function of assimilation, excretion and respiration by bacteria.

7.6.1 Distribution of Biomass

The benthic bacteria have been aggregated into two state variables, the aerobic bacteria [BBAC(I)] and the anaerobic bacteria [ABAC(I)]. The total biomass of the benthic bacteria [TBAC(I)], as determined by epifluorescence microscopy of acridine-stained samples, reaches values that are at least one order of magnitude higher than those in the pelagic. Schröder and Kop (unpublished) found a total biomass of $2–10$ gC $\cdot$ m^{-2} in the Ems estuary (cf. Rublee et al., 1983). The metabolic activity of these populations, as determined by O_2 consumption (Van Es, 1982a) or sulphate reduction (Schröder, unpublished) indicated low metabolic rates per unit biomass, probably caused by a high percentage of dormant bacteria. Therefore we gave low estimates for e.g. specific rates of substrate uptake (Sect. 7.6.2).

The aerobic bacteria [BBAC(I)] were assumed to be fully heterotrophic; the existence of sulphide-oxidizing autotrophic bacteria was neglected. The broad definition of the aerobic zone led to the inclusion of fully aerobic bacteria as well as facultative anaerobes into the state variable BBAC(I). The anaerobic bacteria [ABAC(I)] encompass the sulphate-reducing bacteria that usually dominate the estuarine anaerobic benthos (Jørgensen, 1983; Jørgensen and Sørensen, 1985; Nedwell, 1982). Fermentative bacteria may provide essential substrates for the sulphate-reducing bacteria (Laanbroek and Veldkamp, 1982). Also nitrate reduction or methane formation may contribute to the anaerobic metabolism in the anoxic sediment. In estuarine sediments nitrate concentrations are relatively low, so nitrate is rapidly exhausted as an electron acceptor. Sulphate, on the contrary, is available in large amounts, so it is the quantitatively most important electron acceptor under anaerobic conditions. Methanogenesis is only possible at redox potentials still lower than those needed for sulphate reduction. In practice, in marine sediments methanogenesis will become important when sulphate is exhausted, but substrates such as acetate and hydrogen are still available. The unknown, but probably modest contribution of the latter two groups in the Ems estuary has implicitly been included in the model, since this essentially models the degradation of organic carbon. The comparison of carbon units degraded with units reductant (H_2S, N_2 or CH_4) produced was carried out during the verification of the model (Chap. 11).

7.6.2 Uptake of Labile Organic Carbon

Labile organic carbon in the aerobic and anaerobic layers (BLOC and ALOC respectively) is assumed to be completely assimilable on the day of production (or on the day after arrival in the sediment). This is expressed in the equations for available aerobic and anaerobic labile carbon (RLOC and RLOCA respectively):

$$\text{RLOC} = \text{BLOC(I)} \cdot \text{QLOC (QLOC} = 1 \text{ d}^{-1})$$
$$\text{RLOCA} = \text{ALOC(I)} \cdot \text{QLOC}.$$

The actual uptake of labile organic carbon by benthic bacteria is calculated in a slightly different way than in the pelagic. First, the potential uptake rate is calculated. For the aerobic bacteria:

$$\text{FBAC} = \text{upBAC} \cdot \text{BBAC(I)} \cdot \text{zt3} \cdot \text{zBAC},$$

where zBAC = ksBAC / [ksBAC + BBAC(I) / BAL(I)].
For the anaerobic bacteria:

$$\text{FBACA} = \text{upBACA} \cdot \text{ABAC(I)} \cdot \text{zt3A} \cdot \text{zBACA},$$

where zBACA = ksBACA / [ksBACA + ABAC(I) / BAL(I)].
    The maximum specific uptake rates UPBAC and UPBACA were estimated at a value of 0.75 d^{-1}, much lower than in the bacterioplankton, in view of the large fraction of the benthic bacteria that are usually dormant. The potential uptake rate (FBAC) is temperature-dependent (through the factor zt3) and density-dependent by the Monod-type expressions for zBAC and zBACA. This density dependence is linked to the finite diffusion rate in the neighbourhood of dense colonies of bacteria which become growth-limiting at high bacterial densities. The half-saturation values for this process are 2000 (ksBAC) and 10000 $\text{mgC} \cdot \text{m}^{-2}$ (ksBACA) for aerobic and anaerobic bacteria respectively. Finally, the actual uptake rate by the bacteria (W83 and W83A) is calculated from the available labile carbon and the potential uptake rate by the equations:

$$\text{W83} = \text{RLOC} \cdot \text{FBAC} / (\text{FBAC} + \text{RLOC});$$
$$\text{W83A} = \text{RLOCA} \cdot \text{FBACA} / (\text{FBACA} + \text{RLOCA});$$

expressing some degree of mismatch between substrate supply and bacterial assimilation, caused by spatial heterogeneity in the sediment.
    One transport function should be mentioned here (the others being treated in Sect. 7.5): the available labile organic carbon (RLOC) that is not used by the aero-

bic bacteria is assumed to be available to those anaerobic bacteria that live in the oxygen transition zone. This downward transport from aerobic to anaerobic layers (T8) amounts to:

$$T8 = RLOC - W83.$$

7.6.3 Detritus and Refractive Organic Matter

In compacted layers of organic matter, such as in sediments, the degradation of the more refractive compounds probably depends to some degree on the heterotrophic utilization of simple substrates. This microbial phenomenon, generally known as cometabolism has been included in the formula expressing the bacterial assimilation of detritus (BDET and ADET) and refractive organic matter (BROC). First, the potential uptake from the various sources is calculated:

of aerobic detritus:

$$W'23 = BDET(I) \cdot FRBD \cdot zt3, \text{ with } FRBD = 0.01 \text{ d}^{-1};$$

of anaerobic detritus:

$$W'23A = ADET(I) \cdot FRBDA \cdot zt3a, \text{ with } FRBDA = 0.01 \text{ d}^{-1};$$

of aerobic BROC:

$$W'13 = BROC(I) \cdot FRBR \cdot BAL / TAL \cdot zt3, \text{ with } FRBR = 2 \cdot 10^{-4} \text{ d}^{-1};$$

of anaerobic BROC:

$$W'13A = BROC(I) \cdot FRBRA \cdot [TAL - BAL(I)] / TAL \cdot zt3a, \text{ with}$$
$$FRDRA = 2 \cdot 10^{-5} \text{ d}^{-1},$$

where the FR● constants determine the turnover rates of the carbon sources, by limiting the availability of the detrital substrates to mineralization. Then, to introduce the effect of cometabolism these rates are corrected for the actual rate of uptake of labile organic carbon. The application of the correction factors (zDup to zRupA) introduces a tendency of the bacterial population to utilize increasing amounts of detritus and refractive organic matter with increasing uptake of labile organic carbon. The uptake equations now become for:

aerobic detritus:

$$W23 = W'23 \cdot zDup, \text{ where } zDup = 2 \cdot W83/(W'23 + W83);$$

anaerobic detritus:

$$W23A = W'23A \cdot zDupA, \text{ where } zDupA = 2 \cdot W83A/(W'23A + W83A);$$

aerobic BROC:

$$W13 = W'13 \cdot zRup, \text{ where } zRup = 2 \cdot W83/(W'13 + W83);$$

anaerobic BROC:

$$W13A = W'13A \cdot zRupA, \text{ where } zRupA = 2 \cdot W83A/(W'13A + W83A).$$

Finally, the total uptake of aerobic and anaerobic bacteria is obtained by adding the individual rates calculated for the three classes of organic compounds:

for aerobic bacteria W3 = W83 + W23 + W13,
and for anaerobic bacteria W3A = W83A + W23A + W13A.

7.6.4 Mortality and Excretion

Mortality and excretion in benthic bacteria is assumed to be tightly coupled with the suitability of the organic substrates for growth. The mortality in aerobic bacteria (mor3) is given by the equation:

mor3 = morBAC · BBAC(I) · zt3 · zmor3,

where morBAC was tentatively given a low value of $0.035 \, d^{-1}$. The term zmor3 expresses the dependence of the mortality on an appropriate supply of labile organic matter as substrate:

zmor3 = 1 − MIN(1,W83 / W3).

In comparison to the bacterioplankton, the benthic bacteria are assumed to have higher excretion rates due to the coexistence of several metabolic types that are involved in the degradation of benthic organic matter (cf. Laanbroek and Veldkamp, 1982). The excretion of metabolic intermediates is defined as a fraction of the substrate uptake, but mortality also may be treated as a form of excretion since it regenerates organic substrates. For aerobic bacteria:

W38 = W83 · exlBAC + W23 · exdBAC + W13 · exrBAC
       + mor3 · (1 − pm32),

where exlBAC, exdBAC and exrBAC have values of 0.1, 0.05 and 0.05 respectively. The last term (pm32 = 0.2) specifies that 80% of the cell contents of dying cells are recovered as labile organic carbon, the remaining 20% going to detritus.

Bacterial utilization of detritus is assumed to lead to a modest production rate of refractive carbon:

W31 = W23 · tdrBAC,

where tdrBAC has a conservative value of 0.05. This incomplete stripping of assimilable compounds from detritus may seem insignificant at first glance, but in the long run it is responsible for the accumulation of huge masses of refractive organic carbon in estuaries.

The equations for the mortality and excretion have been given above for aerobic bacteria only; the anaerobic bacteria have been modelled in exactly the same way.

7.6.5 Respiration and Temperature Correction

The respiration of benthic bacteria in the model is composed of activity respiration, taken as a fraction of the substrate uptake, and a biomass-dependent respiration, depending on temperature. For aerobic bacteria the total respiration is:

$$W30 = W83 \cdot arlBAC + W23 \cdot ardBAC + W13 \cdot arrBAC$$
$$+ BBAC(I) \cdot rrsBAC \cdot zt3,$$

where arlBAC, ardBAC, arrBAC and rrsBAC have values of 0.5, 0.65, 0.9 and 0.01 respectively. These parameters stipulate how much of the uptake from a given source is respired away. For anaerobic bacteria the formula is similar:

$$W30A = W83A \cdot arlBACA + W23A \cdot ardBACA + W13A \cdot$$
$$arrBACA + ABAC(I) \cdot rrsBACA \cdot zt3a,$$

where arlBACA, ardBACA, arrBACA and rrsBACA have values of 0.65, 0.75, 0.9 and 0.01 respectively.

The highest respiration of 0.9 associated with the use of refractive organic carbon leaves only 5% of the substrate to be assimilated after subtraction of excretion. The respiration rate of the anaerobic bacteria is set somewhat higher, to reflect the higher energy costs of anaerobic life (Valiela, 1984). The respiration and growth efficiency of aerobic bacteria is similar to that of their planktonic relatives (see Sect. 6.4). Respiration is presented here in terms of carbon loss, but usually this process is measured by the consumption of oxygen or in the case of anaerobic bacteria in estuarine sediments, by the reduction of sulphate.

The conversion from respiratory carbon loss to the uptake in oxygen or sulphate equivalents is discussed in Section 7.5, where the diffusion of these compounds in the sediment is treated and in Chapter 11, where the model results are presented.

The temperature correction terms, zt3 for the aerobic benthic bacteria and zt3A for the anaerobic ones, modify the respiration rates; they increase the respiration for temperatures over the reference value and decrease it for temperatures below the reference temperature. The reference temperature is the annual mean temperature (13.5 °C in the aerobic layer and 11.5 °C in the anaerobic layer).

7.6.6 Conclusion

In modelling the dynamics of benthic bacteria we need nearly 30 parameters to model the flow of organic matter through these organisms. In view of the lack of experimental data on these parameters one might urge for a much simpler model, needing a smaller number of parameters. Yet we felt that the complex submodel presented here accommodated the sequential degradation and recycling of organic matter in the benthos better than could have been achieved in a simpler model.

An obvious simplification in the model would be not to discriminate between aerobic and anaerobic mineralization. This would have eliminated half of the 30 parameters at one time. However, the whole matter how estuarine sediments cope with temporally and spatially varying organic carbon inputs by shifting the predominant mode of mineralization from aerobic to anaerobic and vice versa is too fascinating to simply neglect. The approach we take in this model at least offers the possibility to explore the dynamics of both the aerobic and the anaerobic community in the sediment.

The output generated by this part of the model can only partly be verified. The processes modelled here depend to a large extent on transport processes in the sediment (Sect. 7.5). The interdependence between the functioning of the benthic microbial system and the transport processes determines the depth of the oxygen boundary layer and hence the partitioning between aerobic and anaerobic mineralization. As we will show in Chapter 11, the results indicate that the model reflects the verifiable aspects of the real system.

8 The Construction of the Epibenthic Submodel

J. G. BARETTA-BEKKER and A. STAM

8.1 The State Variables

The epibenthic submodel is a relatively small one, containing only two state variables. Each state variable consists of a potpourri of organisms and is not representative for any one particular species. The reason the epibenthos has been thought worthy of a separate submodel is because it neither fits in the benthic, nor in the pelagic system. It is a link between both submodels, eating mainly from the benthic system and migrating through the pelagic system, according to the state of abiotic factors such as salinity, water temperature and oxygen saturation.

Only the epibenthos that forages on the tidal flats has been modelled. A part of it migrates to the tidal channels at low tide, another part stays on the tidal flats during low tide, buried in the sediment.

Generally, the species that have a tidal and a seasonal migration have been included in the macroepibenthos, whereas the species that stay on the tidal flats and do not generally migrate from the estuary, have been included in the mesoepibenthos.

The property these organisms all have in common is that they can actively migrate through the whole estuary. The epibenthic species are divided over two state variables:

1. *The mesoepibenthos* (EMES), in general the smaller species, staying in the estuary during the whole year;
2. *The macroepibenthos* (EMAC), the species belonging to this group arrive in huge numbers as larvae during spring in the estuary and they return in much reduced numbers, but with a higher total biomass to the sea in autumn. For this group the estuary functions as a nursery area.

Apart from the meso- and macroepibenthos there is a third group of organisms included in this submodel: *the birds*. They also are able to migrate over the estuary and are eating mainly benthic organisms. This group is not modelled as a state variable, but their consumption is included in the model. The carbon flows in the epibenthic submodel have been schematized in Fig. 8.1.

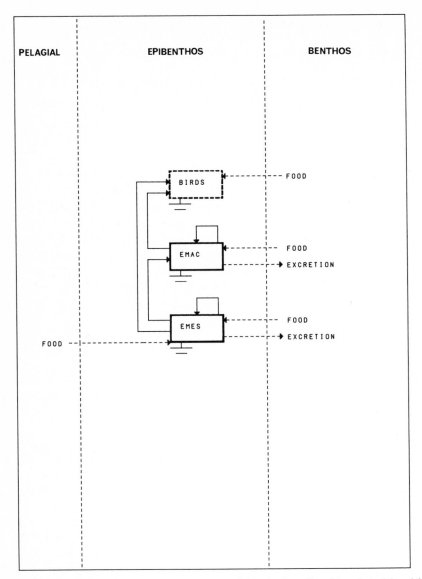

Fig. 8.1. The carbon flows between the state variables in the epibenthic submodel and from and to the other submodels

8.2 The Faunal Components of the Epibenthos

In the Ems estuary the total biomass of the functional group mesoepibenthos (EMES) consists of roughly 25% *Corophium*, of another 25% *Carcinus* and 50% *Nereis* (Van Arkel and Mulder, 1979; Stam, 1982). In the model the mesoepibenthos stays in the estuary during the whole year. It starts increasing in biomass in spring, reaches a maximum in summer and decreases due to predation and lower growth rates in autumn. During the winter the biomass of mesoepibenthos is very low. It forages on the submerged tidal flats and is able to move around in the estuary. Every day a fraction of the total biomass is redistributed over the estuary, depending on biotic and abiotic factors in the different compartments. The size of the migrating fraction depends on the environmental circumstances.

Shrimps (*Crangon crangon*) and plaice *(Pleuronectes platessa)* make up the major part of the macroepibenthos (EMAC). In biomass *Crangon* exceeds *Pleuronectes* by a factor of 5 (Stam, 1982). The macroepibenthos immigrates into the estuary in spring, forages on the tidal flats during high water and has a tidal migration. The assumption is that macroepibenthic organisms do not forage during low tide, when they are in the tidal channels. In autumn the group leaves the estuary, triggered by the time of year and by decreasing water temperature. For the macroepibenthos the same redistribution process is used as for the mesoepibenthos: every day a part of the macroepibenthos biomass redistributes itself over all the compartments. For the macroepibenthos this is a fixed part of the biomass, independent of the environmental circumstances.

The activities of epibenthos have been schematized and aggregated into a number of processes which have been translated into mathematical expressions and incorporated into the model. The processes that are common to all living organisms, such as food uptake, respiration, excretion and mortality have been described in section 4.1 and only deviations in the way processes are formulated will be detailed here.

8.3 Migration

8.3.1 Redistribution

Both macroepibenthos (EMAC) and mesoepibenthos (EMES), having more or less well-developed means of locomotion, can and do move all over the system, concentrating in those compartments where the biotic and abiotic conditions suit them best.

The suitability of the environment in each compartment of the system is expressed in a number of so-called comfort functions (cf. sect. 8.3.2). These comfort functions, together with the amount of food available, determine the preference for the different compartments.

Even for large, actively swimming organisms, such as EMAC and certainly for EMES, the distances to be covered before reaching another compartment are large. Hence, only a part of the total biomass can reach the neighbouring compartment in any one day. From compartment 1 on the inside and compartment 5 on the outside, they can only move in one direction, whereas from the compartments 2, 3 and 4 they can move in either direction. For the macroepibenthos the fraction of the biomass that can reach the next compartment at one side is REDIST. In the model the value of REDIST is estimated to be 0.25. This implies that for compartments 1 and 5 on one side and for the other compartments on both borders of a compartment a quarter of the biomass of macro- and mesoepibenthos may migrate to the neighbouring compartment. For the mesoepibenthos the fraction of the biomass able to move to the adjacent compartment depends on the comfort function $okY(I)$, and is equal to:

$$RDST(I) = 0.1 \cdot REDIST + [1 - okY(I)] \cdot 0.9 \cdot REDIST,$$

where REDIST is the same as for the macroepibenthos. The minimal value of $RDST(I) = 0.1 \cdot REDIST$ and is reached when $okY(I)=1$, while the maximal value $RDST(I) = REDIST$ is reached when $okY(I)=0$.

In the redistribution process no biomass leaves the estuary, neither on the landward side (into fresh water) nor on the seaward side (into the North Sea). Migration across the system boundaries is treated as immigration and emigration (see sects. 8.3.3 and 8.3.4).

Every day a part of the biomass of mesoepibenthos (traEMES) and macroepibenthos (traEMAC) of adjoining compartments equal to:

$$traEMES = RDST(I) \cdot EMES(I) \cdot areaf(I) + RDST(I) \cdot EMES(I+1) \cdot areaf(I+1)$$

and

$$traEMAC = REDIST \cdot EMAC(I) \cdot areaf(I) + REDIST \cdot EMAC(I+1) \cdot areaf(I+1)$$

is redistributed over those compartments. Areaf(I) and areaf(I+1) are the surface areas of compartments I and I+1.

The result of the redistribution over the compartments depends on the preferences (PRMES; PRMAC) for the different compartments. The preferences are calculated in the following equations from the values of the comfort functions (okY; okZ) and the amount of available food (fdY; fdZ), multiplied by the time the flats are submerged (ptWET), since in the model the epibenthos is only foraging during high water!

$$PRMES(I) = okY(I) \cdot areaf(I) \cdot fdY(I) \cdot ptWET(I);$$

and

$$PRMAC(I) = okZ(I) \cdot areaf(I) \cdot fdZ(I) \cdot ptWET(I).$$

The change in biomass of meso- and macroepibenthos per m^2 (TEMES; TEMAC) due to the redistribution process for compartment I is equal to:

$$TEMES(I) = [traEMES \cdot PRMES(I)/ENMES]/areaf(I) + TEMES(I) - RDST(I) \cdot EMES(I)$$

and

$$TEMAC(I) = [traEMAC \cdot PRMAC(I)/ENMAC]/areaf(I) + TEMAC(I) - REDIST \cdot EMAC(I),$$

where $ENMES = \sum_{I=1}^{5} PRMES(I)$ and $ENMAC = \sum_{I=1}^{5} PRMAC(I)$.

8.3.2 Comfort Functions

The suitability to epibenthos of the different compartments in the estuary is defined as being dependent on water temperature, salinity and oxygen saturation of the water. Here the conditions as exist in the pelagic determine the "comfort" value.

For each of these parameters a value between 0 and 1 is calculated, using a Monod-type function (Kremer and Nixon, 1978) where the half-value constant determines the parameter value resulting in a function value of 0.5 and where the exponent in the function determines the steepness of increase or decrease. Increasing the value of the exponent changes this type of function into a switch function, going from 0 to 1 at a small change in the parameter value (Fig. 8.2).

For mesoepibenthos there is only an oxygen comfort function reflecting the suitability, because of the fact that mesoepibenthos is a very heterogeneous group whose component species have widely divergent salinity and temperature preferences. Only low oxygen saturation is a compelling reason for mesoepibenthos as a group to migrate to other parts of the estuary.

$$okY' = [rOSAT(I) / koxY]^{xoxY};$$
$$okY(I) = okY' / [okY(I) + 1];$$

where rOSAT is the relative oxygen saturation, calculated according to:

$$rOSAT(I) = OX(I) / OSAT,$$

where OX(I) is the actual oxygen concentration in compartment I and OSAT is the oxygen saturation value (Sect. 6.5)

The macroepibenthos tries to avoid water of too low or too high temperatures, too low oxygen saturation and too low salinities. The suitability of the environment concerning the temperature is determined in the next equations:

$$oktZmin' = (CTEMP / ktZmin)^{xtZ};$$
$$oktZmin = oktZmin' / (oktZmin' + 1);$$

for the minimum temperature which still is tolerated and

$$oktZmax' = (CTEMP / ktZmax)^{xtZ};$$
$$oktZmax = 1 - oktZmax' / (oktZmax' + 1);$$

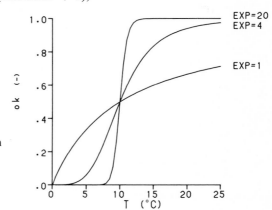

Fig. 8.2. The variation in the function response around the half-value constant (= 10) by varying the exponent in the function

for the upper temperature, where for both cases CTEMP is a check to avoid negative values:

CTEMP = MAX[0,TEMP(I)].

The suitability of the actual salinity is expressed in oksalZ:

$\text{oksalZ}' = (\text{SALT(I)} / \text{ksalZ})^{\text{xsalZ}};$
$\text{oksalZ}' = \text{oksalZ}' / (\text{oksalZ}' + 1);$

and concerning the oxygen saturation in okoxZ:

$\text{okoxZ}' = (\text{rOSAT(I)} / \text{koxZ})^{\text{xoxZ}};$
$\text{okoxZ} = \text{okoxZ}' / (\text{okoxZ}' + 1).$

The product of the various comfort functions reflects the suitability of the abiotic environment of compartment I to the macroepibenthos.

$\text{okZ}'(\text{I}) = \text{okoxZ} \cdot \text{oksalZ} \cdot \text{oktZmin} \cdot \text{oktZmax}.$

To avoid mass migration of macroepibenthos in winter when the water temperature is very low, from the compartments with an okZ value equal to zero to the compartments with an okZ value close to but not equal to zero, an artificial correction for okZ(I) is made:

$\text{okZ}'(\text{I}) = \text{okZ(I)} + \text{I}^2 \cdot 1.\text{E-6}$ (I is the compartment index).

This equation ensures that under identical circumstances the outer compartments have a slightly higher okZ value, and so are more attractive to macroepibenthos.

The parameter values of the comfort function for oxygen saturation for the mesoepibenthos are estimates. The mesoepibenthos seem to be very tolerant to low oxygen (Van Arkel and Mulder, 1979). Only a very low value of oxygen saturation will force them to move to more suitable areas. The parameter values are 0.2 for the half-value constant koxY and 8 for the exponent xoxY (Fig. 8.3). Macroepibenthos is less tolerant to low oxygen saturation. Verwey (1971) stated that very young fish already feel unhappy in water with an oxygen saturation of 90%, while Eggink (1965) mentioned a limit of 50% for older fish. Together with the

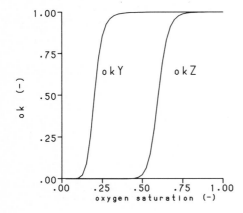

Fig. 8.3. The comfort function (*ok*) of mesoepibenthos (*ok Y*) and macroepibenthos (*okZ*) for oxygen saturation

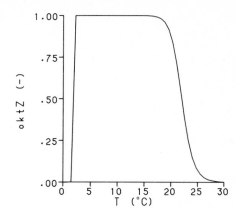

Fig. 8.4. The comfort function of macro-epibenthos (*oktZ*) for temperature, the result of combining the functions oktZmin and oktZmax

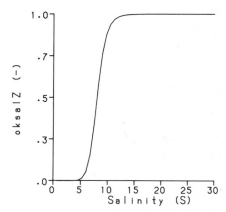

Fig. 8.5. The comfort function of macroepibenthos for salinity

observation on behaviour of shrimps (Stam, 1982), a value of 0.6 equalling an oxygen saturation of 60% is chosen for the half-value constant koxZ. Assuming that a switch function is more realistic than a sigmoid function, a very high value is chosen for xoxZ (Fig. 8.3). The half value constant for low temperatures is set at 2 °C (ktZmin) and for high temperatures at 22 °C (ktZmax). These values are based on investigations of Stam (1979), Fonds (1973) and Van Donk and De Wilde (1981). Here again an abrupt response with xtZ = 20 is chosen (Fig. 8.4). The salinity half-value constant (ksalZ) is set on 8 S (Stam, 1979; Fonds, 1973). The steepness of the function (xsalZ) is set on 10 (Fig. 8.5).

8.3.3 Immigration

Macroepibenthos shows a seasonal migration into the estuary and this migration is primarily governed by temperature (key reference: Stam, 1982). This immigration starts when the sum of daily water temperatures from Jan. 1 exceeds the threshold value TIMM. This kind of triggering is chosen to accomodate the annual differences in the spring increase in water temperatures. It results in a shift in the actual start of immigration (DSTART) depending on the severity of the

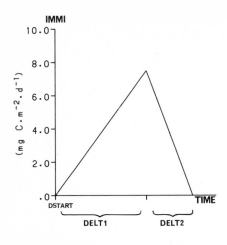

Fig. 8.6. Immigrating biomass of macro-epibenthos, with *DELTI* = 40 days and *DELT* = 20 days. The *area under the curve* is the total biomass immigrating into the system (SETTLE) in mgC · m^{-2} in compartment 5

preceding winter. A mild winter, with fairly high water temperatures will be followed by an early start in immigration, whereas a severe winter will result in a delay in immigration. After a period of DELT1 days, the immigration rate reaches a maximum. From that day on it declines for a period of DELT2 days and then stops (Fig. 8.6). SETTLE is the total biomass of macroepibenthos settling in compartment 5 during one season divided by the size of compartment 5 in m^2.

In the model the temperature sum TIMM is set on 400 °C, DELT1 = 40 days and DELT2 = 20 days, while SETTLE = 225 mgC · m^{-2}. These values are based on data given by Stam (1982). The daily immigrating biomass is calculated with:

TMAXIN = MTIME + DELT1,
TENDIN = TMAXIN + DELT2,
and MAXSETTLE = 2 · SETTLE / (DELT1 + DELT2).

During the first DELT1 days when the immigration is increasing, on day number MTIME:

IMMI = (MTIME – DSTART) · MAXSETTLE / DELT1.

In the following DELT2 days when the immigration is decreasing, on day number MTIME:

IMMI = (TENDIN – MTIME) · MAXSETTLE / DELT2.

The parameter SETTLE now is a constant, but where long-term data on the settlement are available, it should be given as a time series for a number of years.

All settlement takes place in the outer compartment. The redistribution process takes care of further migratory movement between the compartments.

Immigration as modelled here is a passive process. A certain biomass is brought into compartment 5 from outside the estuary and settles on the tidal flats. It does not seem to be realistic for this estuary to couple the total amount of immigrating biomass to the emigrating biomass of the previous year, because the immigrating biomass consists predominantly of larval stages of flatfishes with only low amounts of the older year classes (Stam, 1982). The coupling between the bio-mass emigrating one year and the amount returning the next, is not at all well es-tablished. There may be a coupling between immigrating larval biomass and the density of invertebrate predators during the larval stages (Van der Veer and Berg-man, 1987), implying that regulation of the immigrating biomass takes place out-side the estuary.

8.3.4 Emigration

In autumn, when the water temperature in the estuary falls below the North Sea temperature, the macroepibenthos starts to emigrate from the estuary (key refer-ence: Stam, 1982).

In the model, the timing of this emigration is triggered by time of the year as well as by temperature. After day number DCRIT (day 190), emigration will start once the temperature in a compartment falls below TEMI ($=12\,°C$).

Emigration, in contrast to immigration, is modelled as an active process. Daily, a certain biomass EMAC migrates to the next, more outward compart-ment. The maximum fraction of biomass that emigrates daily is defined by routZ. This maximal fraction is reached over a certain time period. The length of this period, or, in other words, the steepness with which the emigration rate increases, depends on the difference in time between the first possible emigration time DCRIT and the time TCRIT, when the water temperature sinks to the critical level to trigger emigration.

In a warm fall, with high water temperatures, emigration [EXPRT(I)] will be delayed beyond DCRIT and then, once started, will increase sharply to routZ (0.067). In a cold fall, emigration will start early and will increase slowly to routZ (Fig. 8.7). Example:

DEXPRT' $= (MTIME / TCRIT)^{(TCRIT-DCRIT)}$;
DEXPRT' $= DEXPRT' / (DEXPRT + 40)$;
EXPRT(I) $= -DEXPRT \cdot routZ \cdot EMAC(I)$.

Again the data from Stam (1982) are the basis of the chosen parameter values.

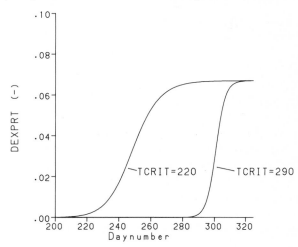

Fig. 8.7. Effect on the emigra-tion rate of macroepibenthos of low water temperatures early (TCRIT) is day number 220) and late respectively (TCRIT = 290) in autumn

8.4 Uptake of Food

8.4.1 Calculation of Available Food

Mesoepibenthos (EMES) has a number of different food sources: benthic bacteria (BBAC), benthic diatoms (BDIA), benthic detritus (BDET), meiobenthos (BMEI), mesozooplankton (PCOP) and EMES itself. Only a fraction of each food source is available for uptake. How large these fractions are, is expressed in availability factors. The final estimate of these availability factors reconciles different properties of the food source:
1. The availability, reflecting the relationship between the food sources and the species composition of the mesoepibenthos.
2. The edibility of the food source. Benthic ditritus has a low edibility, and is thought to be eaten in small quantities by only a small part of the mesoepibenthos and hence has a very low value.

Because of the heterogeneity of EMES and uncertainty about the relative contributions of its constituent parts to the whole, it is very difficult to use experimentally derived values for the availability factors. It should be pointed out that the use of these factors is inevitable, otherwise detritus would automatically become the largest food source because of its high concentration.

Apart from those properties of a food source there is another factor, influencing the amount of available food. It is the accessibility. Only a fraction of a food source may be accessible to mesoepibenthos.

It has been assumed that EMES in searching for food only can reach food particles within the upper half cm of the sediment. Therefore, in determining the total amount of food available to EMES, the fraction of the benthic food sources located in this upper half cm is calculated. Most benthic food sources occur in the aerobic layer, whose extent is determined by the sulphide horizon (BAL). Because the thickness of the aerobic layer is not constant, the accessibility of BBAC, BDIA, BMEI and BDET changes. In the model this is expressed by multiplying the availability factor ($u \bullet Y$; $u \bullet Z$), where $\bullet$ may be any of the food sources, by:

$$F(I) = MIN[1, 0.5 / BAL(I)].$$

Because the concentration of BDIA in the sediment decreases with depth it is not possible to calculate the available biomass of benthic diatoms with the factor $F(I)$. Instead of $F(I) \cdot BDIA$ the state variable CDIA, being the biomass of benthic diatoms in the upper 0.5 cm of the sediment is used (cf. Sect. 7.2).

The amount of mesozooplankton (PCOP) available to EMES is determined by the average water depth above the tidal flats [depfa(I)] in the compartments and the time the tidal flats are submerged (ptWET). Thus the available food for EMES is:

$$\begin{aligned} fdY(I) = \{&[u3Y \cdot F(I) \cdot BBAC(I)]^{xfd} + [u4Y \cdot CDIA(I)]^{xfd} + [u2Y \cdot F(I) \cdot \\ &BDET(I)]^{xfd} + [u5Y \cdot F(I) \cdot BMEI(I)]^{xfd} + [uYY \cdot \\ &EMES(I)]^{xfd} + [uHY \cdot PCOP(I) \cdot depfa(I)]^{xfd}\}^{1/xfd} \end{aligned}$$

EMES (and EMAC) in the model behave as opportunistic feeders, concentrating on the most abundant food source. This behaviour is incorporated in the model by the use of the exponent xfd in the calculation of food availability. When xfd > 1 relatively more food is taken from the abundant food sources than from the less abundant ones. When xfd = 1, uptake from a food source is strictly proportional to its abundance. The values for the parameters concerning the food availability have been calibrated by adjusting their values to obtain plausible fluxes from model runs (see Chap. 12).

Macroepibenthos. The food sources of macroepibenthos are macrobenthos (BBBM and BSF), macroepibenthos (EMAC), mesoepibenthos (EMES) and meiobenthos (BMEI). The calculation of the available food for macroepibenthos is done in the same way as for mesoepibenthos with one important difference: for macroepibenthos the total biomass of the deposit-feeding macrobenthos (BBBM) (polychaetes etc.) is thought to be available as food, since the larger EMAC is able to pull e.g. whole worms from the sediment, making it irrelevant that these deposit feeders have most of their biomass fairly deep in the sediment. For the BMEI, however, only that part that occurs in the upper half cm of the sediment is accessible to EMAC. The available food for EMAC is:

$$fdZ(I) = \{[u5Z \cdot F(I) \cdot BMEI(I)]^{xfd} + [uZZ \cdot EMAC(I)]^{xfd} + [u6Z \cdot BBBM(I)]^{xfd} + [u7Z \cdot BSF(I)]^{xfd} + [uYZ \cdot EMES(I)]^{xfd}\}^{1/xfd}$$

8.4.2 Uptake

The food uptake by mesoepibenthos (upY) and macroepibenthos (upZ) is calculated by first obtaining the relative rate of uptake (rupY and rupZ) and by multiplying this with the standing stock:

$$rupY = ztY \cdot rupYmax \cdot ptWET(I) \cdot fdY(I) / [fdY(I) + kfdY];$$
$$upY = rupY \cdot EMES(I);$$
$$rupZ = ztZ \cdot rupZmax \cdot ptWET(I) \cdot fdZ(I) / [fdZ(I) + kfdZ];$$
$$upZ = rupZ \cdot EMAC(I).$$

The epibenthos is only able to forage when the tidal flats are submerged. When the flats are exposed the macroepibenthos has migrated to the channels and the mesoepibenthos has dug itself into the sediment. Therefore the uptake depends on the time the tidal flats are submerged (ptWET). The uptake is corrected for temperature by ztY and ztZ. When the food supply is very large in comparison to the half-saturation value (kfdY), the fraction fdY(I)/[fdY(I) + kfdY] comes close to 1 and the uptake rate will approach the maximal uptake rate rupYmax. With a smaller food supply the uptake rate will be reduced.

The half-saturation value kfdZ is set on 260 mgC·m^{-2}, though Kuipers (personal communication) gives 3300 mgC·m^{-2} as the optimal value for the food density where the uptake can be maximal. Here a much lower value is chosen because most of the availability factors (u●Z) are much lower than 1. In the model only a small part of the total biomass in the sediment is thought to be available as food, whereas Kuipers includes all the biomass present.

8.5 Loss Factors

8.5.1 Respiration and Excretion

In the epibenthic submodel the losses due to maintenance processes such as rest respiration and rest excretion/mortality are a fixed proportion of the biomass (for EMES: YBILOS and for EMAC: ZBILOS), while the losses due to the uptake-related processes such as activity respiration and activity excretion are a fixed fraction of the uptake (YUPLOS and ZUPLOS). The apportioning of these losses to respiration and excretion is managed by the parameter REFRAC (0.5). The following scheme elucidates this for EMES:

Losses in EMES due to the metabolic processes:

	Respiration	Excretion	Total
Maintenance processes	REFRAC · YBILOS = *rrs Yr*	(1-REFRAC) · YBILOS = *rex Yr*	YBILOS
Activity processes	REFRAC · YUPLOS= *prs Ya*	(1-REFRAC) · YUPLOS= *pex Ya*	YUPLOS

A corresponding scheme can be made for EMAC. It should be pointed out that the activity related losses are dimensionless fractions, while the maintenance losses (BILOS) are defined as rates with the dimension d^{-1}.

Using the calculated proportions gives the following equations for:

Respiration:
Activity respiration of EMES: rsYa = prsYa · upY;
Rest respiration of EMES: rsYr = rrsYr · ztY · EMES(I);
Total respiration of EMES: rsY   = rsYa + rsYr.
For EMAC the corresponding calculations are used.

Excretion:
Activity excretion of EMES: exYa = pexYa · upY;
Rest excretion of EMES: exYr = rexYr · ztY · EMES(I);
Total excretion of EMES: exY   = exYa + exYr.
   In EMAC there is still another form of excretion. Shrimps are moulting several times a year. The moulting products are treated as a form of excretion and that part of the biomass will go directly from EMAC to benthic detritus. The moulting products of the shore crab which is a part of EMES are neglected in the model.
Excretion of EMAC, due to moulting: exZm = pZz · upZ;
Total excretion of EMAC: exZ = exZa + exZr + exZm.

8.5.2 Mortality

Natural mortality is included in the excretion equations. Next to the implicit predation on mesoepibenthos by macroepibenthos and by mesoepibenthos itself, there is another explicit loss, due to predation. Pelagic fishes are eating mesoepibenthos. They prey on the migrating part of the mesoepibenthos. Because the pelagic fishes are not modelled as a state variable, it is necessary to introduce predation on EMES by pelagic fishes indirectly by the equations:

$$\text{rmorY} = \frac{\text{cmorY} \cdot 8^4}{\text{MAX}\,[0, \text{temp(I)}]^4 + 8^4} \cdot \frac{\text{EMES(I)}}{2 + \text{EMES(I)}}$$

and

$$\text{morY} = \text{rmorY} \cdot \text{EMES(I)}.$$

The fraction taken by pelagic fishes (rmory) is dependent on temperature. Especially in winter there is a significant stock of pelagic fishes in the estuary (Stam, 1982). In winter, at a very low water temperature the relative mortality fraction (rmorY) is equal or almost equal to the maximal relative mortality fraction (cmorY). The first term in the calculation of rmorY introduces a temperature-dependent grazing mortality which decreases when the water temperature increases. When the biomass of EMES(I) decreases to low values the relative mortality due to predation by pelagic fishes also decreases. In the equation this is accomplished by the second term.

8.6 Correction for Temperature

Biological processes such as uptake, rest respiration and rest excretion are modified by the water temperature. The values of the parameters have been defined at 12 °C. So it is necessary to calculate a correction factor for the prevailing temperature according to the following equation:

$$\text{ztZ}' = 2^{[\text{TEMP(I)}-12] \cdot Q10Z/10}$$

The temperature/activity curves are flattened at high t:

$$\text{ztZ} = 3 \cdot \text{ztZ}' / (2 + \text{ztZ}').$$

We use the same temperature correction for both macro- and mesoepi-benthos with a Q10Z of 2 (De Vlas, 1979).

8.7 The Role of Foraging Birds

Although birds are not a real state variable in the sense that the state of the model determines their abundance and their growth, they do exert an influence on the system through predation on the benthic and epibenthic state variables.

8.7.1 Estimation of the Number of Birds in the Ems Estuary

The monthly means of the numbers of birds occurring in the Dollard, corresponding with the model compartments 1 + 2, are calculated from Bergman and
Dankers (1978) and have been entered as a time series. The number of birds present in any of the compartments has been calculated from these values in combination with preference factors for each compartment (Smit, personal communication).

The preference factors given for each compartment are annual averages.
Therefore also the food supply for the birds in each compartment has been considered in calculating the actual number of birds present in any compartment. If
the food supply in a given compartment is larger than or equal to the average over
the whole system, then the standard preference is used. If the food supply is smaller than the actual average, the preference factor for the compartment is decreased. The following algorithm accomplishes this:

$$PFOOD(I) = MIN [BFOOD(I) / BFOODM,1] \cdot PREF(I),$$

where BFOOD(I) is the amount of available food for birds in compartment I.
BFOODM is the average amount of food for birds over the whole system and
PREF(I) is the preference for compartment I as given by Smit (personal commu-
nication).

The total number of birds in the whole estuary in a certain month is calculated
from the number of birds in compartment 1 + 2 in that month multiplied by the
sum of all the actual preferences and divided by the sum of the preferences for
compartment 1 + 2:

$$\begin{aligned}
PFOODT &= \Sigma \, PFOOD(I); \\
BIRDS &= BIRD[1 + INT(MTIME/30)] \cdot PFOODT / [PFOOD(1) \\
&\quad + PFOOD(2)].
\end{aligned}$$

From this food-weighted compartment preference the number of birds in compartment I is calculated:

$$BIRD(I) = BIRDS \cdot PFOOD(I) / PFOODT.$$

8.7.2 Calculation of Available Food for Birds

The amount of food available to birds is dependent on the abundance of the different food sources. This in turn also depends on the availability factors (u•X)
for these food sources to the birds. Actually, the u•X factors are viewed as accessibility factors more than availability factors. In determining these factors, the
major consideration has been how much chance a bird would have to detect a certain food source. The average size of the food source and its location (exposed,
submerged, buried) also played a role in establishing these factors.

A special feature of the available food calculation for birds is that when
oxygen saturation in the water drops below 10%, which tends to force EMAC
to the surface, the uZX factor jumps from 0.2 to 1, dramatically increasing the

accessibility of EMAC to birds. UhZX is equal to uZX, unless the oxygen saturation is less than or equal to 0.1. In that case $uhZX = 1$:

$$BFOOD(I) = \{ [uhZX \cdot EMAC(I)]^{xfdX} + [uYX \cdot EMES(I)]^{xfdX} + [u6X \cdot BBBM(I)]^{xfdX} + [u7X \cdot BSF(I)]^{xfdX}\}^{1/xfdX}.$$

The exponent xfdX has the same influence as xfd in the calculation of available food for EMES and EMAC (cf. Sect. 8.5).

8.7.3 Uptake by Birds

In calculating the uptake by birds, it has been assumed that the birds eat their fill and then stop. This implies that the total uptake by birds is equal to the daily food requirement per bird (BNEED) times the number of birds in a compartment, unless this total uptake exceeds the amount of food available. Then the total uptake by birds is equal to the amount available:

$$BIUP = \{MIN[BFOOD(I) \cdot areaf(I), BNEED \cdot BIRDN(I)]\}/ areaf(I).$$

The total uptake thus always is the minimum of the amount required and the amount available. The daily food requirement per bird is calculated from the species composition of the birds in any one month (Bergman and Dankers, 1978; Smit, personal communication) and the food requirements of the individual species (Smit, 1980a). These have been weighted by the number of individuals of the different species present, thus resulting in weighted monthly averages of the daily food requirement.

Part III: Results and Analysis

9 Running the Model

J. W. BARETTA

9.1 The Standard Run

In the following chapters the model results are presented and discussed following the same structure as in Part II, by submodel. In the final chapter, an overview is given of the model as a whole. To ensure that exactly the same model version, identical down to the last parameter value, is presented in each of these chapters, we have adopted the concept of the "standard" run of the model. This is less trivial than it might seem, because each simulation run with the model requires the specification of initial conditions for all state variables (33) in the five compartments, and the setting of hundreds of physiological coefficients, all of them having an extended range of possible values. Added to this is the variability in the annual time series for the boundary conditions, such as fresh water discharges by the rivers Ems and Westerwoldsche Aa. Then there are forcing functions such as irradiance and temperature, which vary considerably from year to year.

We have therefore chosen an arbitrary year as the "standard" year and use the boundary conditions and forcing functions from that year in the standard run. This standard year, in our case 1978, must be a specific year because the boundary conditions and forcing functions not only are time dependent but in varying degrees also interdependent. The whole set of boundary conditions and forcing functions therefore must be derived from the same period. In the standard run all parameters (Appendix A) are fixed at their most probable value.

9.2 The Initial Values of the State Variables

Setting up the standard run also requires establishing the initial concentrations of the state variables. The correct approach would be to obtain accurate field estimates of the average concentrations of all the state variables from all compartments on a certain date of a particular year and to use these data as initial values for a model run for that year. Since it was not possible to obtain the initial values in this way, another approach has been taken to estimate the initial values on January 1, which has been used as the starting date of the standard run. It is especially important to get the initial concentrations right relative to each other, since errors in the relative abundances of consumers versus their food distort the interactions between them in the model.

The approach that has been taken to estimate the initial values for the standard run is based on the following considerations about the internal dynamics of the system: As Hairston et al. (1960) pointed out, the accumulation of organic matter in the environment is negligible in comparison to the amount produced by photosynthesis. This implies that the carbon cycle in the ecosystem is in a dynamic equilibrium between production, consumption, decomposition and transport. This forms the basis for the contention that this equilibrium should also exist in the ecosystem model and that the model system therefore should not significantly accumulate, nor lose, organic matter during a long-term run (6–10 years). This is not to say that the whole system is static, only that when the inputs (as quantified in the forcing functions and boundary conditions) are strictly cyclic on a yearly basis, the organic matter cycling through the system should settle into a stable seasonal pattern. The logical consequence of this is that the concentrations of the state variables also return to the same levels at the end of a year as they had on January 1. For this to happen, the starting values of the state variables, especially of long-lived ones such as the macrobenthos are decisive because their level reflects the carbon fluxes of the previous years. If their levels at the start of a run are far from the steady state value it will take 2–3 years before the steady state value is reached through the internal stabilization within the system.

The concentrations of the state variables at the end of the 6-year run which reflect the steady state values under the current set of boundary conditions and forcing functions have been used as the initial concentrations in the standard run of the model which will form the basis of the following chapters. It should be pointed out that because the ecosystem has to deal with stochastic forcing functions and boundary conditions that vary strongly from year to year, the effect of these varying conditions on the slowly responding state variables such as macrobenthic organisms is carried over to following years, thus damping short-term variations in biotic and abiotic conditions.

The output from this standard run will be used to illustrate and validate the functioning of the model and the interactions between its component variables. As in Part II, the different subsystems will be treated in separate chapters, but the final chapter will present an overview over the whole model.

9.3 Validation of the Model

In Chapters 9–13 the model results will be validated with sets of field data that have not been used in constructing the model; as such they can serve to confirm the model structure if the correspondence between model results and observations is deemed satisfactory. If that is not the case, they serve to point out where there may be major gaps in our understanding of the functioning of the system as expressed in the model formulation.

To take some of the inevitable subjectivity out of the judgment as to how well, or badly, the model results and field observations of a variable do agree, a special validation procedure has been developed (Stroo, 1986). In this routine the correspondence between the model results and the field observations for the same variable is given a value between 0, no correspondence, and 10, perfect correspondence. The calculation of this value is based on the concept of Theil (1961) who used the mean square prediction error (MSE) as his basic quantity.

The MSE is defined as $\text{MSE} \left[\sum_{i=1}^{n} (P_i - A_i)^2 \right] \cdot \frac{1}{n}$:

P_i = simulation value at time t;
A_i = observed value at time t;
n = number of (field) observations.

The MSE is decomposed into three components. Dividing these components by MSE one obtains three relative "inequality" indicators: the Mean Component (MC), the Slope Component (SC) and the Random Component (RC). The Mean Component (MC) establishes whether the mean level of the model result is identical to the mean level of the field data. If both levels are identical, the mean component is zero, signifying that this component contributes no error to the end result. The slope component indicates whether model results and field data show simultaneous changes in abundance in the same direction. When they do, the value of SC becomes zero. Again, this means that there is no error contribution from this component to the total error. All remaining disagreement between model results and observations is expressed in the third component, the random component. In the case where both the mean component and the slope component are nearly zero, the value of RC approaches one, indicating that the remaining error is of a non- systematic nature. In this case it is not possible to obtain a better correspondence between data and model by refining the model; in other words, the optimal correspondence, given these data, has already been obtained, even though the remaining (random) error may be large. The implication of this is that even though there may still be a decided lack of agreement between model results and field data, one can not improve the model formulation on the basis of these data, nor can one reject the model formulation as it is. Generally, of course, both the values of MC and SC are nonzero and the total error is contributed to by all three components, with the fractional value of each indicating its relative contribution to the total error.

To indicate the absolute size of the discrepancy between model results and field observations U2 is calculated.

$$U2 = \frac{\text{MSE}}{\sum_{i=1}^{n} P_i^2} .$$

An U2 $\ll$ 1 occurs when the model results and the field observations generally have the same level. Any U2 value over 1 is very poor.

9.3.1 The Validation Sets

The available data have been taken over a number of years and generally span the period between 1972 and 1983. The standard run of the model covers 1978. Strictly speaking, only field data from 1978 may be used to validate the model. However, the number and the size of the data sets available from this year, or any other 1 year period is much smaller than the total data base. To increase the number of variables and processes that may be validated in order to get an overview of the performance of the model, the assumption is made that the seasonal variability in biomass and rates caused by the annual cycle of the forcing functions dominates the variability from year to year, caused by interannual differences in the forcing functions and boundary conditions. In Chapters 13 and 14 some indications are given that this assumption is not unreasonable. Using data sets from different years for validating the standard run of the model that uses the boundary conditions and forcing functions for 1978 means that the model is not really validated for this specific year, but for the general annual cycle.

The clear advantage of accepting this assumption is that now all sets of field data that have not been used in calibrating the model can be used for validation purposes. A disadvantage is that all interannual differences between variables, as present in the data, will contribute to the various error components discussed in the previous section, and thus will cause lower validation values. The possibility to validate the model with a large number of variables in all the spatial compartments was felt to outweigh this disadvantage.

10 Results and Analysis of the Pelagic Submodel

J. W. BARETTA and W. ADMIRAAL

10.1 Model Results and Validation of the Standing Stocks

When viewing the model results and comparing the five compartments one should realize that all differences in the outcome of the model between the compartments are caused by the differences in the abiotic environment in the compartments. The abiotic differences between compartments in their turn are caused partly by morphological differences (Chap. 2) which determine parameters such as immersion/emersion times of the tidal flats and the relative surfaces of the channels and the tidal flats. For another part the dissimilarity of the compartments is caused by the locations of the fresh water discharges and the seasonal distribution of the river flows. This determines the salinity distribution in the system, the nutrient concentrations and the consequences of waste water discharges.

The abiotic environment thus channels and determines the pathways and sizes of the carbon flows through the system, with the biology adapting to changing conditions.

In the pelagic subsystem the influence of the abiotic environment on the biological variables results in marked differences between the compartments. Figure 10.1 gives an overview of the biomass distributions in all the compartments. To give a first indication of whether the differences reflect reality the available field data are also given in the same plots.

For the variables with sufficient validation data (at least ten) the correspondence value between model results and validation data is determined. These values are given together with their components in Table 10.1.

10.1.1 Primary Producers

In Fig. 10.1 a the biomass in $gC \cdot m^{-3}$ of the phytoplankton is given. The two separately modelled groups of phytoplankton, the pelagic diatoms (PDIA) and flagellate phytoplankton (PFLAG) here have been taken together into phytoplankton (PHYT). This has been done because the validation data are measured chlorophyll-a concentrations that have been converted to phytocarbon by using a C:ChlorA ratio of 35.

The correspondence between model results and validation data is reasonable in compartment 5 (C = 6.9) but insufficient in the other compartments. Especially the spring peak generated by the model is not present in the validation data. A second discrepancy is that the validation data indicate low Aug-Sept values,

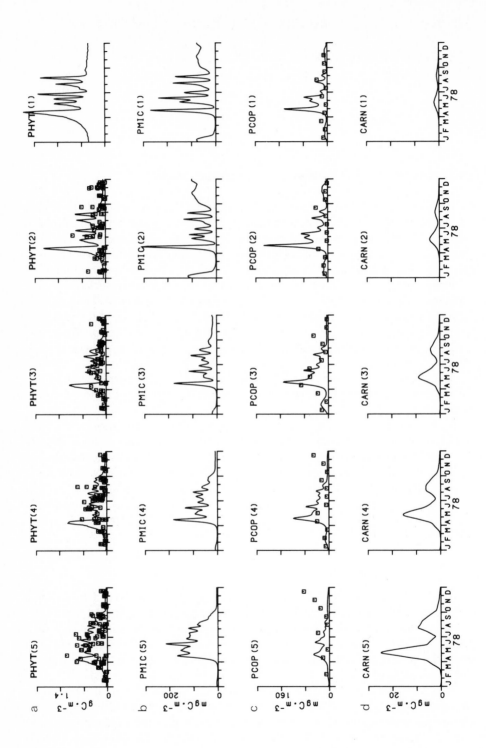

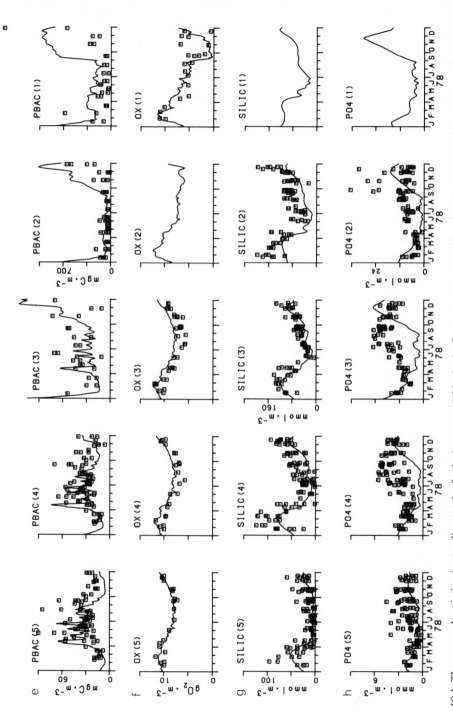

Fig. 10.1. The seasonal variation in standing stock of pelagic state variables in the five compartments with validation data ( ), if available. *a* phytoplankton (*PHYT*); *b* microzooplankton (*PMIC*); *c* mesozooplankton (*PCOP*); *d* carnivorous plankton (*CARN*); *e* pelagic bacteria (*PBAC*); *f* oxygen (*OX*); *g* silicium (*SILIC*); *h* phosphate (*PO4*)

Table 10.1. Correspondence between model results and field data of pelagic state variables and processes C = correspondence value; U2 = measure of variance; RC = fractional contribution to the total error by nonsystematic deviations; MC = idem by the difference between the calculated and observed mean level; SC = idem by nonsimultaneous changes in observed and calculated values (timing error); N = number of observations.

Table 10.1 a.

Variable	Comp.	C	U2	RC	MC	SC	N
Phytoplankton	2	3.9	0.77	0.15	0.14	0.71	73
(PHYT)	3	5.5	0.37	0.27	0.20	0.53	66
	4	5.7	0.48	0.44	0.15	0.41	86
	5	6.9	0.36	0.68	0.16	0.16	92
Mesozooplankton	1	4.4	0.67	0.25	0.01	0.74	11
(PCOP)	2	3.8	0.81	0.22	0.15	0.63	12
	3	7.5	0.42	0.76	0.24	0.00	12
	4	5.6	0.55	0.47	0.01	0.53	12
	5	–	4.44	0.45	0.22	0.33	9
Pelagic bacteria	1	5.9	0.54	0.47	0.36	0.17	28
(PBAC)	2	5.3	0.48	0.15	0.28	0.57	56
	3	7.6	0.41	0.89	0.05	0.05	28
	4	5.7	0.35	0.50	0.03	0.47	71
	5	6.5	0.33	0.62	0.04	0.34	71
Oxygen	1	7.2	0.18	0.58	0.99	0.33	21
(OX)	2	6.6	0.12	0.35	0.63	0.01	21
	3	6.8	0.02	0.20	0.75	0.05	34
	4	7.0	0.01	0.25	0.72	0.03	22
	5	9.3	0.00	0.89	0.11	0.00	20
Silicate	2	6.1	0.40	0.33	0.66	0.01	82
(SILIC)	3	7.6	0.12	0.63	0.31	0.06	73
	4	7.3	0.28	0.62	0.38	0.00	96
	5	7.8	0.44	0.91	0.06	0.02	95
Phosphate	2	5.9	0.71	0.63	0.05	0.32	82
(PO4)	3	4.3	0.71	0.23	0.30	0.47	73
	4	4.3	1.34	0.28	0.50	0.23	96
	5	5.5	0.99	0.56	0.33	0.11	100
Particulate	2	5.7	0.35	0.48	0.40	0.11	18
organic carbon	3	5.5	0.44	0.44	0.48	0.08	14
(POC)	4	4.4	0.45	0.32	0.33	0.35	18
	5	5.4	0.28	0.35	0.43	0.22	35
Dissolved	2	7.8	0.04	0.95	0.00	0.05	18
organic carbon	3	7.8	0.03	0.92	0.02	0.06	16
(DOC)	4	6.8	0.07	0.44	0.51	0.05	18
	5	6.7	0.13	0.55	0.41	0.04	35
Suspended	1	5.8	0.54	0.68	0.08	0.24	28
matter	2	5.7	0.48	0.59	0.23	0.17	82
(SUSM)	3	6.9	0.29	0.88	0.05	0.07	73
	4	7.5	0.37	0.99	0.00	0.00	96
	5	7.2	0.56	0.92	0.07	0.01	100
Vertical	2	5.8	0.55	0.72	0.01	0.27	27
extinction	3	5.7	0.58	0.65	0.29	0.07	32
(EPS)	4	7.2	0.31	0.81	0.19	0.00	46
	5	7.1	0.47	0.84	0.13	0.03	51

Table 10.1 b.

Variable	Comp.	C	U2	RC	MC	SC	N
Primary	1	–	0.89	0.00	0.46	0.54	2
production	2	3.9	0.99	0.25	0.12	0.63	10
(MPPP)	3	6.1	1.16	0.74	0.00	0.25	12
	4	6.4	0.90	0.79	0.00	0.21	11
	5	8.1	0.52	0.99	0.00	0.00	12
Biological oxygen	2	4.0	0.71	0.03	0.28	0.69	17
consumption	3	6.5	0.19	0.45	0.01	0.54	17
(BOC2)	4	5.6	0.32	0.18	0.00	0.82	20
	5	5.1	0.53	0.17	0.00	0.83	16
Bacterial	1	5.0	0.64	0.33	0.43	0.24	16
production	2	6.4	0.43	0.52	0.16	0.33	32
(PRODM)	3	3.9	7.85	0.47	0.49	0.03	16
	4	6.1	0.84	0.52	0.44	0.04	58
	5	7.2	0.42	0.71	0.22	0.07	58

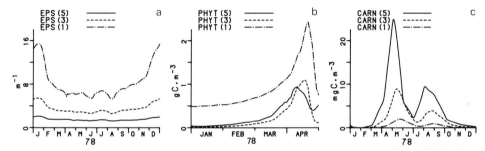

Fig. 10.2. *a* Extinction coefficient (*EPS*) in compartments 5, 3 and 1. *b* The development of the phytoplankton standing stock from January to May in compartments 5, 3 and 1; *c* Standing stock of carnivorous plankton (*CARN*) in compartments 5, 3 and 1 showing the effect of diffusive and advective transport on this state variable

whereas the model suggests a continuation of the summer values, with a tendency to a late-summer peak. The use of a constant C:ChlorA ratio may explain part of this difference, since light adaptation to the high summer light intensities would result in higher C:ChlorA ratios. The field data in summer thus may be underestimates of the phytocarbon present.

As is apparent from Table 10.1 (PHYT), most of the discrepancy or error in the inner compartments is due to timing differences: the field values do not increase or decrease at the same time as the model results. This timing error decreases in importance to 16% (SC=0.16) in compartment 5. The relative error in the average level (MC) is fairly constant in all compartments, varying between 14 and 20%. The correspondence increases from the inner to the outer compartments. This indicates that phytoplankton dynamics over the tidal flats (which decrease in importance from the inner to the outer compartments) probably are

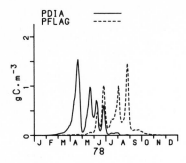

Fig. 10.3. The seasonal succession in abundance of two phytoplankton components: pelagic diatoms (*PDIA*) and flagellate phytoplankton (*PFLAG*) in compartment 1

modified by processes which have not been included in the model, such as the sedimentation and subsequent resuspension of phytoplankton on the tidal flats.

In the model the biomass of primary producers starts increasing earlier in the outer compartments and the increase then proceeds inwards, with a time delay of a few weeks (Fig. 10.2 b). This is caused by the gradient in turbidity, and hence the light extinction, which increases going inward (from high to low compartment numbers) (Fig. 10.2 a). The late winter increase in irradiance is effective first in the less turbid outer compartments (cf. Gieskes and Kraay, 1975).

In compartment 1 large blooms of diatoms occur because of the extreme nature of this compartment: small (only $4 \cdot 10^6$ m^3 average water volume) and so shallow that even at the high prevailing turbidity a large part of the water column over the tidal flats (for data see Chap. 4) participates in primary production. Moreover, the water over the tidal flats here is a large fraction (0.64) of the total water volume and thus the moderating influence of a large, less productive watermass, as in the other compartments, is missing. These blooms are driven by the interaction in the model between the tidal cycle and the daylight period, the so-called light window (Sect. 4.3.2.1), which enhances primary production when the high tide(s) occur during daylight. In the outer compartments this phenomenon is much less noticeable because a much smaller fraction (0.05–0.08) of the total water volume is involved. In compartment 1 the effects from the peculiarities of primary production in a turbid tidal-flat estuary are extreme.

There is a striking difference in the model between the dynamics of the diatom- and the flagellate-phytoplankton in compartment 1 (Fig. 10.3). The diatoms go through three almost equally large blooms in April, May and June, with peak values of up to 1500 mgC $\cdot$ m^{-3}, whereas the flagellates have blooms in July and August, which also reach a peak biomass of 1500 mgC $\cdot$ m^{-3}. The generally lower biomass of the flagellates in compartment 1 in the model is caused by their sensitivity to lower salinities. This decreases their productivity in compartment 1, but less so in August, their period of peak standing stock, because in August the salinity reaches its highest values throughout the system, due to the low fresh water discharges.

10.1.2 Microzooplankton

The seasonal biomass distribution of microzooplankton (PMIC) in the different compartments of the system (Fig. 10.1 b) reflects the different short term variability in the compartments in its main food sources: pelagic bacteria (PBAC) and phytoplankton (PHYT). Proceeding from compartment 5 inwards to compartment 1, the microzooplankton biomass shows increasing fluctuations which are mainly due to the fluctuations in phytoplankton biomass. The microzooplankton, because of its high potential growth rate, shows a fast functional response to changes in food supply, and in summer it is hardly time-lagged relative to the phytoplankton. The lack of validation data on the microzooplankton means the model output has to be treated as speculative. The section on carbon fluxes will establish whether the fluxes through this component agree with literature data. In a sensitivity analysis (Sect. 10.3.4) the role of this state variable in the model will be traced.

10.1.3 Mesozooplankton

The mesozooplankton (PCOP), whose major constituent group is formed by the calanoid copepods, shows a similar seasonal pattern in biomass values in all compartments (Fig. 10.1 c). The absolute biomass values and the details of the seasonal biomass distribution however, are strongly different between the compartments. The correspondence between the biomass as calculated from field data and the model results is only satisfactory in compartment 3. The good correspondence there between model and field results is caused partially because of the influence of the boundary condition with the river Ems, which enters the system in this compartment. The low correspondence between field data and model results for the other parts of the system is mainly caused by the high summer biomasses generated by the model, and by the overshoot of the spring bloom. The biomass is fairly well reproduced in the model from January until April, when the spring increase occurs. This indicates that predation on the mesozooplankton which mainly occurs in summer is underestimated in the model, and that the carnivorous zooplankton, as incorporated in the model does not manage to keep the mesoplankton within its normal limits. The omission from the model of planktivorous fish such as herring and sprat here clearly shows up as a shortcoming.

10.1.4 Carnivorous Zooplankton

This group (CARN), consisting of ctenophores and jellyfishes, shows a biomass distribution in the different compartments which is mainly determined by transport processes (Fig. 10.1 d). Since their growth rates in the inner compartments are very low due to low salinities and other unfavourable circumstances, the loss terms dominate here to such an extent that this group is only maintained in the inner compartments (1 and 2) by replenishment through diffusive transport from the outer compartments.

This is also the explanation for their later (and numerically much reduced) appearance in spring in the inner compartments as compared to the outer compartments (Fig. 10.2 c). Initially they are brought into the estuary across the boundary with the North Sea. In the outer compartments the conditions are such that they can grow, but the transport losses to the inner compartments place a heavy drain on them, which they only can outgrow in compartments 4 and 5, as long as the

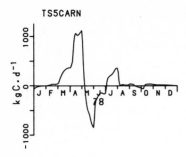

Fig. 10.4. The total amount of carnivorous plankton transported across the seaward boundary into compartment 5 (positive values) or vice versa (negative values)

food conditions are such that they can maintain a high growth rate to counteract the transport losses. In effect, the outer compartment supplies the other compartments with carnivorous zooplankton (Fig. 10.4).

10.1.5 Pelagic Bacteria

The biomass of pelagic bacteria (PBAC) is very different in the various compartments; as shown in Fig. 10.1 e it increases from a mean value of 22 $mgC \cdot m^{-3}$ in compartment 5 to more than 400 $mgC \cdot m^{-3}$ in compartment 1. The mean biomass as calculated by the model for compartments 3–5 agrees rather well with the validation data (Table 10.1).

In compartments 4 and 5, the production and biomass of the pelagic bacteria is closely coupled to the primary production. This is because the exudation by the primary producers of labile organic carbon is the major substrate, utilized almost as rapidly by the pelagic bacteria as it is produced. From compartment 3 inward, the biomass of the pelagic bacteria in autumn is mainly determined by the waste water discharges in compartment 1, increasing sharply at the start of these discharges (note the different scaling of the Y-axes in Fig. 10.1 e). The field data, however, do not show such an increase in biomass until much later in the year. The reasons for this discrepancy are unknown. The scatter in the field data (RC = up to 90%) makes it difficult to say whether the simulated spring peak, coinciding with the spring phytoplankton bloom, is realistic. The large differences between the compartments are reproduced well by the model.

The increase in biomass in autumn, most notable in compartment 2, is caused by the transport of both substrate and bacterial biomass from compartment 1.

10.1.6 Oxygen

As shown in Fig. 10.1 f and Table 10.1 a (OX) the correspondence between model results and field measurements is good in most compartments (C:6.6–9.3). The agreement decreases as the observed oxygen values lie further from the saturation value. This indicates that the reaeration parameter (Sect. 6.5) may be set too high. On the other hand, since the oxygen concentration in the model is the end result of all oxygen producing and -consuming processes, and the model results tend to be higher than the observed oxygen values, it is also possible that the total oxygen consumption in the model is too low. This would imply that the mineralization and respiration rates are too low because of conservative parameter values. The general level of correspondence, however, shows that the total carbon fluxes and the resulting uptake or production of oxygen are consistent with observations. Thus, oxygen serves as a check on the validity of the modelled carbon flows.

Here, it may be concluded that though the correspondence between simulated and observed biomasses leaves much to be desired, the carbon flows between the state variables are modelled correctly.

10.1.7 Nutrients

The modelling of the two different nutrients in the model: silicate (SILIC) and phosphate (PO4) has resulted in a good correspondence with measured concentrations in the case of silicate (Fig. 10.1 g), especially in the seasonal variations expressed in the low SC values in the Table 10.1. This has been accomplished in the model by a simple uptake scheme, which could indicate that silicate in the Ems estuary is not subject to rapid recycling. This is not surprising, since there appears to be no rapid capacity (storage facility) for silicate and its role in the cell is almost entirely structural as opposed to metabolic. Surprisingly, in the model, the low silicate concentrations in summer are not the limiting factor for diatom growth but the limiting nutrient is phosphorus. However, as can be seen in Fig. 10.1 h the modelled phosphate concentrations and the observed concentrations do not agree well. Only from January to April there is a reasonable agreement between them. After that period the simulated concentrations lie far below the measured values. This discrepancy is most likely due to a number of causes. The most important one is that only uptake is modelled and not regeneration.

Nixon et al. (1980) show that the contribution of phosphorus regenerated from sediments in Narragansett Bay is large enough to support half the phytoplankton production in the overlying water. Moreover, Nixon et al. (op. cit.) show the spring/summer increase in phosphate levels in Narragansett Bay to be driven by outputs from the sediments. The observed seasonal distribution of phosphate concentrations in the Ems shows the same increase in spring and summer as reported for Narragansett Bay. This indicates two things: phosphate fluxes from the sediments may play a major role in the pelagic system and second: it is unlikely that phosphate is the limiting nutrient in the Ems estuary. However, for the model to generate the observed phytoplankton distribution and -production, the inclusion of (nutrient) depletion effects is necessary. The constraints imposed by the "phosphate"-uptake on the primary production in the model cannot simply be removed, but we have to conclude that the constraints ascribed to phosphate limitation in the model in reality are not due to phosphate. It is possible that in this system nitrogen is the limiting nutrient.

10.1.8 The Detrital State Variables (PLOC, PDET, PROC, PDROC)

The detrital state variables as they are defined in the model (Sect. 4.4) are well-separated by their different turnover times. The seasonal biomass variation in these components as calculated by the model is given in Fig. 10.5. They cannot be validated directly, since a marine chemist cannot directly sample them as sep-

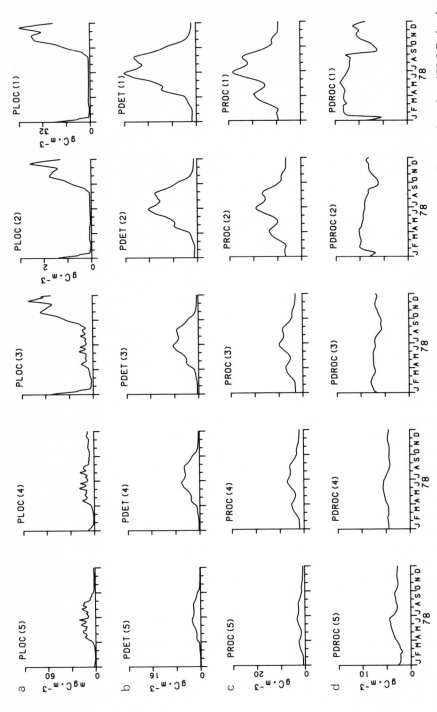

Fig. 10.5. The seasonal variations of standing stock of the detrital state variables in the five compartments. *a* Labile organic carbon (*PLOC*); *b* pelagic detritus (*PDET*); *c* particulate refractory organic carbon (*PROC*); *d* dissolved refractory organic carbon (*PDROC*)

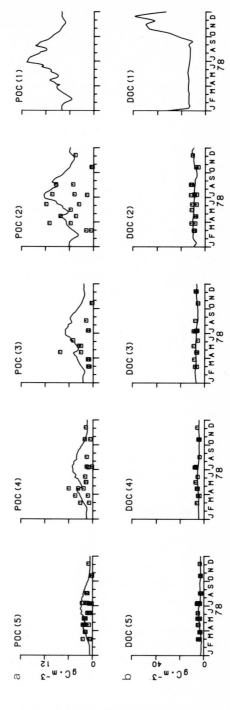

Fig. 10.6. The seasonal variation of (*a*) particulate (*POC*) and (*b*) dissolved organic carbon (*DOC*) with validation data, if available

arate entities in the field. He measures Dissolved Organic Carbon (DOC) and Particulate Organic Carbon (POC). These measurements not only contain the relevant detrital compounds, but also living organic carbon compounds.

Usually, the living and the nonliving fractions are not separated in these samples. For the purpose of presenting validated results for the detrital compounds, which form the bulk of these samples, the concentrations of POC and DOC in the model have been calculated by including the biomasses of the living state variables in POC. The results are presented in Fig. 10.6. There is a clear increase in concentrations from the outer compartment (5) to the Dollard (compartment 1). In the case of particulate organic carbon (POC) this is due to the gradient in suspended matter, with increasing concentrations towards the inner part of the system caused by the silt transport (Chaps. 5 and 13). For the dissolved organic carbon (DOC), it is caused by the discharge of humic compounds by the Ems and the Westerwoldsche Aa in compartments 3 and 1 respectively. These compounds are diluted in the outer compartments by diffusive and advective processes.

The overall concentrations are simulated reasonably well (correspondence values between 4.2 and 7.8), but the model also seems to overestimate the summer concentrations of POC in compartments 3, 4 and 5. However, since this is the sum total of both detrital carbon and living carbon in all 11 pelagic state variables one has to look at the carbon fluxes through all these variables to establish the source(s) of the discrepancies. In the case of the detrital state variables this will be done in Chapter 13.

10.2 Fluxes

In addition to the validation of seasonal biomass distributions there are also a (small) number of carbon fluxes in the pelagic submodel that can directly be validated with field data. The most important of these are primary production (MPPP in the model), bacterial production (PRODM) and biological oxygen consumption (BOC2).

10.2.1 Validated Fluxes

10.2.1.1 Primary Production

The field data give the primary production in $gC \cdot m^{-2} \cdot d^{-1}$ (Colijn, 1982). In the model phytoplankton respiration is subtracted from the calculated gross production and converted from $gC \cdot m^{-3}$ to $gC \cdot m^{-2}$ by multiplying by the channel depth to obtain the comparable figure. Figure 10.7 shows the model results and production measurements. The correspondence between them ranges from $C = 3.9$ in compartment 2 to $C = 8.1$ in compartment 5. Though by no means perfect, the correspondence is fairly good, but with the same too high model values in summer as in the biomass simulation. In compartments 3–5 the correspondence between calculated and observed mean production levels is perfect ($MC = 0.00$) (Table 10.1 gb), but there is still some error in the timing of the increase/decrease

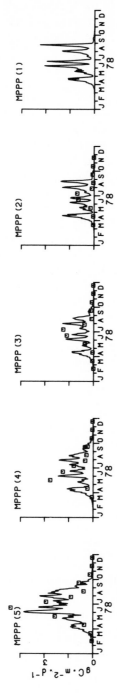

Fig. 10.7. Calculated pelagic primary production in $gC \cdot m^{-2} \cdot d^{-1}$ with validation data, if available

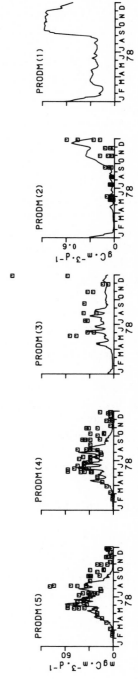

Fig. 10.8. Daily production of pelagic bacteria ($PRODM$) in the five compartments with validation data, if available

of production. The trend of lower primary production in the inner compartments is reproduced faithfully by the model, except for compartment 1. Though there are only a few production data for this compartment, this may well be correct, since compartment 1 is very shallow, with a large productive water layer at high tide.

10.2.1.2 Pelagic Bacterial Production

In 1983 the bacterial production in the water column (PRODM) was measured in the various compartments using the ^{3}H-methylthymidine method of Fuhrman and Azam (1980, 1982). This data set (Admiraal et al., 1985) has been used as validation for the bacterial production as calculated by the model for the standard run of 1978. A perfect correspondence cannot be expected, since the boundary conditions and forcing functions of 1978 as used in the model are different from those of 1983, when bacterial production was measured.

The thymidine method is assumed to reflect the daily gross bacterial production (prodM), which in the model is equivalent to bacterial uptake (uM) minus respiration (rsM): prodM = uM – rsM.

The correspondence between model results and measurements varies rather widely between the compartments (C:3.9–7.2) (Fig. 10.8). In the outer compartments (compartments 4 and 5) the bacterial production closely follows the production of labile exudates (PLOC) by primary producers (Fig. 10.9). This linkage, a central assumption in the model, is completely consistent with the measurements on bacterial production. In the inner compartments the contribution of labile substrate from the waste water discharges dominates the exudate production by primary producers, resulting in high bacterial production in autumn during these discharges. Here again, under rather different circumstances, model results and production measurements agree generally.

In conclusion we can state that the model results of these pelagic bacterial processes are consistent with available observations.

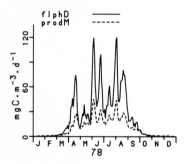

Fig. 10.9. Daily production of labile organic carbon by phytoplankton (*flphD*) together with the daily production of pelagic bacteria (*prodM*), both in compartment 4, showing the coupling between these two processes

10.2.1.3 Biological Oxygen Demand

The sum of all respiratory processes in the pelagic is the biological oxygen consumption (BOC2). This oxygen flux has been measured in the field and the results are given in Fig. 10.10. The correspondence between model results and field data

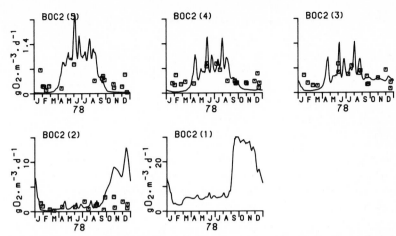

Fig. 10.10. The seasonal variation in biological oxygen consumption ($BOC2$) in $gO_2 \cdot m^{-3} \cdot d^{-1}$ in the five compartments with validation data, if available

varies strongly between the compartments. The mean level is correctly reproduced in compartments 3–5 (MC = 0–1.5%), but the seasonal distribution of the model results is rather different from the field data. Especially the low winter values generated by the model in compartments 5–3 seem to be an underestimate compared to the field data. This indicates that biological activity in the system is higher in winter than the model results indicate. A probable cause might be temperature adaptation, a mechanism not included in the model formulation.

The seasonal distribution of the oxygen consumption can only be corrected by improvements in the formulation of the other state variables. These have to wait for a better understanding of the system.

10.2.2 Nonvalidated Fluxes

The fluxes to be presented here cannot be directly validated and as such should not be considered as proven facts, but, where possible, they will be compared to literature data to see whether their relative magnitude is consistent with reality.

The fluxes are given in the form of tables representing the fluxes in compartment 4. Compartment 4 has been chosen because together with compartment 5 it is the most typical pelagic compartment. Compartment 5, by the influence of the boundary conditions with the North Sea, is a transitional compartment to the North Sea. Therefore compartment 4 is more representative of the dynamics within the pelagic system of the Ems estuary itself.

10.2.2.1 Internal Fluxes and Ratios

The annual carbon flows through the living state variables in $gC \cdot m^{-3} \cdot y^{-1}$, as calculated by the model are given in Table 10.2. The primary producers (PFLAG + PDIA) generate by far the largest carbon flux, with the flagellate phytoplankton (PFLAG) being the largest contributor. Interestingly, the model results indi-

Table 10.2. Internal fluxes in the pelagic submodel in gC · m⁻³ · y⁻¹ in compartment 4

State variable	C uptake	Respiration	Excretion/ Egestion	Net production
PFLAG	29.5	11.4	9.3	8.8
PDIA	11.1	3.3	2.9	5.0
PBAC	18.4	14.7	–	3.7
PMIC	16.1	7.2	6.3	2.6
PCOP	4.6	0.7	2.8	1.1
CARN	0.5	0.03	0.4	0.02
Total	80.2	37.3	21.7	21.2

Table 10.3 Calculated efficiencies and productivities in the pelagic in compartment 4

State variable	Ecological efficiency (%)	$P/B \cdot d^{-1}$	$P/B \cdot y^{-1}$
PFLAG	–	0.09	33
PDIA	–	0.13	48
PBAC	20	0.36	130
PMIC	16	0.10	36
PCOP	23	0.07	24
CARN	4	0.003	0.94

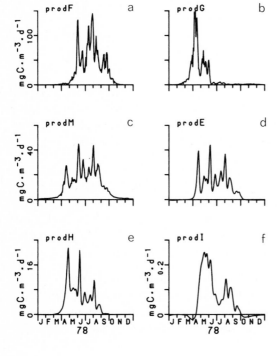

Fig. 10.11. Daily production in compartment 4 of: *a* Flagellate phytoplankton (*prodF*); *b* diatoms (*prodG*); *c* pelagic bacteria (*prodM*); *d* microzooplankton (*prodE*); mesozooplankton (*prodH*); *f* carnivorous plankton (*prodI*)

cate the pelagic diatoms (PDIA) to have a higher specific productivity than the flagellate phytoplankton (Table 10.3, P/B), but still the flagellates dominate the phytoplankton in summer, both in terms of biomass (Fig. 10.3) and production (Fig. 10.11).

Pelagic bacteria (PBAC) and microzooplankton (PMIC) are the major consumers, followed at a large distance by the mesozooplankton (PCOP).

The production of pelagic bacteria has been validated (see Sect. 10.2.1.2); given a consumption of 18.4 $gC \cdot m^{-3} \cdot y^{-1}$ the production of 3.7 $gC \cdot m^{-3} \cdot y^{-1}$ results in an ecological efficiency of 20% (Table 10.3). This value is at the high end of the range given by Valiela (1984), who concludes that in general the ecological efficiency of a trophic link is less than 10%, but some values may reach 25%. The model results for the consumers indicate ecological efficiencies from 4–23% (Table 10.3), reasonably within this range.

Averaged over the pelagic consumers, the respiratory loss is 57% of consumption, with rather large variations between the state variables, ranging from 6% for the pelagic carnivores (CARN) to 80% for pelagic bacteria (PBAC). Literature data, compiled by Valiela (1984) for respiratory losses in marine organisms range from 20–95%. The model thus produces an estimate of the average respiratory loss that is right in the middle of this range.

The excretion/egestion losses as given in Table 10.2 are generally lower than the respiratory losses, except for the mesozooplankton and the planktonic carnivores. In these two groups the loss to egestion is the major factor causing these high values. In the mesozooplankton, unassimilated detritus forms the bulk of the egested material (2.6 $gC \cdot m^{-3}$). In the carnivorous zooplankton in the model, it is the low assimilation efficiency (10%) that causes 90% of the uptake to be egested.

Averaged over all pelagic state variables, both producers and consumers, excretion/egestion is 27% of the carbon (organic or inorganic) uptake, whereas respiration is the major loss factor at 46% of the uptake.

To conclude this section on internal fluxes we have calculated the specific productivities ($mgC \cdot mgC^{-1} \cdot d^{-1}$) (P/B ratios) for the various state variables and summed the daily values to obtain the annual turn- over rates or production over biomass figures (Table 10.3).

The specific productivities, being the outcome of all the internal gains and losses, vary seasonally because of the seasonal variation in irradiance and temperature and over shorter periods because of fluctuations in food supply (Fig. 10.12).

P/B ratios have not been employed to construct the pelagic submodel and therefore may be used as a check on the model results. The specific productivities ($P/B \cdot d^{-1}$) of the primary producers are low (Table 10.3), resulting in correspondingly low annual P/B values of 33 for the flagellate phytoplankton and 48 for the pelagic diatoms. The annual P/B ratio in the model is not identical to the annual P/B ratio, because of the use of the average biomass B as the divisor in the annual P/B ratio. The specific productivity obviously has the actual biomass as the divisor. A common P/B value from literature is 300 for phytoplankton (Parsons et al., 1977). The apparently low productivity of the phytoplankton in the model is caused by the high turbidity of the system: only a small fraction of

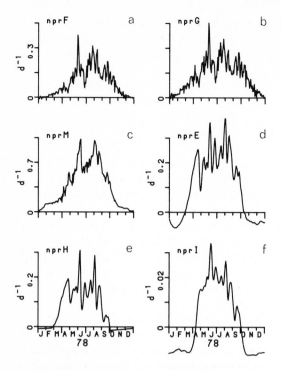

Fig. 10.12. Specific productivity $(mgC \cdot mgC^{-1} \cdot d^{-1})$ in compartment 4 of: *a* Flagellate phytoplankton (*nprF*); *b* diatoms (*nprG*); *c* pelagic bacteria (*nprM*); *d* microzooplankton (*nprE*); *e* mesozooplankton (*nprH*); *f* carnivorous plankton (*nprI*)

the phytoplankton is in the photic zone at any time and this depresses the productivity of the primary producers as a whole.

The P/B values from the model (Table 10.3) for the consumers show the pelagic bacteria to have by far the highest value of 130, as expected. The microzooplankton (PMIC) has a P/B value of 39, which seems reasonable in view of it being in the same size class as the phytoplankton which also has P/B values of the same magnitude. The P/B value of the mesozooplankton (PCOP), 24, is in the range of 10–40 given in the literature for this group (Parsons et al., 1977). The value for the carnivorous zooplankton (CARN), 0.94, seems excessively low, though it should be below 10 (Valiela, 1984).

10.2.2.2 External Fluxes

A main concern in the formulation of the model has been the definition of the various food sources for the functional groups. In this section we are able to view the consequences, in terms of the grazing fluxes between the state variables. It should be stressed that here everything is speculative. About the only thing we can say with confidence is that we have found no demonstrable inconsistencies between the model results and field observations. Thus, the fluxes we discuss here might be more or less correct.

An overview of the grazing fluxes in compartment 4 is given in Table 10.4.

In our model formulation, the microzooplankton (PMIC) clearly is the dominant phytoplankton consumer, with the mesozooplankton share in phytoplankton consumption being more than an order of magnitude less.

Table 10.4. Annual grazing fluxes ($mgC \cdot m^{-3} \cdot y^{-1}$) in compartment 4 between grazers (rows) and grazed (columns)

	PFLAG	PDIA	PBAC	PMIC	PCOP	CARN
PMIC	7 945	4 098	2 162	993	–	–
PCOP	341	496	–	1 553	355	–
CARN	–	–	–	212	269	–

A second surprise is the size of the flux from microzooplankton to meso-zooplankton, which is almost twice as large as the flux from the primary producers. This suggests the microzooplankton to occupy a central place in the transfer of carbon from primary producers (and mineralizers) to the secondary producers. The enormous share of the microzooplankton in the grazing fluxes, as stated before, is speculative. In a sensitivity analysis (Sect. 10.3.4) we will analyze what consequences the removal of this group has to the functioning of the system. The model results underscore the urgent need to obtain more valid data on the microzooplankton, despite the notorious difficulties (Sorokin, 1981) to do quantitative studies on this group.

The grazing fluxes through the mesozooplankton seem fairly innocent; the apportioning of the phytoplankton flux over flagellates and diatoms is debatable and should be viewed as tentative. The existence of cannibalism within the meso-zooplankton has been confirmed experimentally (Daan, 1987; Daan personal communication). The mesozooplankton also ingests a large amount of detrital material, 2171 $mgC \cdot m^{-3} \cdot y^{-1}$, but this contributes little to growth, due to its low assimilation efficiency and is largely egested, thus contributing to the pelagic-benthic flux of detritus, in the form of faecal pellets.

The carnivorous zooplankton (CARN) in this system plays a minor role, with very little impact on the structure of the system. This may be a typical property of the Ems estuary or just a misjudgment on our part. However, also in the pelagic system in the Black Sea the carnivores seem to play a minor role (Parsons et al., 1977).

The microzooplankton thus has a central place in the pelagic model, on the one hand, liberating dissolved organic carbon for bacterial utilization, and on the other hand, indirectly making bacterioplankton available to the larger grazers by grazing on bacterioplankton and subsequently being grazed itself by the meso-zooplankton.

10.3 Sensitivity Analysis

In the following sections analyses will be made of how the model results are modified by changes in parameters considered as central to the performance of the model. This does not imply that the model will not perform differently if other parameter values are varied; it just reflects our view of this estuarine system and its key mechanisms. In fact, a traditional sensitivity analysis, i.e. the systematic varying of all single parameters would be preferable, especially in view of the wide variability of almost all reported parameter values. However, the large number of parameters precludes a full sensitivity analysis of the pelagic submodel, let alone of all submodels. This is one of the reasons why a full listing of the model code is included in this book and why the Fortran-77 source code will be provided on tape to anyone who would like to perform sensitivity runs of the model and thus test the validity of the choices we have made.

10.3.1 Degradation of Algal Production by Bacteria

Phytoplankton is obviously a major source of organic carbon for pelagic heterotrophs. Only in the inner sections of the Ems estuary the riverine input of degradable carbon dominates over the phytoplankton production. The pelagic community is supported directly by the grazing on phytoplankton and indirectly by the utilization of algal exudates and algal detritus. With the model the flux of organic carbon from phytoplankton to dissolved or particulate organic carbon and from there to bacteria will be analyzed.

Many field studies have been executed on this topic. Wolter (1982) and Lancelot (1983) followed the bacterial uptake of algal products in plankton communities labelled with ^{14}C-bicarbonate and indicated a direct coupling between production and consumption of dissolved organic matter. However, Laanbroek et al. (1986) and Lancelot and Billen (1984) found that the bacterial population peaked ca. 5–7 days after a phytoplankton peak. The model predicts a similar time lag

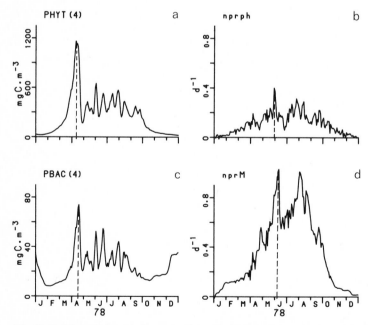

Fig. 10.13. The relationship between phytoplankton standing stock (*a*) and phytoplankton productivity (*b*) as well as between pelagic bacterial biomass (*c*) and bacterial productivity (*d*), showing that both biomass and productivity of the bacteria lag behind those of the phytoplankton

between bacterial and algal biomass (Fig. 10.13 a and c) and between bacterial and algal productivity (Fig. 10.13 b and d).

A delay is to be expected, since the degradation of algal detritus under experimental conditions takes place over periods of a few days or weeks (cf. Newell et al., 1981; cf. Ogura and Gotoh, 1974). Figure 10.13 shows that in the model small peaks of labile organic carbon appear a few days after the production peaks of the phytoplankton. However, detritus (PDET) accumulated slowly in a 3–4 week period following a production peak; furthermore, detritus accumulated during the whole growing season as a result of import from the sea (Chap. 5).

In the standard run the phytoplankton is assumed to release 50% of its excretion/mortality products as labile organic carbon (PLOC) and 50% as slowly degrading detritus (PDET). In two test runs of the model the degradability of algal products was altered: in a first run PLOC amounted to 10% of the total and in a second run the production of PLOC was increased to 90%. Figure 10.14 shows

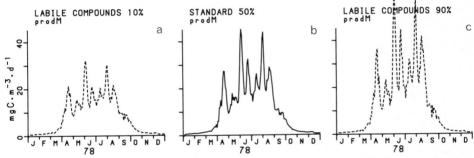

Fig. 10.14. Daily production in compartment 4 of pelagic bacteria (*prodM*) in a run with phytoplankton exudates being apportioned: *a* 10% to labile organic carbon (*PLOC*) and 90% to detritus (*PDET*); *b* 50% – 50% (standard); *c* 90% – 10%

Table 10.5. Yearly budget in $gC \cdot m^{-3} \cdot y^{-1}$ of formation, degradation and transport of labile organic carbon (PLOC) and detritus (PDET) in compartment 4

	PLOC	PDET
Formation:		
Diatoms	2.0	0.9
Flagellates	6.1	3.2
Microzooplankton	3.1	3.1
Bacteria	1.8	–
Copepods	–	0.1
Carnivores	–	0.04
Uptake:		
Microzooplankton	–	0.9
Bacteria	14.0	3.2
Copepods	–	2.2
Import:	1.0	6.3
Sedimentation:	–	7.3

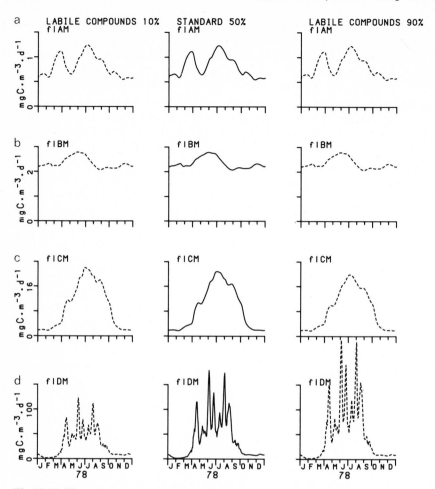

Fig. 10.15. Fluxes in compartment 4 from detrital state variables to pelagic bacteria as modified by the apportioning of phytoplankton exudates over labile and detrital compounds. *a* The flux from particulate refractory organic carbon into bacteria (*flAM*); *b* the flux from dissolved refractory organic carbon into bacteria (*flBM*); *c* the flux from pelagic detritus into bacteria (*flCM*); *d* the flux from labile organic carbon into bacteria (*flDM*)

these strong changes in the degradability of algal products to have only a small effect on the rate of bacterial production and biomass. Apparently, the model is insensitive to the degradability of phytoplankton products, in contrast to the experimental evidence. How can this lack of response by the model be explained? The seasonal development of the algal population was hardly affected (results not shown) so that the observed effect must be a characteristic of the simulated detrital food web. Figure 10.15 gives the simulated yearly menu of bacterial populations. The uptake rate of refractive organic matter (flAM and flBM) was not affected by the composition of the phytoplankton excretion and the uptake of detritus (flCM) was reduced only slightly when phytoplankton released very low amounts. Apparently the seasonal fluctuation of these fractions is not mainly determined by phytoplankton excretion but by other sources, such as the boundary conditions. Furthermore, detritus is also produced in the pelagic by grazers, notably microfauna (Table 10.5). The indirect liberation of algal material by pelagic grazers amounted to 3267 $mgC \cdot m^{-3} \cdot y^{-1}$ of detritus and 3145 $mgC \cdot m^{-3} \cdot y^{-1}$ of labile organic carbon in the standard run and these are significant rates in comparison with the 8114 $mgC \cdot m^{-3} \cdot y^{-1}$ of LOC and the 4056 $mgC \cdot m^{-3} \cdot y^{-1}$ of detritus produced by the phytoplankton.

This test of the model leads to the realization that the liberation of organic compounds by grazers (mainly microfauna) mediates the carbon transfer from phytoplankton to bacteria. Whether or not this aspect of the model is realistic remains to be settled.

10.3.2 Growth Efficiency of Bacteria

The sensitivity of the ecosystem model to the value of effM, the assimilation efficiency of bacteria, will be tested here. Figure 10.16 shows three cases: the standard run (effM = 0.3), a run with a lower efficiency (effM = 0.15) and a run with elevated efficiency (effM = 0.5). The use of an assimilation efficiency of 0.5 produced a simulation of the bacterial production (PRODM) in compartment 4 that gave better results in spring than in the standard run, though with extreme peaks in summer. An efficiency of 0.3 (the value in the standard model) led to a proper

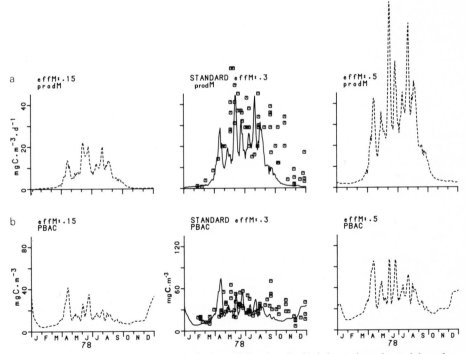

Fig. 10.16. The effect of different assimilation efficiencies of pelagic bacteria on bacterial production (a) and on bacterial biomass (b). In the standard run validation data are given. The plots refer to compartment 4

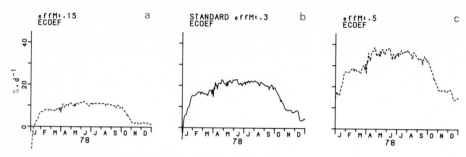

Fig. 10.17. The consequence of different assimilation efficiencies of pelagic bacteria on the ecological efficiency of pelagic bacteria (*ECOEF*) in compartment 4

simulation of bacterial biomass (PBAC), whereas an assimilation efficiency of 0.5 resulted in a too high biomass level (Fig. 10.16). An efficiency of 0.15 resulted in a severe reduction of both the biomass and the production of bacteria. These tests of the model indicate that the assimilation efficiency of bacteria in the Ems estuary is somewhere between 0.3 and 0.5. Indeed, independent calculations on substrate uptake (of small molecules) and productivity (as measured with ^{3}H-methylthymidine) indicate a growth efficiency of 0.5 (Admiraal et al., 1985), which implies an even higher assimilation efficiency.

The simulated bacterial efficiency is given in Fig. 10.17 as a net growth efficiency (including the biomass dependent respiration). In situ measurements on the uptake and assimilation of substrates were carried out with ^{14}C-labelled glucose, leucine and glutamate (for details see Admiraal et al., 1985). Figure 10.17 (prodM/uM) shows that the simulated bacterial population in the growing season showed growth efficiencies near ca. 0.45 in the run with the assimilation efficiency (effM) set to 0.5. Analogously, the ^{14}C-substrates in the growing season showed constant levels of efficiency, varying from 0.4 to 0.75 for the three substrates. Typically low levels of efficiency were observed and simulated by the model during the colder seasons. This indicates that the assimilation efficiency of the pelagic bacteria for labile organic carbon in the model should be nearer to 0.5 than to the standard value of 0.3. However, the assimilation efficiency also should reflect the efficiency with which more refractory compounds are assimilated. This consideration resulted in the setting of effM at 0.3 in the pelagic submodel.

10.3.3 Suspended Phytobenthos as a Food Source for Zooplankton

In the standard run of the model the possible contribution of suspended microphytobenthos, mainly benthic diatoms, to the diet of the zooplankton (both micro- and mesozooplankton) has been ignored. Recent research by De Jonge (1985), however, has shown that in the Ems estuary suspended benthic diatoms can form a considerable fraction of the biomass of primary producers in the water column. Thus it is possible that in the model an important food source is withheld from the zooplankton.

In a test run of the model an amount of 15% of the biomass of benthic diatoms (BDIA), which has been estimated as the suspended fraction (De Jonge, pers.

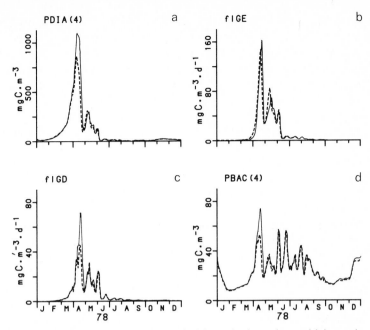

Fig. 10.18. Output from the standard run (——) compared with results from the sensitivity analysis (---) where suspended microphytobenthos (*BDIA*) is an additional food source for planktonic grazers. *a* The standing stock of pelagic diatoms (*PDIA*); *b* the daily carbon flux from pelagic diatoms to microzooplankton (*flGE*); *c* the daily exudate production by pelagic diatoms; *d* the standing stock of pelagic bacteria (*PBAC*)

comm.), was included as an extra food source for the zooplankton. In compartment 4, the reference compartment for the analysis of the pelagic submodel, the direct effects on the carbon fluxes of this extra food source are rather small. The average standing stock of the mesozooplankton (PCOP) increases slightly from 22.6 to 23.3 mgC·m^{-3} and from 36 to 39 mgC·m^{-3} for the microzooplankton (PMIC). Production for both groups also shows a similar increase from 1.06 to 1.11 gC·m^{-3}·y^{-1} for the meso- and from 2.6 to 2.8 gC·m^{-3}·y^{-1} for the microzooplankton. The P/B ratio for the mesozooplankton is virtually unchanged in this compartment as compared to the standard run: 24.7 in the test run with suspended diatoms against 24.3 in the standard run, whereas the P/B ratio of the microzooplankton increases slightly from 36.1 to 36.7. Other state variables, especially the pelagic diatoms (PDIA) and pelagic bacteria, are more affected. The pelagic diatoms show a strongly reduced spring maximum (Fig. 10.18 a). This is caused by the fact that both microzooplankton and mesozooplankton through the increased availability of food during the winter, provided by the suspended benthic diatoms, maintain a higher biomass during this period. This enables the zooplankton, the microplankton the most effectively, to respond quickly to the spring increase in phytoplankton production (Fig. 10.18 b).

Another effect of this change in the biomass distribution of phytoplankton is a shift in the relative contribution to primary production from diatoms to flagellates. The production by diatoms decreases by 4%, while the total primary production decreases less than 2%. Because in this model run the pelagic diatoms are grazed more heavily than in the standard run, they encounter even less incipient nutrient limitation than they otherwise would. This results in a lower excretion of monomeric compounds (PLOC) (Fig. 10.18 c), which in the standard run reaches a maximum at the end of the spring bloom. This reduced exudate flux to the pelagic bacteria lowers the spring biomass maximum, which improves the correspondence with observed bacterial biomass (Fig. 10.18 d) in compartment 4 from C = 5.7 to C = 6.1.

The conclusion from this sensitivity run is that in an estuary with extensive tidal flats and an active microphytobenthos component, the benthic system may, by resuspension of phytobenthos, subsidize the zooplankton. This food subsidy is especially important in winter, when phytoplankton is scarce, leading to a reduced dependence on phytoplankton and thus to smaller fluctuations in zooplankton biomass.

10.3.4 The Role of Microzooplankton in the Model

From the previous sensitivity analysis it is already clear that the microzooplankton is able to profit more from a general increase in food supply than the mesozooplankton. The carbon flux in the model through the microzooplankton is much larger than the carbon flux through the mesozooplankton. Still, we do not have much support from direct observations in the estuary that this situation reflects reality.

To test the relative importance of the microzooplankton to the structure of the model system and the pathways and sizes of the carbon flows through the various components we have made a model run with a potential rate of uptake by the microzooplankton of 0 (zero). In effect the microzooplankton is removed from the model, because it now cannot eat. By removing the most active pelagic grazers, rather conspicuous shifts in standing stocks and carbon fluxes do occur. The average standing stock of phytoplankton more than doubles from 260 to 583 $mgC \cdot m^{-3}$ and primary production increases from 172 to 306 $gC \cdot m^{-2} \cdot y^{-1}$. Pelagic bacteria, which now are released from the grazing pressure of the microzooplankton, also double their standing stock from 24 to 55 $mgC \cdot m^{-3}$, but bacterial production only increases slightly from 3.7 to 4.7 $gC \cdot m^{-2} \cdot y^{-1}$, indicating that the specific productivity is much lower in this run. Specific bacterial productivity is almost halved from 0.36 to 0.17 $\cdot d^{-1}$. The supply of substrate to the larger bacterial population clearly is insufficient to support the same growth rate as in the standard run.

The remaining group of grazers, the mesozooplankton, does profit from the absence of the microzooplankton, increasing its average standing stock from 23 to 41 $mgC \cdot m^{-3}$, while the production increases from 1.4 to 2.6 $gC \cdot m^{-2} \cdot y^{-1}$. A major beneficiary of the absence of microzooplankton is the group of the benthic suspension feeders (BSF) which increases its biomass from 2.1 to 4.4 $gC \cdot m^{-2}$. The carbon fluxes from the pelagic to the benthic system increase by transport and sedimentation processes as a result of the higher standing stocks in the pelagic.

The correspondence between model results and field data is generally worse in this run than in the standard run. The fact that in the model functional groups in the ecosystem whose biomasses and production can be verified in the field, come closer to observed values in the presence of microzooplankton than in its absence, is encouraging. This indicates that the microzooplankton as incorporated in this model plays an essential role in the planktonic food web, linking the production of algae and bacteria to the larger planktonic organisms.

10.3.5 Regulation of the Phytoplankton Succession

The standard run shows a distinct succession in the phytoplankton with diatoms dominating the spring bloom and flagellates persisting throughout the summer. The simplicity of the simulated succession contrasts with observations in the estuary where blooms of diatoms were seen in early spring as well as after vigorous blooms of the flagellate *Phaeocystis pouchetii* in May (Colijn, 1983; Admiraal et al., 1985). The reason for the absence of alternating abundances of diatoms and flagellates in the model seems to be the definition of a single set of capacities for each of the two state variables PDIA and PFLAG, whereas in nature both groups encompass numerous different species. These differences cover various aspects such as temperature response, percentage excretion, edibility by (micro)zooplankton and silicate requirements. All these factors are known to influence phytoplankton succession (Morris and Glover, 1981). The present model offers the possibility to test the relative importance of these factors and to study their interactions.

Figure 10.19 shows the succession of diatoms and flagellates in the standard run together with their net specific growth rates realized over the year. Of course temperature and light, presented here as the light-correction factor CORR, are the driving variables for the growth rate in the two algal populations. It is clear that during the spring and autumn the simulated diatom populations realized higher specific growth rates than the flagellates (Fig. 10.19). This was evidently caused by the low-temperature response attributed to them, but the similar growth rates realized by diatoms and flagellates in summer do not explain the uni-

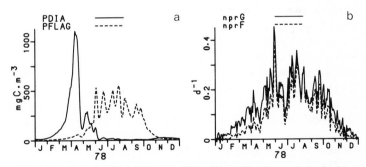

Fig. 10.19. The seasonal succession of pelagic diatoms (*PDIA*) and flagellate phytoplankton (*PFLAG*) in compartment 4 in the standard run. *a* Standing stocks of both phytoplankton components; *b* net productivity of pelagic diatoms (*nprG*) and flagellate phytoplankton (*nprF*)

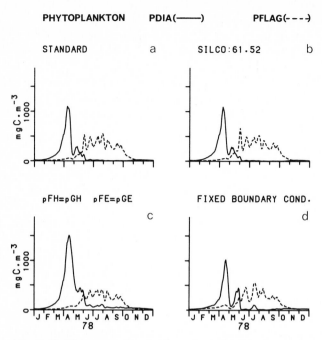

Fig. 10.20. The seasonal succession of pelagic diatoms (*PDIA*) and flagellate phytoplankton (*PFLAG*) in compartment 4 in the standard run and in three sensitivity analyses. *a* Standard run; *b* sensitivity analysis with a doubled silicate requirement of diatoms; *c* sensitivity analysis with equal grazing of both phytoplankton components; *d* sensitivity analysis with fixed boundary conditions at the seaward end for both phytoplankton components

form dominance of flagellates from June onwards, and the impossibility for diatoms to return to dominance in August and September.

    To test the influence on the succession of the various factors, the following three test runs were made in addition to the standard run:

1. The silicate requirement of diatoms was doubled (SILCO = 61.52).
2. The selective feeding of mesozooplankton and microzooplankton on diatoms was replaced by uniform grazing on the two phytoplankton groups (pGE = pFE = 0.4; pGH = pFH = 0.5).
3. The boundary conditions for phytoplankton on the North Sea were given as a constant concentration instead of as a time series with blooms.

Figure 10.20 shows that none of the test runs changes the succession of diatom and flagellates dramatically. Silicate limitation does not seem to play a key role in the succession (see later), whereas non-selective grazing led to a shift in the consumption of phytoplankton towards an increase of diatoms and diminished flagellate populations. Transport of phytoplankton blooms to the sea as stimulated by the low biomass values at the boundary, reinforced strongly intermittent algal growth, an effect probably induced by algae/herbivore oscillations.

    The regulating effects of the nutrients silicate and phosphate was also analyzed for these four cases (Fig. 10.21). In the standard run silicate played obviously no role as we assumed that only the nutrient with the highest limitation,

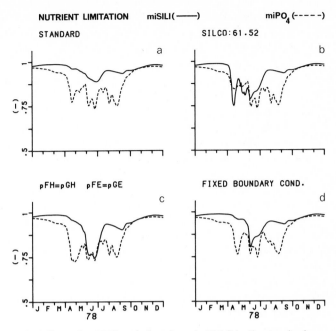

Fig. 10.21. Nutrient limitation for silicate (*miSILI*) and phosphate (*miPO4*) in the standard run (*a*) and for the three sensitivity analyses mentioned in the legend of Fig. 10.20. A value of 1 implies no nutrient limitation. A value of 0.5 reduces the assimilation rate to half the maximum value

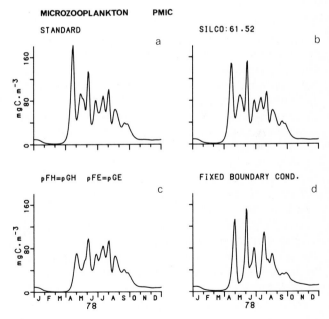

Fig. 10.22. The seasonal variation in standing stock of the pelagic micro-zooplankton for the runs mentioned in the legend of Fig. 10.20

in this case phosphate, modifies algal activity. However, flagellates and diatoms were given the same affinity for this element so that phosphate had no direct selective effect on the phytoplankton.

An increased requirement of silicate by diatoms (SILCO = 61.52) indeed did lead to a distinct silicate limitation (Fig. 10.21), but this had only insignificant effects on the succession. Aselective grazing also resulted in lower grazing rates and thus enhanced the nutrient limitation, but again the effects of silicate limitation were largely overruled by those of phosphate limitation. In this test run diatoms maintained a low but significant population during the summer. In all cases phosphate limitation persists until the end of August and this is unrealistic. This artefact, together with the relatively high temperature of the seawater in autumn and persistent microplankton grazing, may prevent diatoms to form blooms in the second half of the year in the model.

Figure 10.22 shows that decreased grazing by planktonic herbivores on diatoms led to delayed and less dense blooms of microfauna in spring, but to normal concentrations of the grazers in summer. The setting of a low constant phytoplankton biomass as the seaward boundary condition led to a slightly delayed and smaller bloom of the diatoms (Fig. 10.20), also depressing microfauna grazing. Thereby this run produced conspicuous oscillations between the algae and the herbivores.

In summary, the test runs show that various factors of selective value, such as temperature optima, discriminating herbivores, and nutrient limitation are strongly interacting in the succession of the two algal groups: the diatoms and the flagellates.

10.4 Conclusions

The model succeeds in generating the correct average biomass levels in the different compartments of the system, but it is less successful in generating the right seasonal variations in these biomasses.

The carbon fluxes in the model that can be validated reflect the measured fluxes reasonably well. Also, the fact that the simulated biological oxygen consumption is not strongly different from the measured one, indicates that the non-validated fluxes must be of the right order of magnitude.

As is clear from the various sensitivity analyses, the optimal setting of parameter values is not necessarily the standard one.

The agreement between model results and field observations still may be optimized by a repeated calibration of the model. However, a fundamental improvement of the correspondence between model and reality is only to be expected from improvements in the model formulation itself. Some of these improvements have already been suggested. A case in point is the inclusion of nutrient regeneration in the model. Nevertheless, any progress in such ecosystem models is dependent upon the progress in marine ecology itself, in the understanding we have of the key processes and their interactions.

11 Results and Analysis of the Benthic Submodel

W. Ebenhöh

In the first section of this chapter the seasonal cycles of the state variables (standing stocks) are compared with validation values. The knowledge of the standing stocks is the primary type of information on the system. Yet the standing stocks are merely indicators of the internal processes. The analysis of the fluxes between the state variables gives more insight into the functioning of the benthic subsystem, because it directly shows where the major and the minor fluxes are, and how the distortion of the balance between losses and gains affects the development in time of the state variables.

In Sections 11.3 and 11.4 the most important fluxes are discussed. For fluxes in the benthic community validation measurements exist only for respiration rates and primary production. But much more is known qualitatively from the population dynamics of individual species in the state variables. The third level of abstraction (after standing stocks and the fluxes between the state variables) is the level of the regulation of these fluxes. They are regulated by the state variables and by direct interactions between the fluxes, and they are controlled by physical conditions, as described in earlier chapters. The regulations of the fluxes are even harder to study experimentally than the fluxes themselves.

In Section 11.5 a partial sensitivity analysis is carried out. There, control parameters and regulation mechanisms in the model are varied to demonstrate their effect on the model results. Especially the benthic submodel contains many regulation mechanisms with parameters that are only qualitatively known. Our model philosophy, supported by our experience of working with the model, is that the regulation structures are essential but not the absolute values of the regulation parameters.

11.1 The Seasonal Cycle of the State Variables

11.1.1 Total and Anaerobic Bacterial Biomass

In Fig. 11.1 the seasonal cycle of the living state variables is given.

The total bacterial biomass (TBAC) (Fig. 11.1 a) does not show the same sort of seasonal development in all compartments. In the inner compartments (compartments 1 and 2) the input of organic material by the WWA drives up the concentration of anaerobic bacteria (ABAC) from September on. It then remains high during winter, because at the prevailing low temperatures not all available substrate is utilized. Only during summer, after a late spring peak found in all

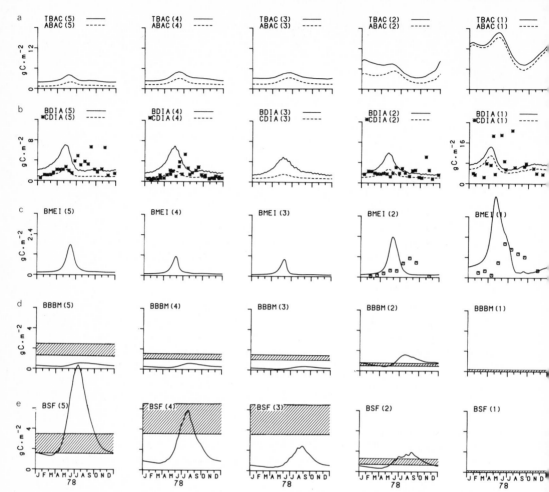

Fig. 11.1. The seasonal biomass distribution of the living state variables in the different compartments of the estuary. a *ABAC* and *TBAC:* anaerobic and total benthic bacterial biomass (the sum of anaerobic and aerobic bacteria). b *BDIA* and *CDIA:* benthic diatom biomass, total and in the upper 0.5 cm of the sediment. c *BMEI:* meiobenthos biomass. d *BBBM:* deposit feeding macrobenthos biomass. e *BSF:* suspension feeding macrobenthos biomass. *Curves:* model results; *squares* or *asterisks:* validation data derived from Colijn and De Jonge (1984); Van Arkel, unpublished data

compartments, the biomass decreases strongly. The waste water effects are superimposed on the seasonal dynamics. The very few biomass determinations available gave average values of 4 gC·m⁻² in the outer compartments and 10 gC·m⁻² in the inner compartments (Schröder, unpublished data).

11.1.2 Phytobenthos

The phytobenthos (CDIA, BDIA) shows a clear spring maximum (Fig. 11.1 b), but the short term fluctuations, characteristic for single-celled primary producers, are missing. The productive phytobenthos in the photic top layer probably shows a dynamic behaviour similar to the phytoplankton, but there are large stocks of phytobenthos deeper in the sediment. The inclusion of these temporarily unproductive stocks damps out any short-term dynamics. In field studies the biomass was always taken from the upper half cm of the sediment (CDIA). In the model the vertical distribution of the benthic diatoms varies between the compartments and with the seasons, hence, a varying part of the simulated biomass (CDIA as a fraction of the total biomass BDIA) has to be compared with the measured biomass. It is quite obvious that such a comparison is problematic due to the influence of wind and wave events and consecutive modifications of the sediment structure. Correspondingly, the observed values scatter strongly in the second half of the year. The densities from December to May are very well reproduced. Due to deviations over the rest of the year the overall agreement is not satisfying (C:4.1–6.8; Table 11.1 a). The inspection of the figures allows the detection of particular deviations, and possibly of their causes. Here the breakdown of the spring peak is too fast. This is a consequence of the too rapid meiobenthos dynamics.

11.1.3 Meiobenthos

The model shows average meiobenthos (BMEI) standing stocks in compartment 1 and 2 similar to the observed biomass (Fig. 11.1 c). The model suggests, analogously to field observations (Bouwman, 1983 b), that the meiofauna in the compartments 3, 4 and 5 is less dense. The peak in the simulated meiofauna follows a peak in the benthic diatoms, indicating the importance of these algae in the diet (cf. Fig. 11.4). The correspondence values in Table 11.1 indicate clearly that the deviations are not due to a wrong average (MC) but due to timing errors (SC). The modelled dynamics of the meiofauna are violent compared with field observations. This reflects the fact that the meiobenthos has been modelled as though consisting mainly of nematodes, whereas in reality it is a complex mixture of nematodes, harpacticoids, turbellaria, etc. with a wide range of physiological properties. Correctly implemented, this would result in much smoother seasonal variations in biomass of the meiobenthos. As it is, the model results conform much better to the seasonal dynamics of individual species of nematodes and oligochaetes than the overall meiofauna. By a future modelling of this functional guild in a more complex way, not analogously to a simple adaptive omnivore species as here, we can gain more insight into the functioning of this subcommunity.

11.1.4 Macrobenthos

Biomass measurements of the macrobenthic fauna (BBBM, BSF) in the estuary have been made during several surveys in spring and summer of different years. Thus there are no continuous series of biomass data available for each of the compartments that may serve for the validation of the seasonal variations. The average biomass (shaded bands) of deposit feeders (Fig. 11.1 d) and suspension feeders (Fig. 11.1 e) is reasonably well simulated in compartment 2 only. In compartment 1 both the simulated and the actual densities are very low. Both the suspension feeders and the deposit feeders become more abundant toward the outer compart-

Table 11.1 Correspondence values of (a) benthic state variables and (b) fluxes

a

State variable	Compart-ment	C	U2	RC	MC	SC	N
Benthic diatoms	1	6.0	0.86	0.81	0.06	0.13	17
in the upper 0.5 cm	2	5.0	0.60	0.52	0.22	0.26	23
of the sediment	4	6.8	0.98	0.88	0.11	0.00	33
(CDIA)	5	4.1	3.55	0.49	0.41	0.10	23
Meiobenthos	1	–	0.61	0.15	0.16	0.69	9
(BMEI)	2	–	0.79	0.21	0.04	0.75	9
Total benthic	1	–	0.00	0.72	0.02	0.27	8
organic carbon	2	–	0.04	0.37	0.57	0.06	8
(BTOC)	4	–	0.08	0.88	0.03	0.09	9
	5	–	0.17	0.14	0.83	0.03	7
Sulphide	1	6.3	0.02	0.32	0.67	0.00	14
(BSUL)	2	6.1	0.35	0.84	0.07	0.09	12
	4	5.3	0.15	0.66	0.00	0.34	16
	5	–	0.27	0.05	0.95	0.00	8
Pyrite	1	–	0.01	0.99	0.01	0.00	8
(BPYR)	2	–	0.24	−0.00	0.00	0.99	5
	4	–	0.12	0.89	0.06	0.05	9
	5	–	0.37	0.05	0.86	0.09	3
Position of	1	5.7	0.70	0.79	0.02	0.19	27
sulphide horizon	2	1.6	>10	0.09	0.85	0.06	27
(BAL)	4	4.5	1.60	0.41	0.59	0.00	66
	5	5.9	0.31	0.76	0.01	0.23	21

C = overall value; U2 = measure of variance; RC = fractional contribution to the error by nonsystematic deviations; MC = idem by the difference between the calculated and observed mean level; SC = idem by nonsimultaneous changes in observed and calculated values (timing error); N = number of observations.

b

State variable	Compart-ment	C	U2	RC	MC	SC	N
Benthic primary	1	6.9	0.25	0.49	0.46	0.05	14
production	2	6.9	0.61	0.82	0.01	0.17	34
(MBPP)	3	7.5	0.33	0.81	0.04	0.15	13
	4	7.6	0.26	0.77	0.08	0.15	17
	5	6.6	0.30	0.56	0.02	0.42	16
Anaerobic	1	–	1.37	0.62	0.37	0.01	7
mineralization	2	–	0.46	0.07	0.77	0.16	9
(W3OA)	4	–	0.42	0.62	0.32	0.07	9
	5	–	0.70	0.68	0.32	0.00	7
Community	1	5.2	0.33	0.24	0.03	0.73	28
respiration	2	6.9	0.18	0.55	0.15	0.31	48
(MCONS)	3	8.4	0.24	0.98	0.01	0.01	32
	4	7.4	0.15	0.63	0.05	0.32	35
	5	5.8	0.27	0.17	0.53	0.30	24

ments; this trend is also present in the model results for the suspension feeders but not for the deposit feeders. The model results for the deposit feeders are too low.

Both deposit feeders and suspension feeders show generally the same seasonal variations despite the different way in which these state variables have been modelled. After an increase in biomass in May/June by a factor of 2 a slow decline follows to minimum values in March/April. This simulated seasonal cycle of macrobenthos biomass agrees well with the seasonal cycle as measured by Beukema (1974) in the western Wadden Sea (Fig. 11.10, data points).

The similarity of the seasonal variations of the two different state variables indicate similarities in the factors which determine these variations: temperature, food availability and predators. Despite their very different feeding strategies and food sources macrobenthic deposit feeders and suspension feeders have a common size range and hence similar physiological rates. Both react much slower than e.g. meiobenthos to external stimuli.

11.1.5 Organic Matter

The seasonal cycle of the nonliving state variables is given in Fig. 11.2.

In sediments the concentration of total organic carbon (BTOC) is easy to measure; because it is dominated by refractory organic carbon, the value does not vary much during the seasons (Fig. 11.2a). On the other hand, the various fractions of organic carbon as used in the model cannot be validated. Yet, the dynamics in the model of the detrital fractions (BDET + ADET) (Fig. 11.2c) and of labile organic carbon (BLOC + ALOC) (Fig. 11.2b) show a typical seasonal behaviour. BLOC and ALOC have a summer maximum due to the excretion of benthic diatoms. This may be a model artefact, because the standing stock of LOC is determined by the rapid balance between production and microbial consumption which both are much higher in summer than in winter. In addition to the summer peak in the inner compartment, the organic waste input can be seen as winter maxima in labile organic carbon (LOC). While LOC has a turnover time of at most a few days, DET is degraded at a much lower rate. Hence, only flat peaks occur, superimposed on a high standing stock.

11.1.6 Sulphide and Pyrite

Due to their long turnover times, sulphide (BSUL) and pyrite (BPYR) do not show much seasonal variation in the model (Fig. 11.2d). The actual concentrations are close to the yearly average concentrations. In an assumed steady state situation the average daily production of pyrite (FLSP + BSUL) must be equal to the average daily pyrite loss due to permanent burial (SDL/TAL · BPYR). Hence, despite the large observed concentration range of pyrite (80–1500), the quotient pyrite/sulphide should be independent of the compartment:

$$\frac{BPYR}{BSUL} = \frac{FLSP}{SDL/TAL} = 5.5 \text{ in the model}$$

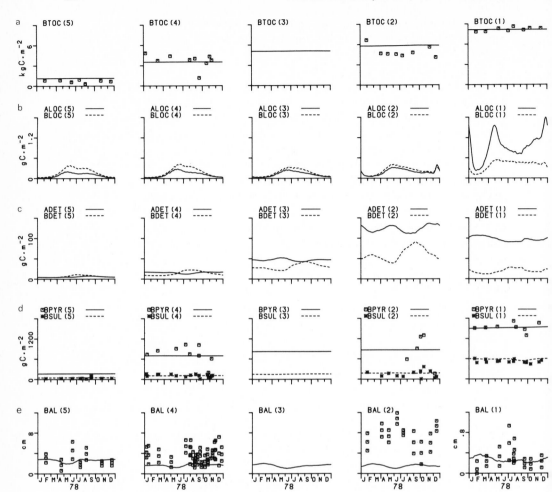

Fig. 11.2. The seasonal abundance of the nonliving state variables in the different compartments of the estuary. a *BTOC:* total benthic organic carbon. b *ALOC* and *BLOC:* labile organic carbon below and above the sulphide horizon. c *ADET* and *BDET:* slowly degradable detritus below and above the sulphide horizon. d *BSUL* and *BPYR:* sulphide (S^{2-}, HS^- and FeS) and pyrite. e *BAL:* position of the sulphide horizon in cm below the sediment surface. Note changed scale in compartment 1. *Curves:* model results; *squares* and *asterisks:* validation data derived from Schröder (unpublished)

In reality this is not fully true (Table 11.2), but the quotients (2.5–6.5) differ only by a factor 2.6 between the compartments compared to 20 for pyrite. Nevertheless, the different quotients indicate deviations from the steady state, or a dependence of the pyrite formation process on the sedimentation rate SDL in the compartment.

The sulphide values of the model agree astonishingly well with the observed values. The correspondence values (Table 11.1) show that the deviations in compartments 2 and 4 are predominantly due to the scatter of the experimental values

Table 11.2 Concentrations of sulphides and pyrites in the sediment

Compartment	Observed sulphide average $(g \cdot m^{-2})$	Observed pyrite average $(g \cdot m^{-2})$	Quotient pyrite/sulphide
1	600	1 500	2.5
2	120	800	6.5
3	–	–	–
4	120	800	6.5
5	25	80	3.3

(RC), while in 1 and 5 the averages (MC) deviate. The general agreement is not due to adjusting the initial values, rather the values are the equilibrium after a simulation run of 6 years with arbitrary initial values. Only the pyrite initial values are fitted to the observations. The small slope of BPYR in the compartments 1 and 5 (Fig. 11.2 d) indicates that the equilibrium is not yet reached there, it takes many years. This equilibrium would be 5.5 times the sulphide values. The observed deviation from the factor 5.5 in compartment 1 can be interpreted in the following way: if the sulphide production in compartment 1 has only been as high as it is now for the last 30 years, the quotient of course is lower, because the pyrite increase lags behind by 40 years. The deviation in compartment 5 cannot be resolved in this manner. A higher (!) sedimentation rate of sand and silt there would reproduce the observed low pyrite/sulphide quotient.

11.1.7 The Location of the Sulphide Horizon

The simulated depth of the sulphide horizon (BAL) (Fig. 11.2 e) corresponds reasonably well with observations in the reduced sediments in compartment 1 and for the oxidized sediments in compartment 5. However, this correspondence as well as the disparities in the other compartments should be regarded skeptically because of the lack of precision in defining the sulphide horizon (BAL). BAL is artificially defined as the sediment layer that is occasionally supplied with molecular oxygen, the permanently anaerobic layer below being characterized by visible accumulation of iron sulphides. The validation data are based on this visible accumulation. In compartment 2 the black subsurface layers lie much deeper for unknown reasons and this results in the disparity shown in Fig. 11.2 e. Oxygen penetrates only millimeter deep into marine sediments as was measured by oxygen microelectrodes. Hence, the simulated mechanisms determining the BAL are highly artificial. BAL is a state variable of a type very dissimilar to the other state variables. It can be considered as an internal variable of the mineralization subsystem. In the model, BAL separates the anaerobic and the aerobic mineralization due to bacteria, hence, in the same way that BAL artificially divides the sediment, also the distinction between the activities of aerobic and anaerobic bacteria is an artificial one. The gap between sediment layers containing oxygen and those containing free sulphide can be significant and may harbour bacterial activities such as fermentation and nitrate reduction (cf. Jørgensen and Sørensen, 1985) that are not explicitly included in the model.

The introduction of BAL nevertheless is a first and important step in the effort to model the interdependence and variability of the various processes, which fall under the heading of "mineralization".

11.2 The Fluxes from and to the State Variables

The importance of a state variable to the system cannot be derived from the standing stocks alone, the fluxes within, from and to the state variables throw much more light on the structure and organization of the system. The fluxes connected with all living state variables are uptake, rest and activity respiration, excretion and mortality, especially due to predation, and finally transport across boundaries. In the following subsections, the most important fluxes are discussed. Validation values exist only for primary production and for total and anaerobic respiration. For these fluxes the model results are compared with the observations (Fig. 11.3). In all other cases the discussion is limited to compartment 2, where the benthic subsystem is of great importance and is not as intensively affected by waste water as compartment 1.

11.2.1 Validated Fluxes

Three series of observations from the Ems estuary are available to validate modelled carbon fluxes: the benthic primary production (MBPP), the anaerobic bacterial respiration (W30A) and the benthic community respiration (MCONS) (Fig. 11.3). The primary production of the benthic diatoms has been measured in all compartments (Colijn and De Jonge, 1984) and is simulated reasonably well in four of the five compartments (Fig. 11.3 a). (C:6.6–7.6 in Table 11.1 b). Field measurements in compartment 2 show an extremely high variability. The variability in this compartment and others is too large to detect the effect of the tidal phase visible as 14-day wiggles in the simulations.

The simulated anaerobic respiration (W30A) (Fig. 11.3 b) agrees in a satisfactory way with the measurements of the sulphate reduction rate. The model seems to overestimate anaerobic respiration in the inner compartment, but this effect has been expected. On the one hand sulphate reduction in compartment 1 tends to be limited by sulphate shortage, which is not included in the model. On the other hand, anaerobic respiration in the field might function to some extent via other processes than sulphate reduction, e.g. methane production or nitrate reduction. Since the contribution by these processes has been implicitly included in the anaerobic metabolism in the model, we may expect discrepancies between simulated anaerobic metabolism and measured rates of sulphate reduction in the direction of overestimation.

The overall oxygen consumption of the benthic community results from a variety of processes such as the respiration of benthic organisms and the chemical oxidation of sulphide. One has to take into account pyrite formation and the buffering action of sulphide, which may lead to a partial temporal uncoupling between anaerobic respiration and total oxygen consumption. Furthermore, the measurements were carried out with bell jars (Van Es, 1982 a) that alter the dif-

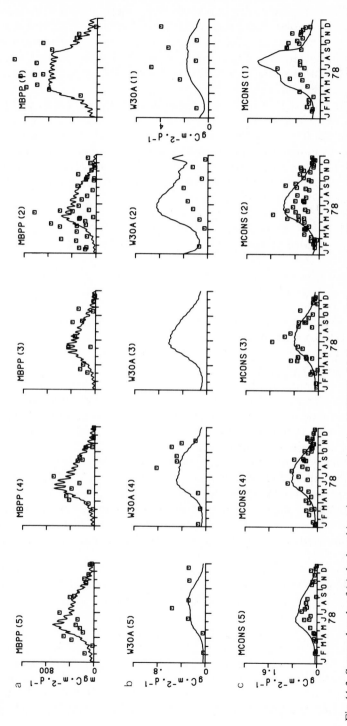

Fig. 11.3. Seasonal cycle of (a) the benthic primary production (*MBPP*), (b) the anaerobic respiration (*W30A*, note changed scale in compartment 1) and (c) the total oxygen consumption (*MCONS*). *Curves:* model results. *Squares:* validation data derived from Schröder (unpublished), Van Es (1982a) and Colijn and De Jonge (1984)

fusion of oxygen across boundary layers and apply to high tide only. Hence, to make the simulated oxygen consumption comparable to the measured values we needed to formulate a model variable MCONS to simulate measurement of the community respiration. MCONS contains the aerobic respiration and the oxidation of the upward diffusing sulphide. Figure 11.3 c shows that the model variable MCONS simulated the measured values of oxygen consumption in an acceptable way (C:5.2–8.4 in Table 11.1 b).

In summary, the validation of the simulated primary production (MBPP), the anaerobic respiration (W30A) and the oxygen consumption (MCONS) shows that the benthic submodel approximates the overall carbon flow in the natural benthic community reasonably well. This is more valuable and encouraging than the validation of the seasonal developments in biomass (see Sect 11.2). Together these results justify a closer examination of the carbon flow through the various state variables.

11.3 Unvalidated Fluxes

The benthic submodel, partly validated now with ca. ten data sets, can be used to demonstrate a variety of quantitative and qualitative aspects of the benthic ecosystem. The generated results serve in the first place to indicate the capabilities of the model and secondly allow an intuitive check of the model's functioning. The following examples were selected.

11.3.1 Diet of Meiofauna

In winter the simulated meiobenthos mainly feeds on bacteria, whereas in summer the feeding on diatoms dominates. During the short periods of high densities of meiobenthos, "cannibalistic" feeding occurs (Fig. 11.4). With some caution the feeding components can be interpreted as groups of food-specialized species. Considering the seasonal variation of the meiobenthos (Fig. 11.1 c) the bacterial feeding component of meiobenthos also peaks in early summer, but this peak is not so pronounced as the peaks of the other components.

In support, Bouwman (1983 b) provided evidence for the predominance of diatom-feeding nematode species in summer, whereas feeding on bacteria seems to be restricted to a few less abundant species. However, there is no evidence from

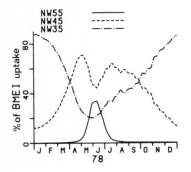

Fig. 11.4. The relative contribution from different food sources to the total uptake of meiobenthos in the model. *NW35* = Uptake of bacteria; *NW45* = uptake of diatoms; *NW55* = cannibalistic feeding

the Ems estuary that cannibalistic feeding is as strictly density-dependent as the model indicates. It is a very attractive feature of the model that it indirectly describes the varying species composition of meiobenthos by its food adaptation, even if this adaptation in the model is immediate and not delayed by a month or so due to the internal population dynamics. This causes the artificially violent dynamics of the cannibalistic component of the meiobenthos in the model.

11.3.2 Grazing and Predation in the Benthos

Table 11.3 gives a survey of the annual carbon fluxes through the simulated food web. Table 11.3, together with the data in Table 11.4, shows that only about 25% of bacterial production is removed by grazers. Excretion losses, respiration and mortality by other causes than grazing seem to be much more important in absorbing the production of these microorganisms.

Benthic diatoms constitute by far the largest food source for the benthic community. In the meiofauna, cannibalism is the largest single loss factor; nevertheless the consumption by macrobenthos (BBBM) and epifauna (EMES and EMAC) is also significant. The epifauna and the birds together consume almost all of the macrofaunal production in compartment 2. In this averaging consideration all the information about seasonal variations is lost. As exemplified in Section 11.3.1 and Figure 11.4 the model also provides this very important additional information.

The total food production in the benthos is indeed dominated by unicellular organisms, but the next trophic level, the fauna, is highly productive also. This implies that the consumption by benthic and epibenthic fauna is a quantitatively important regulator for the populations of unicellular organisms (cf. Sects. 11.4.3, 11.4.4 and 12.2.2).

Table 11.3. Yearly food uptake of benthic and epibenthic fauna as simulated in compartment 2 (unit: $gC \cdot m^{-2} \cdot y^{-1}$)

Food sources					
Consumer	BBAC+ABAC	BDIA	BMEI	BBBM	BSF
BMEI	8.3	16.4	5.6	–	–
BBBM	7.3	7.7	1.9	–	–
BSF	–	7.4	–	–	–
EMES	1.2	5.9	2.1	–	–
EMAC	–	–	0.7	2.1	0.8
BIRDS	–	–	–	0.7	0.7
Total	16.8	37.4	10.3	2.8	1.5

11.3.3 Carbon Budgets of Organisms

Food uptake is distributed over various metabolic processes that differ widely in the functional groups of benthic organisms. Figure 11.5 shows that the food (in case of algae: inorganic carbon assimilated) is only partly converted to new bio-

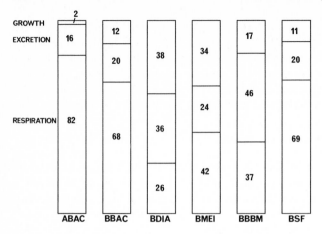

Fig. 11.5. Partitioning of total uptake into respiration, excretion and growth for anaerobic bacteria (*ABAC*), aerobic bacteria (*BBAC*), benthic diatoms (*BDIA*), meiobenthos (*BMEI*), deposit-feeding macrobenthos (*BBBM*) and suspension-feeding macrobenthos (*BSF*) in the model. The values are calculated for compartment 2

mass. These losses determine the ecological efficiency; here they are calculated via the model, thereby including variations in food limitation, stress excretion, etc. Benthic bacteria are calculated to have an ecological efficiency of only 2–12%, macrobenthos converts 11–17% of their food into tissue, whereas meiofauna shows a highly efficient conversion with 34% of the food intake being converted into biomass. Of course the benthic diatoms, as autotrophs, show a high efficiency of 38%. Using the model to calculate parameters that have an ecological meaning, such as the ecological efficiency of a functional group, gives an indication of the performance of this group in the model. Moreover, since parameters such as ecological efficiencies, P/B ratios, etc. are output parameters, the end results of the model calculations can serve to check whether the performance of the model is reasonable or not. Also, because the choice of parameter values for the physiological processes in the state variables determines the final values of the ecological parameters, the model offers a useful vehicle for discussion about the ecological importance of physiological parameters.

11.3.4 Net Production and Productivity

In Table 11.4 the calculated production and productivity values for benthic organisms are given. We use the symbol $\bar{B}$ for the yearly average standing stock ($gC \cdot m^{-2}$), P for the total yearly production ($gC \cdot m^{-2} \cdot y^{-1}$), and the quotient $P/\bar{B}$ is the yearly productivity of the organisms. The $P/\bar{B}$ ratio is of course affected by the parameters defining the feeding and growth characteristics described earlier; here the resulting yearly productivities include the impact of competing functional groups and seasonal changes in food availability.

Schwinghamer et al. (1986) surveyed the literature on annual $P/\bar{B}$ ratios of estuarine benthic organisms measured in situ. The simulated data for macrofauna are similar to those data from many field studies (0.8–5.2). The productivities of benthic diatoms and meiofauna are also in the range of reported values (1.9–123), but that of aerobic bacteria in the model is below the reported range of productivities (105–593). Little is known about the in situ productivity of anaerobic bacteria.

Table 11.4. Average standing stocks and annual production in gC · m^{-2} and resultant P/B̄ ratios of benthic state variables in compartment 2

State variable	B̄ (gC · m^{-2})	P (gC · m^{-2} · y^{-1})	P/B̄ (y^{-1})
Aerobic bacteria BBAC	2.44	44.8	18.4
Anaerobic bacteria ABAC	3.55	22.0	5.6
Benthic diatoms BDIA	2.65	37.5	14.2
		73.3[a]	27.6[a]
Meiofauna BMEI	0.50	10.3	20.6
Deposit-feeding macrobenthos BBBM	0.90	3.0	3.3
Filter-feeding macrobenthos BSF	0.91	1.6	1.8

[a] Including excretion products

In summary, we may conclude that the productivities of benthic organisms resulting from the model seem to match in situ observations. The model allows the extraction of P/B̄ ratios for a part of the year; this would simplify the comparison with in situ observations, and this would possibly resolve discrepancies like that in the state variable aerobic bacteria which may partly be the result of the averaging process.

11.3.5 Dynamics of Refractory Organic Carbon in the Sediment

The continuous spectrum of detritus in the model is divided into three classes ROC, DET and LOC. These classes differ by their turnover times. In addition, in the benthic submodel it is assumed that LOC is mainly dissolved and hence diffuses rapidly into the sediment while the particulate DET and ROC fractions can only be transported vertically by sediment reworking.

Table 11.5 shows the balance of import, burial and mineralization of ROC in compartment 2.

In equilibrium there should be a balance of the gain terms deposition and production and the loss terms of mineralization and burial of ROC. Due to the fact that the mineralization rate of ROC is much lower than the deposition rate, ROC accumulates to large concentrations in the sediment. The amount buried (149 gC · m^{-2}) corresponds to the ROC content of the 0.8-cm-thick layer which leaves the system per year across its lower boundary as a consequence of sand and silt sedimentation. The ROC concentration (5600 gC · m^{-2}) is very high com-

Table 11.5. Annual budget of refractive organic carbon in the benthos in compartment 2

Deposited in the benthic system	297 gC · m^{-2} · y^{-1}
Produced in the benthic system	10
Decomposed by aerobic bacteria	35
Decomposed by anaerobic bacteria	50
Buried	149
Mass increase	73

pared to the mean values for DET (185 gC·m⁻²) and LOC (0.42 gC·m⁻²). Because all the produced and imported DET and LOC is decomposed in the long run, one can easily calculate the turnover times (in the model) to be months for detritus and a few days for LOC. In contrast, only ca 1% of ROC is degraded per year according to the model, and 70% of the deposited ROC is never degraded at all. In the model the ROC concentration will slowly increase to about 8300 gC·m⁻². Then burial and decomposition balance deposition and production, and the mass increase vanishes.

11.4 Sensitivity Analysis of the Benthic Subsystem

A full sensitivity analysis of the benthic submodel is practically impossible due to the number of parameters. But during the modelling process the most sensitive aspects of the model have clearly manifested themselves. These aspects have been addressed in the following sensitivity analyses.

In a sensitivity analysis a regulation parameter or regulation mechanism is changed to demonstrate its effect on the model results. Insensitivity to change of a single parameter does not mean that the model is too robust, compared to nature. The reason is often that most processes are regulated by several mechanisms, such that in case of a failure of one mechanism another will take over. In such cases a single limiting factor cannot be identified. Not all regulations and controls in a real system are known.

In the following sections some variations of parameters and regulation mechanisms are discussed. It is a small selection of a great number of computer experiments performed with the model.

11.4.1 Reduced Detritus

To clarify the role of detritus (DET) in the benthic system, the quantity of DET has been reduced. This has been done by adding an artificial "detritus aging" by transition of DET into ROC with a half-life time of 14 days. The detritus standing stock is thereby strongly depleted (Fig. 11.6a and b; ADET, BDET) and its degradation by bacteria is decimated.

In this run the bacterial biomass decreases to one-half of its value in the standard run (ABAC, TBAC). Meiobenthos, deposit feeders and epibenthic organisms decrease slightly. The sulphate reduction rate (W30A, anaerobic respiration) is also reduced to one half of its standard value, hence the sulphide concentration (BSUL) itself drops over the year and will reach a lower equilibrium in a few years. As a consequence of the reduced sulphide concentration the position of the sulphide horizon (BAL) moves slowly downward. The total respiration (MCONS) is also slightly reduced. This reduction as well as other changes will be larger in later years when the sulphide reservoir is oxidized. Other state variables are essentially unaffected.

A diminished role of detritus has only a limited impact on the model results because detritus in the standard model is mainly oxidized by bacteria and only a few percent of the detrital carbon is transferred into the food web. Therefore,

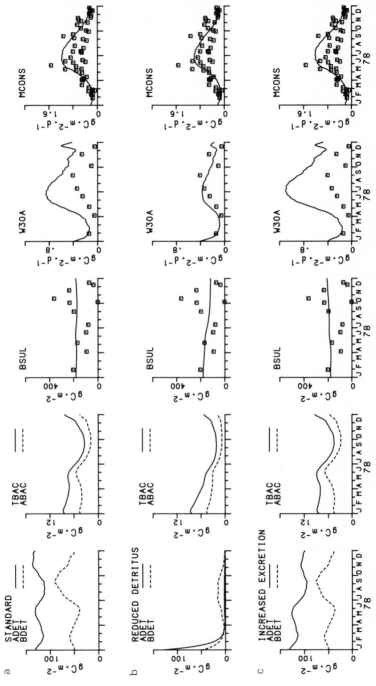

Fig. 11.6. Results from sensitivity analyses compared to results from the standard run. a Results from the standard run. b Results from a run with reduced detritus concentrations (see Sect. 11.4.1). c Results from a run with increased excretion rates (see Sect. 11.4.2; Table 11.6)

we may conclude that (with respect to the carbon cycle) sedimentation of detritus on the tidal flats predominantly is a carbon sink. But it would be wrong to conclude that sedimentation of detritus is an ecologically unimportant process. The sedimentated detritus is certainly very important as a carrier of nutrients, which are recycled during degradation (Höpner et al., 1983). This recycling is not contained in the version of the model presented here, but it will be an essential feature of a new model of the western Wadden Sea.

11.4.2 Bacterial Excretion and Efficiency

Because it is not very well known which percentage of the organic matter taken up by bacteria is respired, excreted (as LOC, DET or ROC) or assimilated as cellular material, we tested the model to find out how important an accurate estimate of these coefficients is. We have used the set of parameters shown in Table 11.6 (test run coefficients in parentheses). In the test run we have reduced the percentage respired with 5% and increased the percentage excreted as LOC by the same percentage, keeping the sum of the loss terms constant.

In the test run the LOC excretion by bacteria is nearly twice as high as in the standard run, whereas the excretion of DET is reduced. The standing stocks, however, change less than expected (Fig. 11.6c), because in the model, the total LOC production by bacteria also includes a "natural" mortality of bacteria. On average, 2.5% of the bacterial biomass per day is transformed into detritus which consists of 80% LOC and 20% DET. Considering this, the use of a modified set of parameters (Table 11.6) increases the total LOC production by bacteria by only 20%. Accordingly, the test run resulted in a 20% increase in bacterial biomass. Benthic detritus is reduced somewhat more, whereas the anaerobic respiration (W30A) is increased by 30% (Fig. 11.6c). In contrast, the overall oxygen consumption (MCONS) remains essentially unchanged. It seems that the cometabolism of DET and ROC with LOC accelerates the degradation cycle without affecting the overall dynamics of the benthos. This test run indicates that the rather complicated formulation of bacterial metabolism and the extensive list of parameters still needs to be clarified further in order to simulate the degradation of organic carbon in the benthos correctly (cf. Sect. 7.5).

Table 11.6. Excretion as a fraction of uptake in the standard run and in a sensitivity analysis run (in parentheses)

State variable	Food source	Assimilable per day	Fraction respired	Fraction excreted as LOC	Fraction excreted as DET	Fraction excreted as ROC	Fraction assimilated
BBAC	LOC	0.40	0.50(0.45)	0.10(0.15)			0.40
(aerobic	DET	0.25	0.65(0.60)	0.05(0.10)	0.05	0.05	0.30
bacteria)	ROC	0.05	0.90(0.85)	0.05(0.10)			0.05
ABAC	LOC	0.25	0.65(0.60)	0.10(0.15)			0.25
(anaerobic	DET	0.15	0.75(0.70)	0.05(0.10)	0.05	0.05	0.20
bacteria)	ROC	0.05	0.90(0.85)	0.05(0.10)			0.05

11.4.3 Phytobenthos Regulation

In the model the primary production by the phytobenthos is calculated in the following way: first a "potential primary production" PPR is determined proportional to the illuminated biomass. In the following calculation step PPR is subjected to feedback in order to obtain the actual primary production PP (cf. Sect. 7.2).

The presence of the diatoms in the photic layer depends of course on the irradiance, but it was also assumed that at low biomass levels the diatoms tend to occur more superficially. There is no satisfactory justification for this assumption.

The second regulation is based on observations on diffusion limitation in diatom mats imposing an estimated upper limit of $1500 \, \text{mgC} \cdot \text{m}^{-2} \cdot \text{d}^{-1}$ on the photosynthetic rate. This feedback of photosynthesis is tested in various modifications of the model.

In a series of test runs we altered the formulation of photosynthetic feedback from the gradual correction in the standard model (Sect. 7.2) to a simple Michaelis-Menten formula:

$$PP = \frac{PPR \cdot PPH}{PPR + PPH},$$

where PPH as before is the upper level of photosynthetic production, but here PPH serves also as a half-saturation value.

The test run with a Michaelis-Menten-type regulation was run twice, once with the original value for PPH of $1500 \, \text{mgC} \cdot \text{m}^{-2} \cdot \text{d}^{-1}$ and once with an increased value of $5000 \, \text{mgC} \cdot \text{m}^{-2} \cdot \text{d}^{-1}$. The rationale is that the upper level of photosynthetic production as determined by bell-jar experiments is about $1500 \, \text{mgC} \cdot \text{m}^{-2} \cdot \text{d}^{-1}$ (Colijn and De Jonge, 1984; Admiraal et al., 1982), but recent measurements with O_2 micro-electrodes indicate much higher potential productivities (Revsbeck et al., 1980; De Jonge and Sandée, personal communication).

Figure 11.7 shows that the conservative estimate for PPH of $1500 \, \text{mgC} \cdot \text{m}^{-2} \cdot \text{d}^{-1}$ leads to model results similar to the standard run for the less productive compartment 5, but for the densely populated compartment 1 the photosynthetic rates predicted by the model are too low. This is serious since we think that the actual measurements in this compartment are low because of the restrictions of the bell-jar technique. Using a PPH value of $5000 \, \text{mgC} \cdot \text{m}^{-2} \cdot \text{d}^{-1}$, we see that the calculated biomass distribution is more dynamic, with a possibly too pronounced spring bloom, but with an appropriate autumn bloom. The production peaks generated are obviously too high compared with the validation data, but this is to be expected. From the results presented in Fig. 11.7 we may conclude that feedback regulation of the benthic photosynthesis is a key factor for the benthic system. It seems likely that the maximum production, given by PPH, is somewhere between 1500 and $5000 \, \text{mgC} \cdot \text{m}^{-2} \cdot \text{d}^{-1}$. The simultaneous reproduction of the observed values for phytobenthos densities and primary production is possible only for very few combinations of parameter values.

Benthic diatoms are an essential food source for grazers and their excretion plays an important role for the benthic bacteria (see above). The test runs in

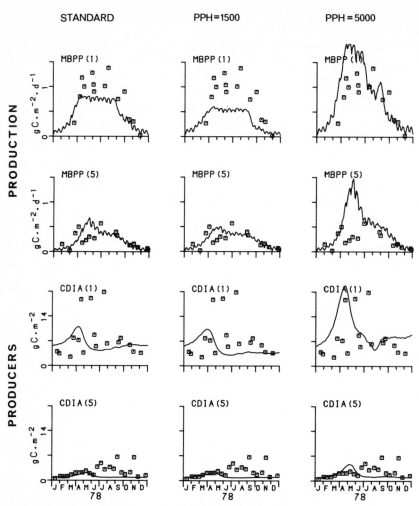

Fig. 11.7. Results of a sensitivity analysis with Michaelis-Menten regulation of photosynthesis compared with the standard run (see Sect. 11.4.3). *PPH* = 1500 and *PPH* = 5000: runs with maximum photosynthetic rates of 1500 and 5000 mgC $\cdot$ m^{-2} $\cdot$ d^{-1}, respectively. Production rates (*MBPP*) and biomasses of producers: diatoms (*CDIA*) are shown for compartments 1 and 5. *Squares:* validation data

Fig. 11.7, showing a modified diatom growth and production, were also used to demonstrate the effect on benthic grazers. The minor variations in the test run PPH 1500 produced no significant alterations in the three animal groups shown in Fig. 11.8. In contrast, the increased biomass and production of diatoms in the run with PPH of 5000 led to explosions in the benthic communities: meiofauna and even the slowly growing suspension feeders increased to unrealistically dense populations. Furthermore the benthic bacteria reached very high activities (results not shown). Thus, moderate changes in the production regulation of dia-

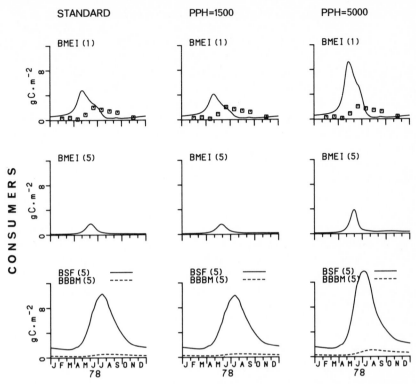

Fig. 11.8. As Fig. 11.7. The biomass of consumers: meiofauna (*BMEI*), deposit-feeding (*BBBM*) and filter-feeding (*BSF*) macrofauna. The extremely low values for macrofauna in compartment 1 were omitted

toms induce increased grazing activity. The grazers in turn keep the effect on the diatoms themselves within limits.

We may learn from this observation that it is not possible to deduce fluxes (rates) from such indirect evidence as standing stocks, because generally many fluxes and feedback mechanisms cooperate. This is not only true in nature, but it is also the case in the model even though this is much simpler than nature.

Another illuminating aspect of these model runs is the coherence of the model: a fairly moderate change in the potential benthic primary productivity leads to large changes in other components of the system that damp the effect on the primary producers themselves. The changes in the other components cannot simply be undone by changing a few other parameters, without having these parameters take very unlikely values. The implication of this is that the use of one extreme parameter value in a biological state variable shows up in a reduced correspondence between model results and observations of other state variables. It appears that in the present model structure the freedom to change parameter values in order to improve the performance of a state variable is rather restricted because the response of other state variables to this change may very well lead to a degradation of the model performance as a whole.

11.4.4 The Role of Meiobenthos in the Benthic Community

The partitioning of the available food among benthic bacteria, meiofauna and macrofauna is, among numerous other things, dependent on the response time of the respective faunal elements to changes in food supply. Obviously, the shorter the generation time, the faster the response. This would tend to favour the meiofauna over the macrofauna in any competition for the same food resource.

However, since the macrofauna also acts as a predator on the micro- and meiofauna the situation is a bit more complex, especially when we take into account that the epifauna (ranging from mysids to flounders) also competes for the same food resources and acts as a predator on almost the whole benthic community. Several mechanisms act together to allow coexistence between meiobenthos and sediment-feeding macrobenthos: food structure, predator-mediated coexistence, direct interactions and strong temporal variations due to the seasons (Chesson, 1986). Especially the temporal fluctuations or oscillations strongly enhance the possibilities for coexistence (Ebenhöh, 1988).

To visualize the role of meiobenthos in this complex whole, a model run was made where the meiobenthos feeding activity was reduced to zero. In effect, meiobenthos thus was removed from the system. The results of this run were compared to the results of the standard run (Fig. 11.9). The consequences on the seasonal biomass distributions of the other benthic fauna elements are considerable: the benthic diatoms now show a strongly pronounced spring maximum, being released from meiobenthic grazing pressure (Fig. 11.9). Benthic bacteria show the same effects (Fig. 11.9). Both groups have lowered standing stocks in the second half of the year, due to the delayed but increased grazing pressure by macrobenthos, which by then has caught up with the increased food supply.

Both macrobenthos groups, deposit feeders and suspension feeders show increased standing stocks in summer, and are depressing the standing stocks of benthic diatoms and benthic bacteria below the levels found in the presence of

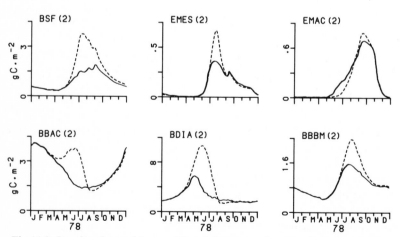

Fig. 11.9. Results of a model run with meiofauna uptake set at zero. *Continuous curves:* results from the standard run. *Dashed curves:* results from the sensitivity analysis run

meiobenthos. The effects of the removal of meiobenthos are also clearly visible in the epifauna (Fig. 11.9). The mesoepibenthos directly profits from the enhanced phytobenthos stocks, whereas the macroepibenthos, as a predator on meiobenthos, has a lower abundance than in the standard run until July/August, when the increased stock of macrobenthos outweighs the loss of meiobenthos as a food source.

The major role of meiobenthos in this model thus is that of a direct competitor for food with the other benthic consumers, with the additional role of prey to macroepibenthos which generally cannot use the primary production directly. It is of special interest to observe the readjustment of the system after the meiobenthos removal. The weblike structure of the carbon flows makes the system much more resistant against major distortions than a chainlike structure.

11.4.5 The Activity Pattern of Macrobenthos

The mussel, *Mytilus edulis*, one of the major species among the benthic suspension feeders (BSF), has an enhanced activity in early spring as measured by their O_2-uptake, compared to its activity at the same temperature in late summer (De Vooys, 1976). The annual activity curve has the same shape as the temperature curve, but it precedes the temperature curve by about 2 months. The high activity in spring is connected with spawning. It is obvious that the activity of this organism is not dependent on the actual temperature alone, but is modified by another factor, possibly temperature adaptation.

The effect of the time lag between activity and temperature was analyzed in a model run. The temperature curve for benthic suspension feeders, which determines their activity in the model, in this run was time-shifted such that it coincides with the activity curve observed in situ. It was applied to benthic deposit feeders as well.

Figure 11.10 shows the relative biomass value with the validation points. In the time-shifted run, the fit in late winter and spring is improved, but in summer the biomass reaches too high values, as it does in the standard run too. Compared to the standard run, the annual mean biomasses are slightly higher but by the end of the year they have lower values.

As a consequence of this activity shift the spring bloom of benthic diatoms (Fig. 11.11, BDIA) is reduced, and the sharp peak of meiobenthos (BMEI) is reduced (Fig. 11.11). Deposit feeders (BBBM) and meiobenthos interact in two

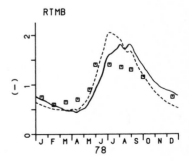

Fig. 11.10. Seasonal cycle of relative biomass of macrobenthos (*RTMB*) in compartment 2. Standard run (*continuous curve*) and test run with macrobenthos activity shifted 2 months backward in time (*dashed curve*). Squares: validation data (scaled) taken from the Balgzand (Beukema, 1974). *RTMB:* total macrobenthic infaunal biomass, divided by the annual average biomass

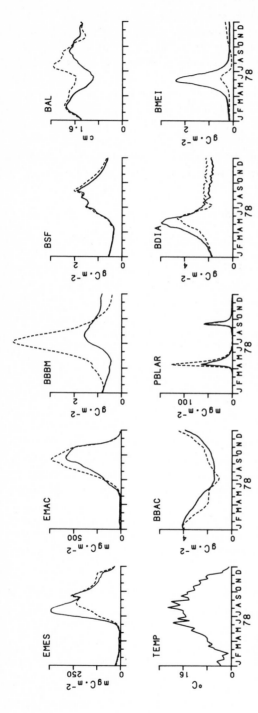

Fig. 11.11. Results (*dashed curves*) from a test run with early macrobenthic activity (see Sect. 11.4.5) and the model results from the standard run (*continuous curves*)

ways: as food competitors and as prey-predator. The meiobenthos peak occurs in April and May. During this period the carbon flux from meiobenthos to deposit feeders did not change but the flux from benthic diatoms to deposit feeders increased five-fold, whereas the flux from benthic diatoms to meiobenthos decreased to one-fifth. So the interaction between meiobenthos and deposit feeders ran via food competition.

11.5 Conclusions

The benthic submodel cannot be separated from the pelagic and the epibenthic submodels, because sedimentation from the pelagic and predation by the epibenthic predators are essential processes in the benthic system. As demonstrated in Chapter 12, modifications in the epibenthic submodel strongly influence the size and the seasonal dynamics of the benthic state variables.

In combination with the standard version of the other submodels the benthic submodel reproduces the basic features of the observations on the state variables. But according to present knowledge there are still several possibilities for the control mechanisms of the main benthic fluxes, hence further attempts at fine calibration of the parameter values have to wait. During the test phase and the sensitivity analysis the urgency of several improvements to the model structure became obvious. Cases in point are the nutrient recycling in connection with the mineralization, a more realistic control of phytobenthos growth and a better treatment of the reworking activity of the macrobenthic deposit feeders. Also the assumption of meiobenthos as a single omnivore species is unsatisfying. But most important in an improved benthic model is probably a better modelling of the biochemical processes connected with the border or the gap between the oxygen-rich and the sulphide-rich layer of the sediment.

12 Results and Analysis of the Epibenthos Submodel

J. G. BARETTA-BEKKER

12.1 Model Results

There are several ways to validate the results of the epibenthic submodel. The correspondence between the simulated standing stocks and the field values is a measure for the goodness of fit but also the fluxes, both internal and between state variables, can reveal something about this part of the ecosystem.

12.1.1 The Standing Stocks in the Five Compartments

The lack of calibration data for the epibenthic state variables manifests itself clearly in the low correspondence values (C:3.9–4.7) between the model results and the validation data (Table 12.1) for macroepibenthos (Stam, 1979, 1982, 1984). Relatively the largest part of the error is in the slope component (SC) (50–70%) and almost all of the remaining error is due to the mean component (MC). The RC error component in all the compartments is very small.

There are several reasons for the discrepancies. A major reason is that in the model the epibenthic and the benthic systems are confined to the area of the tidal flats, and that the subtidal channel area is ignored. For the macroepibenthos this probably is a severe distortion of reality, since it is hardly possible that it does not feed at all during the time (the low water period) that it is denied access to the tidal flats. This modelling approach has been taken because the subtidal area in the inner three compartments is virtually identical with the area of the tidal channels and the tidal channels have been shown to be very poor in benthic organisms

Table 12.1. Correspondence values of epibenthic state variables

State variable	Compart-ment	C	U2	RC	MC	SC	N
Macroepibenthos	2	4.5	0.69	0.05	0.25	0.70	19
(EMAC)	3	4.0	0.78	0.07	0.23	0.71	22*
	4	3.9	0.83	0.03	0.31	0.67	22*
	5	4.7	0.66	0.03	0.47	0.50	12

C = overall correspondence value; U2 = measure of variance; RC = fractional contribution to the error by nonsystematic deviations; MC = idem by the difference between the calculated and observed mean level; SC = idem by nonsimultaneous changes in observed and calulated values (timing error); N = number of observations; * = the same observations are used for compartments 3 and 4.

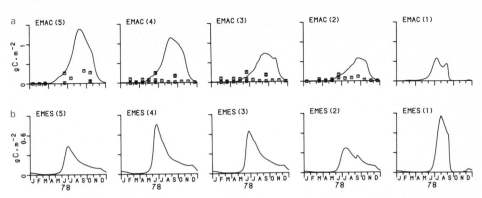

Fig. 12.1. Simulated biomasses (in mgC · m⁻²) of macroepibenthos (a) and mesoepibenthos (b) in the five compartments with the available field data. In the field measurements no distinction between compartment 3 and compartment 4 has been made

(Van Arkel, unpublished data). The two outer compartments, however, have more extensive subtidal areas with the potential for a well-developed benthic fauna. The consequences of ignoring this part of the benthic and epibenthic system in the model are difficult to estimate. Kuipers and Dapper (1981) have shown that the brown shrimp (*Crangon crangon*), the main constituent of the macroepibenthos, shifts its behaviour gradually from staying on the tidal flats all the time when small, to migrating to and from the flats with the tides, to always staying in the subtidal zone when over 30 mm in size. With samples on the subtidal parts of the flats being underrepresented, the field measurements may be underestimates of the mean standing stock on the flats, especially for compartments 4 and 5.

Other reasons can be found in the model formulation itself. It is very difficult to estimate parameter values for this group which may change quite considerably in mean size during the season, with a concomitant change in actual parameter value. For example, the value for the parameter rupZmax, the maximum relative uptake rate of macroepibenthos, has been calculated for the generally small immigrating organisms in spring which makes it probably too high later in the year when the mean size has increased. This is one of the typical drawbacks of modelling an unstructured population. Though the simulation of the biomass level seems doubtful, the seasonal dynamics and the differences between the compartments of macroepibenthos are reproduced reasonably well (Fig. 12.1 a). Note the migration from compartment 1 to more outward compartments in September, when the waste water discharges result in oxygen saturations which are too low for the macroepibenthos to tolerate.

The mesoepibenthos (Fig. 12.1 b) cannot be validated with field data because the group of small epibenthic omnivores has not been studied as such. Some species in this functional group, such as *Nereis*, have been sampled as benthic components, others, such as mysids, as pelagic components. The different sampling procedures prohibit simply adding the different components to obtain an estimate of the standing stock, since converting concentrations expressed per m³ to concentrations per m² is a dubious enterprise. Indications whether this group behaves in the model as in reality or not may be found in the interaction between this group and other components of the system. The sections on sensitivity analysis will return to these interactions.

12.1.2 Internal Fluxes and Productivity

The lack of direct validation values for epibenthic processes makes it necessary to resort directly to a comparison of the internal fluxes generated by the model with literature data.

In Fig. 12.2 the realized relative uptake of macroepibenthos (rupZ) and mesoepibenthos (rupY) is given for compartment 2 through the year. The maximal value for rupZ is around 0.16 while rupY has a maximum of 0.40. Both groups realize at most half their maximum potential rate of uptake (rupZmax = 0.35; rupYmax = 0.70). The uptake is regulated by temperature and the amount of available food. This combined regulation results in the relative uptake rate reaching a maximum in June (Fig. 12.2). The loss factors respiration and excretion in the model are equal to each other. The average respiration loss for macroepibenthos in the model (43.6%) is close to literature data. Kuipers and Dapper (1981) calculated a value of 40% for plaice, Jansen (1980) 65% for shrimps and Peters (1983) 48% for poikilotherms generally. The excretion in the model is probably too high, reducing the remainder of the uptake that can be used for growth (13%) to a lower percentage than the values (average 35%) mentioned by De Vlas (1979), Kuipers and Dapper (1981), Jansen (1980), Redant (1980) and Peters (1983).

The net productivity (the daily growth divided by the biomass) of macroepibenthos (Fig. 12.3 a) is highest in the two innermost compartments from January to June. From June on the highest productivity is found in the outer compartments with compartment 4 as the most productive. In compartment 1 the productivity decreases rapidly after August, due to the waste water discharges. Thus in spring, conditions for macroepibenthos are best in the inner part and in autumn in the outer part of the estuary.

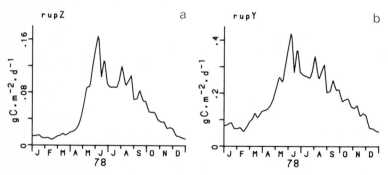

Fig. 12.2. Relative rate of uptake of macroepibenthos (*rupZ*) (a) and of mesoepibenthos (*rupY*) (b) in compartment 2

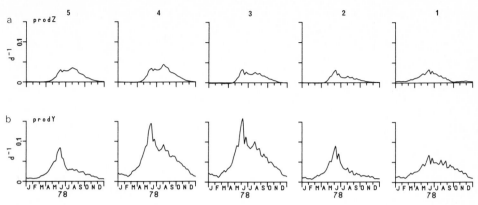

Fig. 12.3. The productivity for macroepibenthos (*prodZ*) (a) and mesoepibenthos (*prodY*) (b)

The mesoepibenthos (EMES) productivity (Fig. 12.3 b) on average is about five times higher than the macroepibenthos (EMAC) productivity. The shift in productivity during the year from the inner compartment to the outer compartments as found in EMAC does not occur in EMES. In this group productivity is highest in compartments 3 and 4, while productivity is only half as much in the compartments 1, 2 and 5.

In summary, for EMAC the excretion losses are too high and hence the growth rate is too low. Still, even with a low productivity, the standing stocks are too high, indicating that predation pressure either by pelagic predators or by external predators (birds) is too low in the model. This is certainly plausible, since predation of the macroepibenthos by pelagic fishes is not included in the model.

The macroepibenthos is a biological component suitable to quantify the so-called nursery area function of the estuary. In spring the immigrating biomass in compartment 5 is 225 mgC·m^{-2} (cf. Sect. 8.3.3). According to the model, 1391 mgC·m^{-2} emigrates from compartment 5 to the sea in autumn. Thus, the nursery area returns biomass immigrating in spring multiplied by a factor 6.2 to the sea in fall. From data of Kuipers and Dapper (1981), collected on the Balgzand between 1976 and 1979, it can be calculated that the ratio of output to input of shrimps for that area is 4.5. Taking into account that the macroepibenthos in the model contains not only shrimps but also larger, less productive species such as flatfishes, the ratio of output to input of 6.2, found for this system in the standard run seems too high.

12.1.3 Carbon Fluxes to the Epibenthic Submodel

The total consumption by macroepibenthos in the different compartments (Table 12.2) is more or less in agreement with literature data for the compartments 1, 2 and 3. Kuipers et al. (1981) mentioned a consumption of 4.3 gC · m^{-2} · y^{-1} by different species belonging to the state variable macroepibenthos (EMAC). In the compartments 4 and 5 the total amount of food consumed by macroepibenthos seems too high compared to literature data. Especially the con-

Table 12.2. The annual fluxes ($mgC \cdot m^{-2} \cdot y^{-1}$) in the standard run from the different food sources to the epibenthic state variables, birds and pelagic fishes

	PCOP	BDET	BBAC	CDIA	BMEI	EMES	EMAC	BBBM	BSF	Total
Compartment 1										
EMES	1	10	1 022	10 306	4 435	329				16 103
EMAC					2 280	1 005	23			3 308
Birds						1 149	162	1	1	1 313
Pel. fish						96				96
Total	1	10	1 022	10 306	6 715	2 579	185	1	1	20 820
Compartment 2										
EMES	21	69	1 152	6 103	2 110	364				9 819
EMAC					407	580	270	2 045	663	3 965
Birds						166	65	754	719	1 704
Pel. fish						120				120
Total	21	69	1 152	6 103	2 517	1 230	335	2 799	1 382	15 608
Compartment 3										
EMES	325	56	4 161	11 667	2 269	2 069				20 547
EMAC					298	2 209	1 262	399	2 771	6 939
Birds						288	113	240	945	1 586
Pel. fish						160				160
Total	325	56	4 161	11 667	2 567	4 726	1 375	638	3 716	29 232
Compartment 4										
EMES	322	11	3 229	13 590	2 746	2 562				22 460
EMAC					392	2 097	2 696	764	9 261	15 210
Birds						177	86	191	1 017	1 471
Pel. fish						192				192
Total	322	11	3 229	13 590	3 138	5 028	2 782	955	10 278	39 333
Compartment 5										
EMES	21	1	556	6 742	2 152	650				10 122
EMAC					304	516	1 865	413	13 363	16 461
Birds						57	50	105	945	1 157
Pel. fish						128				128
Total	21	1	556	6 742	2 456	1 351	1 915	518	14 308	27 868

tribution of the benthic suspension feeders to the total consumption by macroepibenthos in compartments 4 and 5 results in consumption values which are higher than the literature data from the Balgzand area. Kuipers and Dapper (1981) estimated from different literature sources (Afman, 1980; De Vlas, 1979; Van Beek, 1976) predation by species such as *Carcinus maenas*, *Pleuronectes platessa*, *P. flesus and Gobies spec*, in the model all belonging to the state variable macroepibenthos, on *Crangon crangon* (also belonging to EMAC) to be 60% of the *Crangon* production on the tidal flats. In compartment 2 the production of macroepibenthos is 461 $mgC \cdot m^{-2} \cdot y^{-1}$, while the internal predation of EMAC is 270 $mgC \cdot m^{-2} \cdot y^{-1}$. This is 58% of the production of EMAC, which agrees well with the estimate of Kuipers and Dapper (1981).

The total simulated uptake by birds as given in Table 12.2 is almost the same as estimates by Smit (1980b) of 1640 $mgC \cdot m^{-2} \cdot y^{-1}$ and Swennen (1976) of

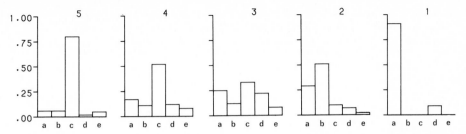

Fig. 12.4. Relative food composition of macroepibenthos in compartments 1 to 5. The relative uptake is defined as the fractional uptake divided by the total uptake. a Fractional uptake of meiobenthos; b fractional uptake of deposit feeders; c fractional uptake of suspension feeders; d fractional uptake of mesoepibenthos; e fractional uptake of macroepibenthos

$1480 \ mgC \cdot m^{-2} \cdot y^{-1}$. Even taking into account that the model results and the literature data on the total uptake by birds from the tidal flats of the Wadden Sea are based on the same estimates of bird abundance and species composition, it is gratifying to note the close agreement between the two approaches.

The food composition of the macroepibenthos is quite different in the five compartments. There is a steady shift in the food composition of EMAC from compartment 1 to 5 (Fig. 12.4). Meiobenthos dominates the diet in compartment 1 but has a decreasing importance going from compartment 1 to compartment 5, whereas the importance of the suspension feeders increases going from the inner to the outer compartments. The uptake by meso- and macroepibenthos from the different food sources has a strong seasonal variation related to the relative abundance of the various food sources (Fig. 10.1).

The most important food source for EMES is CDIA, the benthic diatoms in the upper 0.5 cm of the sediment, in June supplemented by the meiobenthos. Macroepibenthos in compartment 2 mainly feeds on meiobenthos in the beginning of the year, shifting to the other benthic organisms in the second half of the year with deposit-feeding macrobenthos as main component. The birds in compartment 2 mainly feed on macrobenthos. In the second half of the year mesoepibenthos and macroepibenthos form a minor food source for birds.

12.2 Sensitivity Analysis

12.2.1 Varying the Immigrating Biomass

The production of macroepibenthos in the whole system expressed as the ratio of the emigrating biomass to the immigrating biomass in the standard run is ≈ 6, as discussed before. When the immigrating biomass (SETTLE) into compartment 5 is increased from the standard value of $225 \ mgC \cdot m^{-2} \cdot y^{-1}$ to $500 \ mgC \cdot m^{-2} \cdot y^{-1}$ the emigrating biomass [EXPRT(5)] decreases slightly from $1391 \ mgC \cdot m^{-2} \cdot y^{-1}$ in the standard run to $1274 \ mgC \cdot m^{-2} \cdot y^{-1}$ (Table 12.3). The ratio of output to input thus decreases from 6.2 to 2.6. The cause is the enhanced internal predation (flZZ) in the macroepibenthos because of the higher

Table 12.3. The emigrating biomass [EXPRT(5)] and the internal predation of macroepibenthos (flZZ) in compartment 2 for model runs with a different immigrating biomass (SETTLE) in compartment 5, together with the resultant ratio between the immigrating and the emigrating biomass

SETTLE (mgC · m^{-2})	140	225	500
EXPRT(5)(mgC · m^{-2})	1 449	1 391	1 274
flZZ (mgC · m^{-2})	236	270	440
Ratio IN/EX	9.7	6.2	2.6

spring biomass, combined with the fact that the macroepibenthos species in the model behave as opportunistic feeders, eating relatively more from more abundant food sources. The internal predation flux in compartment 2 from EMAC to EMAC is 270 mgC·m^{-2}·y^{-1} in the standard run, while this flux is 440 mgC·m^{-2}·y^{-1} in the model run with increased immigration.

A decrease in SETTLE from 225 mgC·m^{-2}·y^{-1} to 150 mgC·m^{-2}·y^{-1} results in a smaller internal predation flux of 236 mgC·m^{-2}·y^{-1}. Despite the lower spring biomass, visible in a somewhat delayed increase in macroepibenthos, we still get a total emigrating biomass of 1449 mgC·m^{-2}·y^{-1} in autumn, which gives a production ratio of 9.7.

The net production, as expressed in emigrating biomass in fall, is almost constant, despite the variations in immigrating biomass in spring. This dynamic stability is caused by the strongly density dependent predation within the macroepibenthos in the model. The level of this internal predation is determined by the carrying capacity of the area in terms of the abundance of the various food sources for EMAC. It is the outcome of the dynamic interactions between the various submodels, as manifested in the abundances of the various food sources. If the predation pressure of macroepibenthos is such that it starts to deplete its food sources in the benthos, its own standing stock will become the most abundant food source, thus increasing internal predation in the macroepibenthos until the other food sources become relatively more abundant again.

12.2.2 Model Runs, Partially or Wholly Excluding Epibenthos

The effects of small (EMES) and large epibenthic organisms (EMAC) on the performance of the other components of the system can be made visible by setting the potential uptake rate of the state variable(s) to be tested to zero. This effectively removes this state variable from the model, since the initial standing stock is low and the various loss processes (respiration, excretion and mortality) continue, reducing the standing stock quickly to virtually zero. In this way we can mimic exclusion experiments such as have been performed by Reise (1985) on the tidal flats in the German Wadden Sea.

In view of the lack of direct validation values it is of interest to compare the model results from such exclusion runs with the results of Reise's field experiments. A similarity between them would indicate that at least the interactions between the benthic and epibenthic groups are modelled realistically.

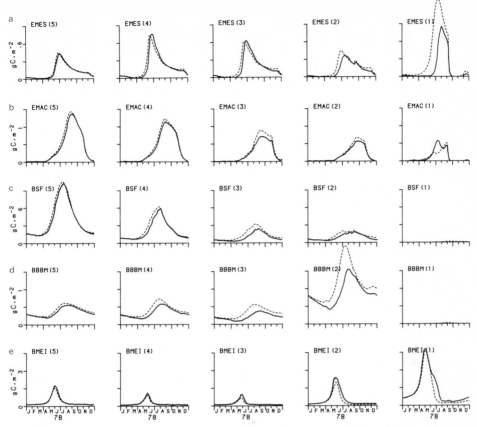

Fig. 12.5. Simulated biomass in gC·m⁻² of a mesoepibenthos (*EMES*); b macroepibenthos (*EMAC*); c suspension feeders (*BSF*); d deposit feeders (*BBBM*); and e meiobenthos (*BMEI*) in a simulation run without birds (--) together with the simulated biomass of the standard run (–)

In one of Reise's experiments large carnivorous fishes and birds were excluded from the benthic community on the tidal flats. Since in the model large carnivorous fishes are only a small part of the macroepibenthos we could not really duplicate Reise's experiment in the model, because we cannot exclude a part of a state variable. Therefore only birds, which strictly speaking are not a state variable (cf. Sect. 8.7), were excluded from the model.

The results are given in Fig. 12.5. The seasonal dynamics of the state variables are all changed to some extent. The strongest changes can be seen in the two inner compartments, where the predation pressure by birds in the standard run is heaviest (cf. Table 12.2). The nature of the changes is rather different for each state variable. The mesoepibenthos (Fig. 12.5a) only shows slight changes, except in compartment 1. Apparently, in the standard run it is limited by bird predation in compartment 1.

The macroepibenthos (Fig. 12.5b) reaches somewhat higher late summer values in most compartments. In compartment 1 it now has a much lower concen-

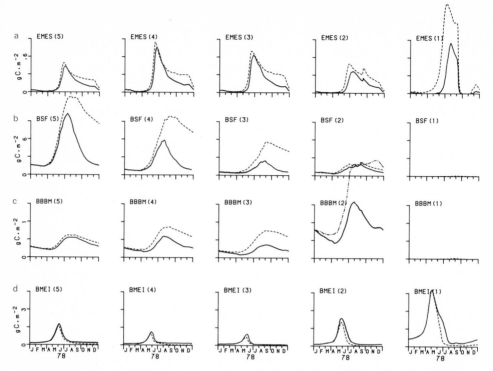

Fig. 12.6. Simulated biomass in $gC \cdot m^{-2}$ of a mesoepibenthos (*EMES*); b suspension feeders (*BSF*); c deposit feeders (*BBBM*); and d meiobenthos (*BMEI*) in a simulation run excluding both birds and macroepibenthos (--) together with the simulated biomass of the standard run (—)

tration than in the standard run, except for a short period just before the waste water discharges force it from the compartment. The reason is that the food conditions in the neighbouring compartment 2 are better than in compartment 1, thanks to the higher biomass of the macrobenthos (Fig. 12.5 c and d).

Both the benthic suspension feeders (BSF) and the deposit feeders (BBBM) increase their biomass in the absence of birds. However, for BSF this increase almost disappears in the second half of the test run, indicating that EMAC is the main predator in the second half of the year (cf. Fig. 12.6 b). The deposit-feeding macrobenthos (BBBM) increase their standing stock more strongly than the suspension feeders, especially in compartment 2, now that the predation by birds is removed.

The main loser from the exclusion of bird predation is the meiobenthos (BMEI), which is reduced in biomass by the deposit-feeding macrobenthos (BBBM). This effect is clearest in compartment 2, where the increase of deposit feeders is largest.

Reise's conclusion from his experiment was that excluding this group had only a small effect on the benthic community. He observed a shift from small to larger individuals in the macrobenthic community, which is in agreement with the results of the sensitivity run without birds.

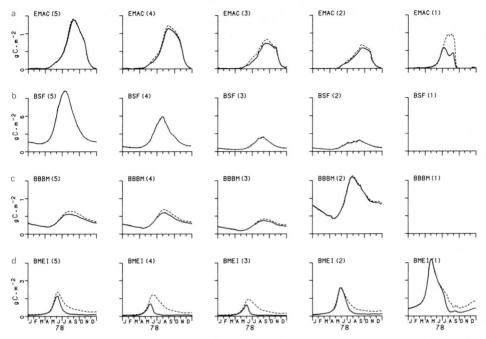

Fig. 12.7. Simulated biomass in gC·m^{-2} of a macroepibenthos (*EMAC*); b suspension feeders (*BSF*); c deposit feeders (*BBBM*); and d meiobenthos (*BMEI*) in a simulation run without mesoepibenthos (--) together with the simulated biomass of the standard run (–)

In a second sensitivity run both predation by birds and by macro-epibenthos (EMAC) was excluded. This leads to large changes in the seasonal biomass distribution of the benthic and the epibenthic state variables (Fig. 12.6). The trend of the change is the same as in the run without birds, but the changes are larger. Especially in the second half year the standing stocks are higher in the sensitivity run. In the compartments 2–5 meiobenthos (Fig.12.6 d) seems to be grazed nearly to extinction, but it will probably stabilize at low biomass. In the response of the suspension-feeding macrobenthos (Fig. 12.6 b) to the release of predation pressure by macroepibenthos and birds one can see the effect of turbidity on the filtration rate. In compartment 2 the high turbidity limits the maximum filtration rate, thus preventing the suspension feeders from realizing their maximum growth rate and profiting optimally from the release of predation pressure to raise the standing stock. From compartment 2 to 5 the turbidity decreases, allowing gradually higher filtration rates and, thus, higher growth rates.

The response of the deposit-feeding macrobenthos (Fig.12.6 c) shows an opposite trend: a relatively slight increase in standing stock in compartment 5, becoming larger towards compartment 2. This is caused by the larger food supply in these compartments, both of phytobenthos (due to the longer light exposure periods of the tidal flats) and detritus (due to the gradient in suspended matter).

In conclusion, the removal of the major predators of macrobenthos shows that the macrobenthos in the standard run is controlled by these predators,

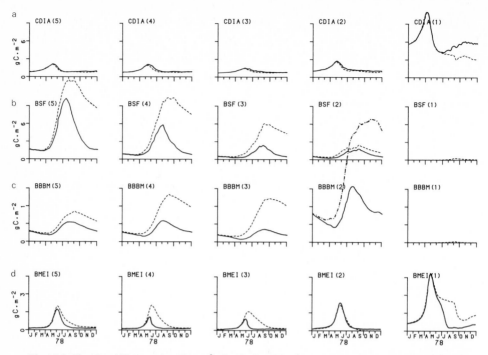

Fig. 12.8. Simulated biomass in gC · m⁻² of a the benthic diatoms in the upper 0.5 cm of the sediment (*CDIA*); b suspension feeders (*BSF*); c deposit feeders (*BBBM*); and d meiobenthos (*BMEI*) in a simulation run excluding the whole epibenthos (--) together with the simulated biomass of the standard run (–)

whereas in the sensitivity analysis it is regulated by food supply and abiotic conditions.

One of Reise's conclusions, based on experiments excluding small predators such as small fish and decapod crustaceans (in our model belonging to the state variable EMAC), states that the small predators have much stronger effects on the benthos than the larger ones. This agrees with the results of the simulation run of the model excluding both birds and macroepibenthos (EMAC) as described above: the biomass of the benthic deposit feeders and benthic suspension feeders increases dramatically, with as much as a threefold increase in compartment 2.

To clarify the role of the mesoepibenthos (EMES) a model run has been made without EMES. The removal of EMES has quite different effects on the performance of the system (Fig. 12.7) than the removal of the large predators, EMAC and birds. In the second half of the year EMAC (Fig. 12.7a) profits from slightly increased standing stocks of BBBM and the higher biomass of meiobenthos (BMEI). The main effect of the exclusion of EMES is on the meiobenthos; this group increases its standing stock considerably from May on in the sensitivity run. At the end of the run it has double the biomass of the standard run. This leads to a moderate increase in the biomass of deposit feeders (Fig. 12.7c). The

EMES in the model thus functions as a predator on meiobenthos and thereby as a food competitor of macroepibenthos and deposit-feeding macrobenthos. Still, its effect on the benthic system in terms of changing the dominant functional group, seems to be slight.

A final sensitivity run has been made without the epibenthic submodel altogether, to test for synergistic effects. The results (Fig. 12.8) are about as expected: greatly enhanced standing stocks of both macrobenthos groups (BSF and BBBM) and a larger standing stock of meiobenthos (BMEI).

In this run it is very clear that in the absence of predators the benthic macrofauna tends to be dominated by suspension feeders in the outer compartments and by deposit feeders in compartment 2.

12.3 Conclusions

In this submodel the disadvantages of aggregating various species into one variable are painfully manifest in the low correspondence value between model results and field data in the case of macroepibenthos. In the case of mesoepibenthos the different sampling techniques applied to the various components of this functional group do not allow a reliable estimate of the biomass, making direct validation of this state variable impossible.

The species in the functional groups have generation times ranging from a few weeks to 3 or more years. The rate constants especially of the species with a generation time of years should decrease with increasing size and age. However, since the state variables are modelled as unstructured populations, this cannot be included in the model in a straightforward way. Future versions should introduce the dominant species of the macroepibenthos (*Crangon crangon* and *Pleuronectes platessa*) as state variables and thus disaggregate the macroepibenthos and at the same time allow for a population structure in these new state variables.

The macroepibenthos functions as the top predator in the model, being only preyed upon by birds, who do not regulate them, and from within the group. The internal predation level corresponds with literature data, making it unlikely that a too low internal predation is responsible for the elevated biomass levels. In reality, shrimps (macroepibenthos) that reach a certain size, migrate from the tidal flats to deeper water. This type of emigration is not incorporated in the model because of the unstructured population approach. This type of emigration creates a steady drain of biomass from the system, thus lowering the average standing stock of the macroepibenthos.

The nursery function of the system is shown to be independent of the immigrating biomass in spring; though the ratio of output to input varies strongly with the input, the output is virtually constant. The size of the output is largely determined by the carrying capacity of the area, which may vary from year to year with the primary production.

Sensitivity analyses of the model, excluding components of the epibenthos, induce changes in the benthic state variables that correspond well with exclusion experiments in the field. The conclusion thus may be that the interaction of the epibenthos with the other parts of the system is qualitatively realistic.

13 Results and Validation of the Transport Submodel

P. RUARDIJ

13.1 Introduction

In principle all pelagic state variables treated in Chapter 10, being subject to transport processes, could be discussed again, but now with the emphasis on transport. However, we will limit ourselves to those variables that are either determined exclusively by transport processes (salinity: SALT and suspended matter: SUSM), or to a large extent (dissolved refractory organic carbon: DROC and particulate organic carbon: POC).

The validation data have not been corrected for the tidal phase at the time of sampling. For our tidally averaged model, this would have removed the random noise from the data introduced by the tidal excursion. However, in view of the fact that this noise probably is one of the smaller sources of error in the model, it was decided to use the data as they are.

The validation data in this chapter are only used for validation of the model run of the same year as they were taken. This has two reasons: first, there are sufficient validation data for each of the specific years (1978–1980) and, second, the transport processes are strongly modified by the boundary conditions which vary considerably from year to year. The latter makes it imperative to use data only for the model run of the same year.

In the transport model organic and inorganic particulate matter is aggregated into one compound variable. This implies that the components are transported in an identical way. To find out whether this is correct, in Section 13.5 the carbon fluxes generated by the biological processes will be compared to other budget studies.

13.2 Transport of Dissolved Compounds

13.2.1 Salinity

The model has been calibrated with salinity data from the years 1972–1977. The validation data are from 1978–1979.

Figure 13.1 shows the simulated salinity in compartments 1–5 during 1978. The lowest salinities occur in compartment 1 and the highest in compartment 5, reflecting the actual salinity gradient in the estuary. The seasonal variation in the salinity for compartments 2,3,4 and 5 is plotted in Fig. 13.2 together with validation data which have been taken at various stages in the tidal cycle. Taking into

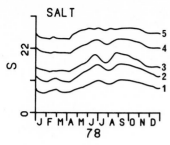

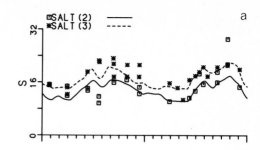

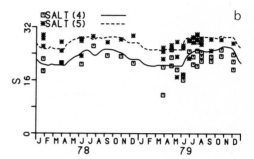

Fig. 13.1. The calculated salinity in
compartments *1–5*

Fig. 13.2. The calculated salinity
for compartments *2* and *3* (a) and compartments *4* and *5* (b) with validation data

account that the model is tidally averaged, the correspondence between model results and field data (Table 13.1) is very good (C = 7.9–8.7).

13.2.2 Dissolved Organic Carbon

The calculated concentration of dissolved organic carbon (DOC) for 1979
(Fig. 13.3) is in agreement with earlier work of Laane (1980). The high average
concentration in compartment 1 is caused by the waste water discharges. The calculated seasonal variation, together with validation data, is given in Fig. 13.4. The
agreement between model results and measurements is better in the inner compartments (Fig. 13.4a) than in the outer compartments (Fig. 13.4b). This is due
to the fact that the field observations in compartments 4 and 5 were made around
low tide. As Fig. 13.4b and Table 13.1 (MC ≈ 0.3) show there is a systematic difference between the field data and the model results in that the latter are consis-
tently too low. This is because the field data, taken at low tide, represent the DOC
concentrations in the water column some 8 km further towards the inner estuary
at mid tide, where the DOC concentrations (cf. Fig. 13.4a) are higher.

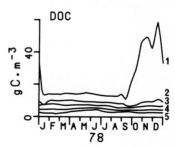

Fig. 13.3. The concentration of dissolved organic carbon
(*DOC*) in gC · m⁻³ as computed by the model in compartments *1–5*

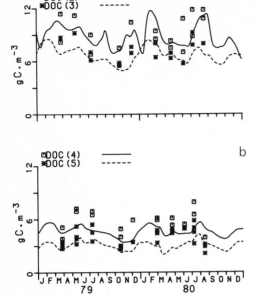

Fig. 13.4. The computed concentration of dissolved organic carbon (*DOC*) in gC·m⁻³ in the compartments *2–3* (a) and *4–5* (b) with validation data

Table 13.1. Correspondence values for those state variables that are mainly determined by transport processes

	Compart-ment	C	U2	RC	MC	SC	N
Salinity	2	8.3	0.06	0.90	0.10	0.00	22
(SALT)	3	8.7	0.02	0.88	0.12	0.00	24
	4	7.9	0.02	0.73	0.24	0.03	33
	5	8.2	0.01	0.86	0.12	0.02	40
Dissolved	2	8.6	0.03	0.95	0.00	0.04	18
organic	3	7.6	0.03	0.65	0.32	0.02	16
carbon (DOC)	4	6.3	0.12	0.38	0.62	0.00	18
(mgC · m⁻³)	5	6.4	0.15	0.49	0.51	0.00	35
Silt (SILT)	2	6.2	0.95	0.86	0.00	0.13	131
(mgC · m⁻³)	3	6.9	0.32	0.87	0.02	0.11	129
	5	6.3	1.20	0.77	0.23	0.01	120
Particulate	2	5.6	0.36	0.48	0.38	0.14	18
organic	3	5.6	0.43	0.43	0.51	0.06	14
carbon (POC)	4	4.6	0.43	0.34	0.37	0.29	18
(mgC · m⁻³)	5	5.5	0.26	0.34	0.48	0.18	35
Suspended	2	6.4	0.99	0.91	0.00	0.09	131
matter (SUSM)	3	7.1	0.32	0.93	0.01	0.07	131
(mgC · m⁻³)	5	6.0	0.37	0.69	0.30	0.02	123

C = correspondence value; U2 = measure of variance; RC = fractional contribution to the error by nonsystematic deviations; MC = idem by the difference between the calculated and observed mean level; SC = idem by nonsimultaneous changes in observed and calculated values (timing error); N = number of observations.

13.3 Transport of Particulate Compounds

13.3.1 Suspended Sediment

The seasonal variation of the suspended sediment (SILT) concentration is given in Fig. 13.5 a. Gradients are shown with high concentrations in compartment 1 and low concentrations in compartment 5. This is in agreement with the field results of De Jonge (1983). For the winter period relatively high concentrations of suspended sediment are simulated in each compartment. This is due to wind-driven resuspension (see Chap. 5). Note that the wind speed is given as a time series of monthly average wind speeds, and that this time series is invariant from year to year. The small year to year difference (Fig. 13.6) in the simulated suspended sediment concentration is caused by differences in fresh water discharge.

In view of the large variation in the suspended sediment data (SILT) (Fig. 13.6; note the different scales) the agreement between model results and field measurements is satisfactory.

The model uses a second state variable, the inorganic fraction of the "fluid mud" layer on the channel bed (CSILT) (see Chap. 5) to model the transport of particulates. This state variable shows a seasonal distribution which is opposite to that of suspended sediment. In Fig. 13.5 b CSILT is presented as a layer by conversion from $mg \cdot m^{-3}$ water column to $mg \cdot m^{-2}$ channel bed. This layer is thinner in winter than in summer because the equilibrium between SILT and CSILT is shifted towards SILT by high wind speeds. Average wind speeds are higher in winter than in summer and thus the model generates these seasonal differences between suspended sediment (SILT) and "fluid mud" (CSILT) which are in agreement with field observations (Eisma et al., 1985). From Fig. 13.6 it is clear, however, that the use of a monthly averaged wind speed to generate the seasonal distribution of suspended sediment is too coarse since the very high SILT concentrations occurring during stormy periods, visible in the field observations, are not simulated by the model. Using an invariant time series of wind speeds worsens the situation. Future versions of the model should at least incorporate averaged wind speed time series on the appropriate time scale of wind events of a few days to a week.

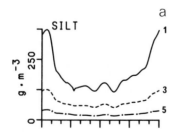
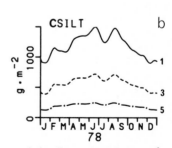

Fig. 13.5. The calculated variation in the concentration of suspended sediment (*SILT*) $(g \cdot m^{-3})$ in compartments *1, 3* and *5* (a); idem near the channel bottom (*CSILT*) $(g \cdot m^{-2}$ channel bottom) (b)

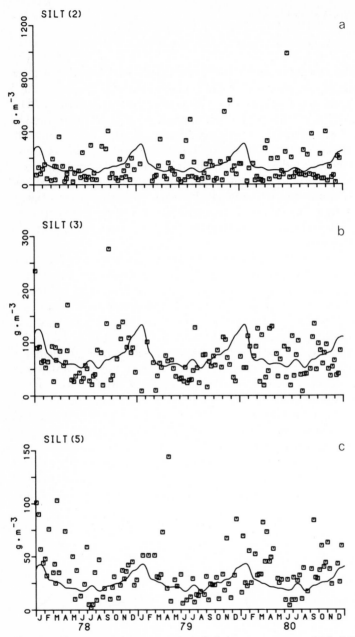

Fig. 13.6. The computed concentration of silt in compartments *2* (a), *3* (b) and *5* (c). Note the different scaling of the y-axes

13.3.2 Particulate Organic Carbon

As with suspended sediment, the concentration of particulate organic (POC) carbon decreases going from the river to the sea (Fig. 13.7). This is in agreement with
the gradients found by Laane et al. (1986) for the Ems estuary. In each compartment the highest concentrations are found in summer. The seasonal distribution
thus is opposite to the one of SILT (Fig. 13.6). The difference is mainly caused
by local production of POC in the form of phytoplankton and detritus and to
some extent by seasonal variations in the boundary conditions for particulate organic carbon and suspended sediment.

The correspondence between the observed and calculated concentrations of
particulate organic carbon (Fig. 13.8 a, b) is not very good (Table 13.1). The simulated values of POC are generally too high. The cause of this deviation is unknown. The sensitivity analysis in Section 13.5.1 will address this problem.

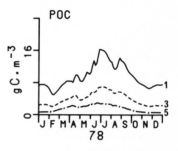

Fig. 13.7. The concentration of particulate organic carbon
(*POC*) in gC·m^{-3} as computed for compartments *1, 3*
and *5*

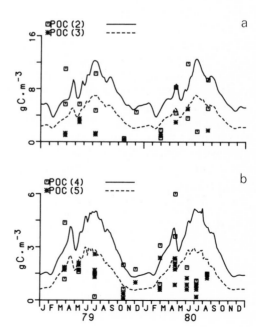

Fig. 13.8. The computed concentration in
gC·m^{-3} of particulate organic carbon
(*POC*) in the compartments *2* and *3* (a)
and in the compartments *4* and *5* (b) with
validation data

13.3.3 Suspended Matter

Suspended matter (SUSM) consists of silt (SILT) and particulate organic carbon (POC). The silt fraction is by far the largest constituent ($> 90\%$). It is therefore not really surprising that the correspondence values for silt and suspended matter in the different compartments (Table 13.1) are not very different. However, since neither the method of measurement of both variables nor the modelling of them is identical, it was felt to be informative to present them both.

13.4 Flows and Annual Budgets

13.4.1 Suspended Sediment Transport

The tidal asymmetry in the Ems estuary is responsible for the transport of suspended sediment in a landward direction and the strong gradient in suspended matter, with very high concentrations in the Dollard (compartments 1 and 2). The gradient in suspended matter corresponds to a similar gradient in the net sedimentation rate of fine particulates (Fig. 13.9) which shows up in the composition of the sediments: the clay content of the sediment increases towards the inner compartments (BOEDE, unpublished data).

Figure 13.10a gives the calculated amount of suspended sediment entering the system from the sea (TS5SILT). The amount is high during summer but much lower in winter and the import of the suspended fraction (the difference between the two curves in Fig. 13.10a) even becomes negative for a short period in winter. During that period suspended matter is flushed out of the system due to the fact that diffusive and advective transport processes, which also operate on the suspended sediment, dominate over the transport processes driving suspended sediment into the system. Figure 13.10a makes it clear that most of the suspended sediment entering the system enters as CSILT (TS5CSILT) or fluid mud. Small annual variations are only due to differences in river flow; the invariant wind speed time series produces the same effect each year. The total amount of suspended sediment imported into the system each year corresponds to the observed sedimentation in the estuary because the parameters for sedimentation and resuspension have been estimated from field observations on the clay ($\approx$ silt) content of the upper sediment layer. The amounts of suspended sediment imported into the system from the Ems and the Westerwoldsche Aa (Fig. 13.10b) vary with the

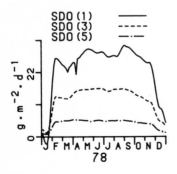

Fig. 13.9. The seasonal variation of the sedimentation rate in $g \cdot m^{-2} \cdot d^{-1}$ of inorganic silt (SDO) in compartments 1, 3 and 5

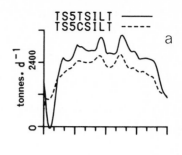

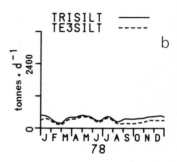

Fig. 13.10. Silt fluxes across the boundaries of the estuary in metric tonnes. In a: *TS5TSILT:* the total flux of silt from the sea into compartment 5. *TS5CSILT:* the total flux of silt near the bottom (CSILT) from the sea into compartment 5. In b: *TRISILT:* the total flux of silt from the rivers into the system. *TE3SILT:* the total flux of silt from the Ems into compartment 3

Table 13.2. The annual silt budget (10^3 metric tonnes $\cdot$ y^{-1}) for the Ems estuary in 1978

Import		Export from the pelagic	
Westerwoldsche Aa	34.7	Sedimentation	1 004.7
Ems	97.5	Mass increase	1.7
Sea (suspended)	135.3	(deviation from steady	
Sea („fluid mud")	739.0	state in the model)	
	1 006.4		1 006.4

fresh water discharge, but are otherwise relatively constant. The riverine contribution to the suspended sediment load is low in comparison to the marine contribution as calculated by the model. This is in agreement with data from Favejee (1960), Salomons (1975) and Rudert and Müller (1981). They all found the suspended sediment fraction < 55 µm diameter in the sediments of the Ems estuary to be predominantly of marine orgin. Summing these rates over all compartments for a whole year results in the budget presented in Table 13.2. This budget shows that the annual sedimentation in the Ems estuary, according to the model, is slightly over a million tonnes of sediment, corresponding to 0.7 cm $\cdot$ y^{-1}.

13.4.2 Organic Carbon Transport

13.4.2.1 Particulate Organic Carbon Budget and Flows in the Pelagic

Particulate Organic Carbon (POC) consists of a number of state variables: phytoplankton (PFLAG + PDIA), mesozooplankton (PCOP), microzooplankton

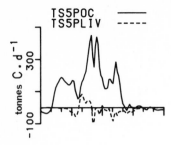

Fig. 13.11. The particulate organic carbon flux (*TS5POC*) and the flux of living particulate organic carbon (*TS5PLIV*) across the seaward boundary in tonnes $\cdot$ C $\cdot$ d^{-1}

Table 13.3. The annual POC budget (10^3 metric tonnes $\cdot$ y^{-1}) for the Ems estuary in 1978

Import/production		Export/consumption	
Westerwoldsche Aa	1.9	Sedimentation	63.3
Ems	10.6	Mineralization	77.0
Sea	37.9	Uptake by filterfeeders	5.8
DOC uptake by bacteria	43.5	Excretion as DOC (by	24.6
Primary production	77.2	plankton and bacteria)	
	171.1		171.1

(PMIC), bacteria (PBAC) and the suspended fraction of the detritus (PDET and PROC).

The detrital material forms the bulk of POC. Because this component is transported in the same way as silt the transport is landwards (Fig. 13.11: TS5POC). However, the living fraction in POC (Fig. 13.11: TS5PLIV) is transported seaward (negative values in Fig. 13.11), because this fraction is predominantly in the water column and consequently is mainly controlled by the diffusive and advective transport. As a consequence of these opposite transport directions, easily degradable carbon (the living fraction of POC) is replaced in the estuary by more refractory material (PDET and PROC).

The budget calculated from the model (Table 13.3) shows that more particulate matter is imported from the sea than from the rivers; the contribution by the sea to the total import of carbon is 74%, slightly lower than for silt (87%). An explanation for this difference is that the carbon content of particulate matter from the rivers is much higher than at the seaward boundary. The local production of POC in the estuary itself is not of the same order as the import.

13.4.2.2 Dissolved Organic Carbon Budget and Flows in the Pelagic

The high dissolved organic carbon (DOC) concentration at the river boundaries and the low concentration at the sea boundary result in an import from the river Ems (Fig. 13.12: TE3DOC) and the Westerwoldsche Aa (Fig. 13.12: TW1DOC), a transport through the estuary and an export to the sea (Fig. 13.12: TS5DOC). Because DOC consists mostly of refractory compounds (Laane, 1980), to a large extent it behaves as a conservative variable. Thus the amounts transported to the

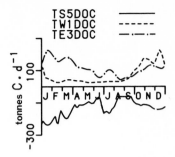

Fig. 13.12. DOC fluxes across the boundaries of the system in metric tonnes $\cdot C \cdot d^{-1}$. *TS5DOC* = import from sea; *TW1DOC* = import from the WWA; *TE3DOC* = import from the Ems

Table 13.4. The annual DOC budget (10^3 metric tonnes $\cdot y^{-1}$) for the Ems estuary in 1978

Import/production		Export/consumption	
Westerwoldsche Aa	25.8	Sea	47.5
Ems	41.5	Consumption	43.5
Production (=excretion)	24.6	(=uptake by organisms)	
		Sedimentation	0.9
	91.9		91.9

sea are mainly controlled by the riverine input. Only the labile organic carbon which is brought into the estuary from the Westerwoldsche Aa in autumn in the waste water is completely mineralized in the estuary.

Just as with POC an annual budget can be made. The local production of DOC is nearly exclusively in the form of labile organic carbon which is very rapidly consumed by bacteria. Therefore local production of DOC does not lead to higher concentrations of DOC.

The riverine fluxes (Fig. 13.12: TE3DOC,TW1DOC) are derived directly from the boundary conditions to the model which are based on field observations (van Es and Laane, 1982). However, the other contributions to the budget (Table 13.4) were largely unknown, mainly because they are nearly impossible to measure.

The budget in Table 13.4, calculated with the model, indicates that about half the amount of DOC entering the system by transport or local production is removed by biological processes in the system. The other half is transported out of the system to the sea.

However, viewed in another way, the refractory DOC from the Ems is transported through the system unchanged and discharged to the sea, while the labile DOC (=LOC), mainly originating from the primary producers and the waste water discharges from the Westerwoldsche Aa is mineralized in the estuary. From this view, the DOC budget has a biologically active component (LOC), produced and consumed within the system and an almost inert component (PDROC) that is transported unchanged through the system.

The small amount of DOC going to the sediments is the fraction of the dissolved organic carbon in compartment 1 that flocculates and sediments (see Sect. 6.5). Another conclusion from this budget is that the bacterial loop in the ecosystem plays an important role in recycling dissolved organic carbon into particulate organic carbon and thus returning it to the food web. Excretion products (mainly from the phytoplankton) and the labile organic carbon from the Westerwoldsche Aa are rapidly consumed by bacteria ($43.5 \cdot 10^3$ tonnes $\cdot$ y^{-1}) and converted into biomass.

13.4.2.3 Annual Carbon Budget for the Whole System

Earlier studies show that it is difficult to establish a carbon budget from field observations on the flux of organic matter to and from the sea and the flux of organic matter between the sediments and the overlying water column (Kjerfve et al., 1981; Cadée, 1982).

With an ecosystem model, however, one can derive such a budget simply by adding up the daily fluxes, assuming that the model is a correct description of the ecosystem. Such carbon budget calculations (Table 13.5) at first glance, looking at the carbon flux across the sea boundary, show this system to be a net carbon exporting system. The annual import of $37.9 \cdot 10^3$ metric tonnes of particulate organic carbon from the sea is offset by an export of $47.5 \cdot 10^3$ metric tonnes of dissolved organic carbon.

However, when we take into account that $48.6 \cdot 10^3$ metric tonnes of dissolved refractory organic carbon (PDROC) enters the system from the rivers and that the bulk of this amount passes through the estuary unchanged, we see that the export across the seaward boundary mainly consists of this material which is imported from the rivers. The system thus turns out to be a carbonimporting system, in other words a netmineralizing system.

The only potentially important flux of carbon, which is not included in the model is the carbon flow from and to the marshes in the Dollard. Dankers et al. (1984) calculated both the export of DOC ($0.1 \cdot 10^3$ tonnes $\cdot$ y^{-1}) from the marshes and the import of POC ($1 \cdot 10^3$ tonnes $\cdot$ y^{-1}) to the marshes. These flows are small compared to most of the fluxes in Table 13.5 and are therefore neglected in the model.

In Table 13.5 the pelagic and benthic system are taken together. In Fig. 13.13 a diagram is given of the budget for the pelagic and benthic systems separately. It shows a carbon flux from the pelagic to the benthic system of $63.3 \cdot 10^3$ tonnes C $\cdot$ y^{-1}.

The import of organic carbon from the Ems and Westerwoldsche Aa has already been given in Figs. 13.11 and 13.12. The amount of organic carbon produced locally by benthic and pelagic primary production is of the same order as the amount introduced by the rivers. In other systems carbon import may dominate the local production as has been found in the Nanaimo estuary (Sibert et al., 1977) and the Cumberland Basin (Gordon et al., 1986).

Only a small part of the organic carbon imported and produced ($181.4 \cdot 10^3$ tonnes $\cdot$ y^{-1}) is exported to the North Sea. Most (60%) is consumed (mineralized) in the estuary itself and 10% is buried in the sediments.

A carbon budget of the Ems estuary was published by Colijn (1983), who estimated the amount of organic carbon imported by the Westerwoldsche Aa at $30.3 \cdot 10^3$, the primary production at $89.7 \cdot 10^3$ tonnes $\cdot$ y^{-1} and the carbon buried

Table 13.5. The annual carbon budget (10^3 metric tonnes $\cdot$ y^{-1}) for the Ems estuary in 1978

Import/production		Export/consumption	
Westerwoldsche Aa	26.8	Sea	9.8
Ems	52.1	Mineralization	145.3
Primary production	102.5	Birds	0.3
		Buried	21.2
		Biomass increase	4.8
	181.4		181.4

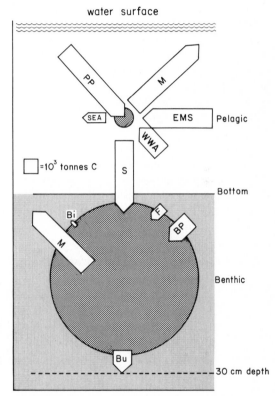

Fig. 13.13. The schematized annual carbon budget as calculated from the model. The *circles* represent the average carbon mass (including detritus) in the estuary, the *arrowed rectangles* represent the flows. *PP* pelagic primary production; *M* mineralization; *BP* benthic primary production; *S* sedimentation; *Bu* buried; *Bi* birds; *EMS* input of Ems; *WWA* input from Westerwoldsche Aa; *Sea* output of sea; *F* filtered by suspension feeders

in the sediment at $22.1 \cdot 10^3$ tonnes $\cdot$ y^{-1}. The primary production calculated from the model (Table 13.5) is higher than his calculation, because the model takes account of the enhanced primary production in the water overlying the tidal flats, which was computationally impossible for Colijn. Otherwise, the results in Table 13.5 are in good agreement with the values, with the exception of a deviation between the mineralization from Colijn and the mineralization in the model budget: $109 \cdot 10^3$ tonnes $\cdot$ y^{-1} against $145.3 \cdot 10^3$ tonnes $\cdot$ y^{-1} respectively. Colijn (1983) already indicated that his carbon consumption estimate is probably low due to an underestimate of the annual oxygen consumption from which the mineralization is calculated. This underestimate is caused by the fact that oxygen consumption is usually measured in the dark, thus underestimating the oxygen consumption which is caused by the bacterial uptake of excretion products of primary producers.

There is a small mass increase in the amounts of refractory organic carbon in the benthic system in the inner compartments [BROC(1), BROC(2)] and consequently in the total carbon mass in the benthos from one year to the next. For these state variables the model is not in a dynamically steady state. A sensitivity analysis will show the possible cause of this imbalance (see Sect. 13.5.1B).

13.5 Sensitivity Analysis

From the discrepancy between model results and field observations (Sect. 13.3.2) it can be concluded that the modelling of particulate organic carbon dynamics (POC) is problematic. The simulated POC concentrations are too high and the sedimentation of particulate carbon, which is closely coupled to the concentration in the water column, generates an annual increase in the state variable BROC in compartments 1 and 2.

This observed imbalance has inspired a sensitivity analysis on the validity of assuming a complete coupling of organic and inorganic compounds in modelling the silt transport processes. Other sensitivity analyses will deal with the validity of the silt transport process itself. The consequences of the import of particulate organic and inorganic material will be discussed both for the pelagic and the benthic system.

13.5.1 Behaviour of Particulate Matter During Transport

A. Net sedimentation of POC and Silt in Relation to Transport. The assumption of identical behaviour of organic and inorganic particulate material when transported by the residual silt transport has to be rejected if we compare the correspondence values between model results and field data for silt and particulate organic carbon (Table 13.1).

Assuming the field observations for both variables to be representative, the agreement between simulated and observed values is much better for silt than for particulate carbon and an explanation might be that the overall behaviour of both types of material is different during transport. The sensitivity analysis is limited to the process of resuspension and sedimentation on the tidal flats.

To obtain a lower POC concentration, one of the parameters that define the residual transport is changed. The parameter controlling resuspension is changed only for particulate carbon: REASIL is set to 25% of its original value. This causes POC, once sedimented, to be less susceptible to resuspension than the inorganic silt particles. This modification results in a better agreement between model results and field observations for POC (Table 13.6, cf. Fig. 13.8 a). The correspondence value is higher due to the fact that the mean simulated level of POC is lower, which is in better agreement with the field observations. The same is true

Table 13.6. Comparison of some important factors between a standard run and the run with a lower resuspension of POC

Unit: 10^3 tonnes C $\cdot$ y^{-1}		Standard run	Sensitivity analysis
Sedimentation of PDET		24.7	33.6
Sedimentation of PROC		+37.6	+53.9
Total sedimentation of particulate material		62.3	87.5
Correspondence value	Comp.	C	C
Particular organic	2	5.6	7.3
carbon (mgC $\cdot$ m^{-3})	3	5.6	6.6
	4	4.6	5.1
	5	5.5	5.8
Community	1	5.2	5.2
metabolism (MCONS)	2	6.9	6.8
(mgC $\cdot$ m^{-2})	3	8.4	8.4
	4	7.4	7.0
	5	5.8	5.6
Annual mass	1	0.02%	1.70%
increase of BROC	2	1.30%	2.91%

C: The correspondence value between model results and validation data

for the POC concentrations in compartments 4 and 5. This improvement is caused by a higher net sedimentation (see Table 13.6), which lowers the concentration of POC in the water.

A consequence of this higher sedimentation is that more detritus is added to the benthic system which causes a higher community metabolism (MCONS) in the benthic system. The correspondence values for MCONS deterioriate slightly. The increased sedimentation causes a slow rise to a higher equilibrium of BROC(1) and BROC(2) to values which are about a factor 2 higher than in the standard run.

The conclusion of this sensitivity analysis therefore is that the assumption of a similar sedimentation/resuspension behaviour of organic and inorganic particulates is the best one at this moment.

B. The Relation Between Import of POC and the Concentration of POC at Sea. According to the model a large amount of POC is imported from the sea (Fig. 13.11). Combining this model result with the observation that the simulated POC values are too high in comparison to the field data (Fig. 13.8) and with the accumulation in the model of refractory organic carbon in the sediment(BROC) in compartments 1 and 2, we would like to know whether a decrease in the sea boundary concentration of POC affects the POC dynamics in the whole system, and how the sedimentation rate of POC would change.

An additional reason to make such a sensitivity analysis is that the concentrations of POC and suspended sediment at sea, both important boundary conditions in the model, have been measured in different ways. For the determination of the silt concentration at sea, data were collected from an area with the same surface as compartment 5. However, for the determination of the POC concentration at sea only data were available which had been taken on the boundary between compartment 5 and the sea. Because the POC concentration at sea decreases with increasing distance from the estuary, this sampling method leads to an overestimate of the POC concentration at sea. For this sensitivity analysis the probable extent of this overestimate is inferred from the silt gradient which the model generates: the mean silt concentration at sea is calculated to be 87% of the mean silt concentration on the boundary between compartment 5 and the sea.

Assuming the POC concentration gradient to be identical to the silt gradient the sensitivity analysis is run with a boundary concentration of POC at sea of 87% of the value used in the standard run. From a comparison between the standard run with this run the following points emerge:

1. The correspondence between field observations and model results of POC is improved. The average POC level in the system decreases in the same way as the POC concentration at sea (Table 13.7).
2. The annual increase of BROC(1) and BROC(2) is terminated and this state variable now is in equilibrium.
3. The import of POC from the sea decreases from 37.9 to $25.7 \cdot 10^3$ metric tonnes $\cdot y^{-1}$.

Apart from the conclusion that this small change in the model leads to an improvement of the model for a number of variables, it is remarkable that such a slight decrease of the boundary condition of POC results in a decrease in the import of POC from the sea of more than 30%.

The effects throughout the system of such a small change in a boundary condition illustrate the importance of an accurate determination of the boundary conditions. In our case, the importance of obtaining good data sets for the boundary conditions was only realized long after the field programs were ended, since the construction of the model was only started after field investigations were terminated.

Table 13.7. POC concentrations in a standard run and in a sensitivity run with POC concentrations at the seaward boundary reduced to 87% of the standard values

	Standard run		Sensitivity run	
Compart-ment	Average POC conc. in 1978 $(gC \cdot m^{-3})$	C	Average POC conc. in 1978 $(gC \cdot m^{-3})$	C
1	9.7	–	8.5	–
2	7.3	5.6	6.3	6.4
3	4.0	5.6	3.4	6.2
4	2.9	4.6	2.5	5.2
5	1.6	5.6	1.4	6.2

C: The correspondence values between model results and validation data.

A lesson to be learned from this is that field investigations and model development should go hand in hand to allow for modifications in the sampling programs.

13.5.2 The Role of Imported POC in the System

The input of organic particulate material is an important additional food source for some estuaries such as the western Wadden Sea (De Wilde and Beukema, 1984). The imported material is expected to particularly benefit the intertidal benthic community. For the Ems estuary Van Es (1982 b) and Colijn (1983) suggested that import from the sea could also play an important role. Particulate organic carbon (POC) is a summation of different state variables which can be separated into two groups: living material (phytoplankton, zooplankton, bacteria) and detritus. In Section 13.2 it has been shown that the living fraction of the POC is transported in the direction of the sea, and that it does not form a part of the imported POC.

This sensitivity analysis will deal with the question whether the imported particulate detritus (PDET, PROC) is an essential supplement to the local production of organic carbon. The emphasis of the sensitivity analysis is on the effect of this material on the benthic system and whether or not it is an essential part of the total food supply.

In Table 13.8 the total production, sedimentation and consumption in the benthic system is given for the three detritus fractions in the standard run. Table 13.8 illustrates that almost all LOC is produced in the benthic system, while the other detritus components are mainly brought in by sedimentation.

To test the sensitivity of the benthic system to changes in the detritus supply a simulation is run in which the external input and the initial values of detrital pelagic POC (PDET and PROC) are set at zero.

At first sight the consequences are large: the sedimentation of carbon (Table 13.9) decreases with 78% compared to the standard run. Hence, sedimentation of detrital POC depends largely on imported material. The decrease in sedimentation leads to a 48% lower total input of carbon in the benthos.

In the absence of an allochtonous source of particulate organic carbon, the size of the autochtonous input (Table 13.9) becomes visible which is a nice byproduct of this sensitivity analysis. The autochtonous input in the sensitivity run is assumed to be the same as in the standard run. In the standard run it contributes 40% to the total carbon input, slightly lower than the estimate of De Wilde and Beukema (1984) of 48% for the western Wadden Sea. The consequences of the

Table 13.8. The main detritus fluxes (in 10^3 metric tonnes $C \cdot y^{-1}$) in the benthic system

	Sedimentation	Production	Consumption	Burial
LOC	1.5	23.2	24.7	–
DET	24.6	6.0	30.8	–
ROC	37.5	1.5	13.0	21.1

Table 13.9. Annual rates of benthic processes concerning total carbon flows in the whole estuary in the standard run and in a sensitivity analysis without POC import

Total carbon (10^3 tonnes C $\cdot$ y^{-1})	Standard	No detrital POC import
Phytoplankton filtered by macrobenthos	5.3	5.7
Sedimentation	63.6	14.2
Production in benthic	+25.3	+25.2
„Fresh" input in benthic	94.2	45.1
Carbon mass decrease	+−	+38.7
Total input in benthic	94.2	83.8
Burial (only BROC)	−21.1	−20.7
Mineralization	73.1	63.1

lower input in the sensitivity run only become fully evident when the whole benthic system is in equilibrium again with this reduced carbon input, which, due to the long turnover time of benthic detrital carbon in combination with the very high concentration of this compound, will take about 150 years.

The large amounts of detrital carbon available for mineralization in the benthic system still can fuel the mineralization for a long time, despite the decrease in input, thus masking the effects in this 1-year sensitivity run. The lower input in the long term leads to a slow decline in the concentration of organic carbon in the benthic system. In this analysis, the decrease in organic carbon in the benthos is calculated to be 4.7% of the total organic carbon mass in the first year. This decrease is an extra source which in the first year contributes 46% of the total input (Table 13.9). The withdrawal of detrital organic carbon from the large buffer in the benthic system thus suffices to maintain the mineralization at 86% of its original level.

13.5.3 The Role of Silt in the Estuary

Silt, as incorporated in the model, influences the biological system of the estuary in three ways:

1. It serves as a carrier of particulate carbon, which is transported into and through the estuary;
2. The silt concentration determines the light penetration into the water, and thereby modifies the primary production to a large extent;
3. The silt concentration modifies the food uptake of benthic suspension feeders.

De Jonge (1983) argued that the average silt concentration in the Ems estuary over the last 30 years has increased steadily, possibly caused by dredging.

Dredging removes mainly sand from the bottom of the channels, but silt also is removed. Partly this silt becomes suspended in the water and so contributes to the actual silt concentration. To study the effects on the system of a lower silt concentration, like the situation of 30 years ago, a model run was made with a 33% lower silt concentration at the seaward boundary. According to the model such a change results in an almost proportional decrease in all compartments (Table 13.10). The silt transported from the river masks this decrease somewhat

Table 13.10. Silt concentrations and sedimentation rates in the standard run and in a run with lower silt concentration at the seaward boundary

Compartment	Average silt concentration $(g \cdot m^{-3})$		Sedimentation rate of silt $(kg \cdot m^{-2} \cdot y^{-1})$	
	Standard	Sensitivity analysis	Standard	Sensitivity analysis
1	191	139	8.5	6.1
2	144	103	7.9	5.5
3	75	53	4.7	3.2
4	53	37	3.4	2.3
5	26	17	1.7	1.1

Table 13.11. Some major process rates in the pelagic in the whole system in the standard run with a run with a lower silt concentration at the seaward boundary

Unit: 10^3 metric tonnes $C \cdot y^{-1}$	Standard	Sensitivity analysis
Pelagic primary production	77.1	91.0
Sedimentation of organic carbon	63.3	64.0
Export to sea	9.8	15.3
Filtered by macrobenthos	5.6	6.7

Table 13.12. The average biomass in compartment 2 in the standard run and in a run with a lower silt concentration at the seaward boundary

State variables	Standard	Sensitivity analysis	Unit
Flagellates	112	104	$mgC \cdot m^{-3}$
Diatoms	126	178	$mgC \cdot m^{-3}$
Bacteria	238	244	$mgC \cdot m^{-3}$
Microplankton	58	58	$mgC \cdot m^{-3}$
Mesozooplankton	27	34	$mgC \cdot m^{-3}$
Suspension feeders	918	1 278	$mgC \cdot m^{-2}$

in the compartments 1 and 3 which have a river boundary. A more or less proportional change was also found by De Jonge (1983) from a comparison of the situations in the period 1950–1960 and 1975–1985.

Together with silt also particulate carbon is resuspended by dredging. To avoid confusing the issue, the changes in the POC concentration due to dredging are neglected in this run.

A major consequence of a lower silt concentration in the water is a lower sedimentation rate of silt (Table 13.10) on the tidal flats. The most important direct effect on the biology of the system is an increase in primary production. This increase (Table 13.11) leads to a higher biomass of the pelagic diatoms, but not of the flagellates (Table 13.12). The effects on the other groups in the pelagic except for the mesozooplankton are small (Table 13.12).

Another indirect effect is the higher production of detritus in the estuary by excretion and/or mortality of phytoplankton. This increases the export of organic carbon to the sea (Table 13.11).

Benthic suspension feeders are stimulated in two ways by the lower silt content of the water (Table 13.12): the filtration rate can be higher because of the lower silt concentration (Winter, 1976; Widdows et al., 1979), and they get a higher supply of phytoplankton as food. This results in an increase of the suspension feeders by 38% in compartment 2.

A reduction in the average silt concentration according to the model will enhance the primary production in the system and benefit the suspension feeders. At the same time it will reduce the importance of allochthonous carbon to the functioning of the system.

13.6 Conclusions

The model results together with the sensitivity analysis lead to the following conclusions:

1. The concentration of dissolved organic material in the system is completely determined by the rivers and the sea despite a large production and consumption of DOC locally in the system.
2. The concentration of particulate material depends mainly on the concentration at the seaward boundary. Inorganic particulate matter concentrations (silt) in the estuary are fully controlled by the boundaries, while for particulate organic carbon there is a small but important input by local production. The autochthonous input is more important than the allochthonous input due to its much easier degradability.
3. The complete coupling of POC transport to the transport of silt (against the concentration gradient) for the moment seems to give the best overall results. However, in view of the conflicting results of the sensitivity analysis in Section 13.5.1A, the transport mechanisms operating on particulate matter should be further elucidated.
4. POC in the form of living matter, mainly phytoplankton, is exported from the estuary and replaced by POC of a more refractory nature.

14 Model Applications and Limitations

J. W. Baretta and P. Ruardij

14.1 Introduction

This chapter will discuss the potential of the model as a research and management tool. As a research tool it already has proven its worth, since in Chapters 9–13 it has been shown, especially in the sensitivity analyses, which aspects of the Ems ecosystem are understood well enough to model correctly and which are not, thus defining new lines of required investigation.

When the model is used as a management tool, the experiments to confirm or reject the model results or predictions would be exactly those perturbations of the system whose effects the model should have predicted. The model results from such applications can only be verified or falsified after the fact. The problem here is that if a far-reaching management decision is taken on the basis of model results that predict this decision to have no adverse impact on the system, and the model results are falsified after the fact by the occurrence of adverse effects, the "experiment" cannot easily be undone. To minimize the chance that the model will produce false-positive predictions when used in shaping management decisions, the restrictions to which its use as a predictive tool is subject have to be clearly understood. These restrictions are the subject of the next section.

The model can be applied to some of the questions that are relevant to the management of the Ems estuary; an example is given in Section 14.3.1. The model is sufficiently general to be applied to other estuaries in the temperate zone, too. This is illustrated in Section 14.3.2, where the model is applied to a completely different system, the Bristol Channel and Severn estuary. The results are compared to the data gathered in that area by IMER and to the model results from GEMBASE (Radford and Joint, 1980; Radford and Uncles, 1980; Radford, 1981).

Nevertheless, if the model is used as a management tool, its users should be fully conversant with its strengths and weaknesses in order to reliably translate model results into management recommendations.

14.2 Limitations of the Model

The first and obvious limitation of the model is that it can be used only for modelling the response of system variables and processes that have been included in the model. The other restrictions are somewhat more subtle than this. They will be treated in increasing order of importance to the functioning of the system.

The sensitivity of the system to changes in parameter values as observed through sensitivity analysis depends on the category the parameter belongs to. Any parameter in the model may be considered to belong to one of the following three categories:

1. Parameters defining the internal dynamics of the state variables (category I);
2. Parameters defining the structure of the system (category II);
3. Parameters defining the interaction between the abiotic environment and the system (category III).

The properties of the various categories and the consequences of varying the parameter values of each category are treated in the following section.

14.2.1 Category I: "Physiological" Parameters

To this class belong all parameters defining the uptake, respiration and excretion of the state variables. A number of sensitivity analyses dealing with parameter changes in this class are discussed in Chapters 10 and 11. A general conclusion from these analyses is that the system is not very sensitive to changes in these values, if kept within the limits found in the literature. Obviously, as soon as one assigns unrealistic values to these parameters, the internal dynamics of the state variables will also become unrealistic. The restrictions in varying parameter values of this class thus are that they have to be within the known range for these parameters.

Generally, applications of the model based on variations in this parameter class are not very relevant to its use as a management tool, because this category is unaffected by human intervention in the system.

14.2.2 Category II: "Structural" Parameters

The structure of the model, expressed in the size of the carbon flows between the state variables, can be affected and changed in various ways. In all cases, however, the structural changes are caused by the response of state variables to shifts in their environment, both biotic and abiotic.

This category, which determines the structure of the simulated ecosystem, has three subclasses:

1. Parameters defining the structure of the food web in the model. A number of parameters quantifies the extent to which different food sources may be used by the biological state variables. The values chosen define the "menu" of the state variables. Including more possible food sources into this menu, by setting the relevant parameter value to a nonzero value, can have large consequences for the seasonal dynamics of the other food sources. This is illustrated by the sensitivity analysis of Section 10.3.3 where the inclusion of suspended benthic diatoms into the diet of micro- and mesozooplankton results in a much lower spring peak of pelagic diatoms.

   A more exact quantification of such parameters is both necessary and possible through extensive analysis of gut contents and faecal pellets. This will, moreover, provide a check on the soundness of the parameter estimates in the model.

The restrictions in this category lie partly in the constant values chosen for them. The correct values may in reality be quite variable during the season, reflecting seasonal changes in food preference. Also it is entirely possible that potential food sources have been omitted from the menu. These restrictions can be removed by additional research as outlined above.

2. "Comfort" parameters: These parameters determine the response of functional groups to their abiotic environment, i.e. to variables such as oxygen saturation, salinity, temperature, nutrients and suspended matter. The usual response to an unfavourable or "uncomfortable" abiotic environment in the model is increased mortality or excretion. However, the response of the epibenthic functional groups is to migrate away from an uncomfortable environment. The influence the comfort parameters thus have on the various biological groups is large, but by no means uniform or even parallel for all groups. The comfort functions describing the response of primary producers to low nutrient levels determine to a large extent the phytoplankton production and thus the food available to planktonic herbivores. The distribution and the level of the standing stocks in the different functional groups are coupled in this way to the specifications of comfort parameters. Changes in the comfort parameters will not only cause shifts in the functional group directly modified by this parameter but also in the groups that use this group as a food source. The values of comfort parameters are known from literature for separate species but not for whole functional groups. This poses a restriction on the reliability of such parameters, since the correct value may be expected to depend on the species composition of the functional group, and to vary with changes, seasonal or otherwise, in the species composition.

3. Boundary conditions: To attain the necessary closure of the system, boundary conditions must be given for all the state variables subject to transport. Though the boundary conditions are not parameters in that they usually do not have a constant value during a model run, neither are they variables which are modified by the performance of the system. They exert a strong influence on the system, especially the boundary conditions for detrital carbon. Transported all over the system from the boundaries by the transport model, these boundary conditions determine to a large extent the sedimentation and mineralization of detritus and refractory organic carbon in the benthic system.

Quite small shifts in the boundary conditions can mean the difference between a system which is in dynamic equilibrium and a system where detritus keeps accumulating in some compartments. A sensitivity analysis of some boundary conditions (Sect. 13.5.2) elucidates the consequences.

The sensitivity of the system to quite small errors in the boundary conditions is due to the large water volume transported across the boundaries. A small error, multiplied by a very large number (the water volume transported), creates a large erroneous carbon source or carbon sink in the model.

14.2.3 Category III: "Dimensional" Parameters

This category defines the morphological and physical properties of the estuary to which the model is to be applied (Sect. 4.2 and Chap. 5). Only this category will be very different from estuary to estuary, to account for the morphological differences between systems. When the values for the morphological and physical characteristics of another estuary are assigned to these parameters, together with the boundary conditions of that system, a model run should describe the dynamics of this other system (Sect. 14.3.2).

The morphological characteristics and hydrodynamics of the Ems system as defined by the relevant parameters are well known (Helder and Ruardij, 1982) and impose no restrictions on the use of the model for the system as it is, i.e. as long as the estuary is not physically modified by diking or other changes in its size and shape. If, however, the morphology of the system is modified, the hydrodynamic parameter values will change, too. Using this model to simulate the consequences of changes in size and shape of the estuary will therefore only be possible if one also can obtain the "new" values of the hydrodynamic parameters, such as diffusion constants, sedimentation rates, etc. This restriction on the use of the model only can be removed by first calculating the new values for these parameters with a hydrodynamic model. The effects of changes in morphology on the performance of the ecosystem thus might be calculated by using a two-stage approach: first, the calculation of the hydrodynamic effects of the morphological changes and then the estimation of the ecological effects, using the new set of parameter values.

14.3 Applications of the Model

Subject to the limitations given in the previous section, the model can be used to trace the consequences of human activities. Depending on what aspects of the system are affected by these activities, the limitations as given in the previous section will determine to what extent the model may be expected to predict the consequences.

The applications described here directly modify structural or dimensional parameters. The first case details the consequences of changed boundary conditions at the fresh water end of the system. The second case treats the performance of the model with the dimensional parameters (category III) defining a completely different system: the Bristol Channel and Severn estuary.

14.3.1 Case 1: Reduced Organic Loading

For this case, a 3-year run of the model has been made, with the first year having the "normal" discharge of waste water as measured by the RIZA (Anonymus, 1979). The discharge of waste water rich in labile organic carbon (LOC) was terminated on January 1 of the second year. Figure 14.1 shows the sudden decrease in the amount of labile organic material (LOC) entering the system in compartment 1 (WWAPLOC in Fig. 14.1). This decrease removes the largest source of easily degradable organic carbon from the system.

Figure 14.2 shows the effect on the oxygen concentration in the water in compartment 1: The oxygen concentration in the water drops almost to zero in year 1 when the waste water discharges (Fig. 14.1) start, but this does not happen again

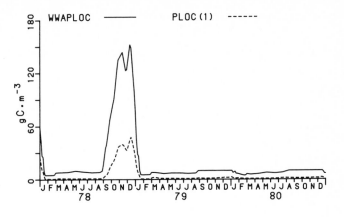

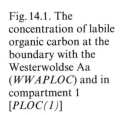

Fig. 14.1. The concentration of labile organic carbon at the boundary with the Westerwoldse Aa (*WWAPLOC*) and in compartment 1 [*PLOC(1)*]

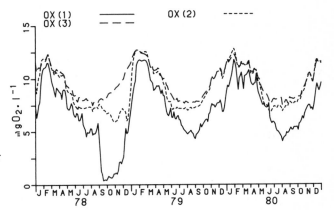

Fig. 14.2. The oxygen concentration in the water of compartments 1–3 [*OX(1–3)*]

in years 2 and 3, in the absence of the discharges. Notable is, however, that oxygen consumption due to mineralization in compartment 1 is still high enough in years 2 and 3 to cause significantly lower oxygen saturations than occur in compartment 5 which stays close to the saturation value throughout the year.

The change in boundary conditions causes large shifts in some of the biological state variables. The primary producers (benthic diatoms, pelagic diatoms and flagellate phytoplankton) are taken together as PRIM in Fig. 14.3. This group itself is not very sensitive to changes in waste water discharge, but in combination with the effects on the consumer variables, which changes the grazing pressure on this group, the seasonal biomass distribution clearly changes. In autumn of the first year one can see a slight increase in the standing stock of primary producers in compartment 1. This is caused by the release in grazing pressure, because of the mortality/migration of the consumers out of compartment 1 under the almost anoxic conditions. This recovery in autumn of primary producers is much larger in the second year but absent in the third year, indicating a slow shift to a new equilibrium, with grazing on the primary producers being dominated by the benthic suspension feeders (BSF). In compartments 3 and 5 the effects on the primary producers are much smaller.

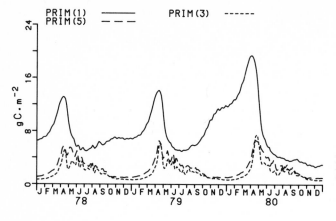

Fig. 14.3. The biomass in gC·m⁻² of benthic and pelagic primary producers (*PRIM*) in compartments 1, 3 and 5

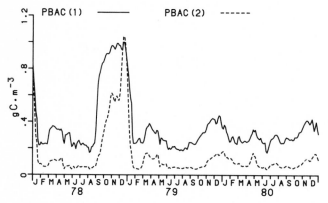

Fig. 14.4. The biomass in mgC·m⁻³ of bacterioplankton (*PBAC*) in compartments 1 and 2

Another biological component which reacts almost instantaneously to changes in the concentration of labile organic carbon (cf. Sect. 10.3), is the bacterioplankton (PBAC) (Fig. 14.4). The enormous increase in the biomass of this group in compartment 1 and to a lesser extent in compartment 2, during the waste water discharges in the autumn of year 1 disappears immediately when the discharge is terminated. Any change in the concentration of labile organic carbon, caused by fluctuations in excretion and mortality of other biological variables, leads to a rapid response of the bacterioplankton. The bacterial biomass thus keeps fluctuating, on a much lower level, after the termination of the waste water discharges.

A group that responds strongly to the changes in oxygen saturation is the mesoepibenthos (EMES) (Fig. 14.5) that flees from compartment 1 as soon as the oxygen saturation declines to low levels, to return, strongly depleted, in December, when oxygen content increases again. This mass emigration does not recur in the following years. Otherwise, the concentrations of mesoepibenthos do not increase very much. The long-lived group of benthic suspension feeders (BSF), containing among others the species *Macoma balthica*, *Mya arenaria* and *Cardium edule*, are very low in biomass in the first year in compartment 1 due to

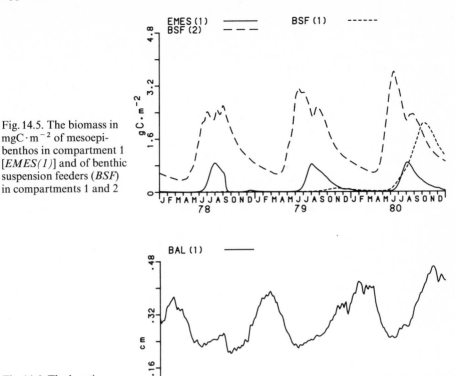

Fig. 14.5. The biomass in mgC·m^{-2} of mesoepibenthos in compartment 1 [*EMES(1)*] and of benthic suspension feeders (*BSF*) in compartments 1 and 2

Fig. 14.6. The location below the surface in cm of the sulphide horizon (*BAL*) in compartment 1

the long period of oxygen depletion during the waste water discharges. In years 2 and 3, this condition no longer prevails. In year 3, the benthic suspension feeders show a strong increase in biomass in compartment 1. In compartment 2, this group also profits from an end to the waste water discharges shown by a biomass increase in year 2. The increase is not as dramatic as in compartment 1, reflecting the fact that the conditions in compartment 2 do not improve as drastically as in compartment 1.

Another typical example of slow changes in the system is the change in the thickness of the aerobic layer (BAL) (Fig. 14.6). The thickness of this layer always shows strong seasonal dynamics under the influence of the oxygen demand from the benthic system, which is maximal in late spring. At that time the thickness of the aerobic layer reaches a minimum, whereas the seasonal maximum occurs in winter due to the low oxygen demand at that time of the year. This seasonal pattern is only slightly different after terminating the waste water discharges, but the thickness of the layer starts to trend upward slowly in the second and third year.

The results from this application may be treated with confidence to reflect what would happen if the waste water discharges really were stopped, because this is an interpolating type of application: it shifts the conditions in compartment 1 to a state comparable to the normal conditions in compartments 2 and 3.

Independent support for these model results that predict a stronger year-round presence of mesoepibenthos (*Nereis* sp.) if the waste water discharges were reduced or terminated may be found in Esselink and Van Belkum (1986). This publication is based on field observations in the Dollard (model compartments 1 and 2) after a reduction in waste water discharge was effected.

14.3.2 Case 2: Applying the Ems Model to the Severn Estuary

The Institute of Marine Environmental Research (IMER) has developed GEM-BASE (Radford and Joint, 1980; Radford, 1981), an ecosystem model of the Bristol Channel and Severn estuary. A major application of this model has been a simulation of the impact which a barrage in the Severn estuary would have on the estuarine ecosystem (Fig. 14.7). The impact predicted by GEMBASE obviously cannot be verified until this barrage really has been built. To get a second opinion we have, together with Radford, adapted the Ems model to the Bristol Channel-Severn estuary area. This entailed providing our model with the morphological and physical parameter values for the Severn area, as well as with the relevant boundary conditions. Also, hydrodynamic parameters such as diffusion constants for the Severn area had to be included.

In effect, the transport model was adapted to the area to be modelled, while the biological models remained unchanged. Then this modified model was run for a 3-year period, with a barrage between the compartments 2 and 3 (Fig. 14.7).

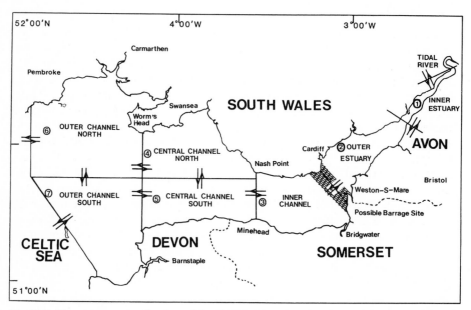

Fig. 14.7. The compartmentalization of the Bristol Channel and Severn estuary in the GEM-BASE model showing the location of a hypothetical barrage (*shaded*) in the model

The major change, predicted by both models, was a large decrease in turbidity, caused by the decrease in diffusive exchange between compartments 2 and 3 because of the barrage. All other changes, notably the strong phytoplankton blooms from May to July, can be traced to the increased transparency of the water in compartment 2. Here, again, both models predict a large bloom in the summer of the first year after the barrage is completed, followed by lesser blooms in the following years. The reduced phytoplankton blooms in the following years are due to high grazing pressure by the benthic suspension feeders which already increase strongly during the first year in GEMBASE, with a slow increase in the following years. In the modified Ems model the first-year increase, though considerable, is dwarfed by the second-year growth. The biomass level reached during the second year seems to be an overshoot, since it starts to decline again during the third year. The Ems model predicts a larger increase in suspension feeders behind the barrage than GEMBASE does: a 50-fold increase versus a 30-fold increase for GEMBASE. The deposit feeders show a transient response to the barrage: in the Ems model, the biomass returns to its starting level, after a strong increase in the second year, and in GEMBASE the response is not quite clear: the deposit feeders trend slowly upwards in one run and slowly downwards in another.

The zooplankton in both models tracks the change in phytoplankton occurrence: a strong bloom in the first year, followed by lesser blooms in the following years.

The differences between the results from both models mostly stem from a shortcut mistakenly made in the assignation of the nutrient concentrations at the fresh water boundary. In GEMBASE, the concentrations are correctly given at the boundary between the tidal river and compartment 1 (the inner estuary) (Fig. 14.7). In the case study these concentrations were assigned to the boundary of compartments 1 and 2, leaving out compartment 1 altogether. This results in much too high nutrient concentrations in the outer estuary, because the dilution in the inner estuary now does not occur.

This results in exorbitantly high phytoplankton levels since the reduced turbidity effectively removes irradiance as the factor limiting primary production. The other trophic levels respond to this exaggerated supply of food and also reach very high levels.

On the other hand, this error does not change the fact that both models, though conceived, structured and implemented independently predict the same qualitative consequences of a barrage to the ecosystem. The exercise described here, the exploratory application of the Ems model to a quite different, but well studied estuary, where most necessary parameter values were known and available, indicates that the biological model developed for the Ems estuary can describe the functioning of a quite different type of estuary in the temperate zone.

14.4 Conclusions

The results from the model, both from the standard run and especially from the sensitivity analyses, as presented in Chapters 10–13 show the model to be dynamically stable and still responsive to perturbations. Especially the sensitivity analyses where biological state variables were in effect removed from the model (for instance Sects. 10.3.4 and 12.2.2) show the model behaviour to be adaptive to these changes in its energy-flow structure. None of the sensitivity analysis runs resulted in a breakdown of the model, thus indicating a fair degree of robustness. This is of course not miraculous: because the model has to represent an estuarine system where large-scale perturbations in the form of waste water discharges are a regular feature, as well as to represent a more or less undisturbed area of the Wadden Sea, it was designed to be resilient and adaptive to the wide range of conditions found in the Ems estuary.

In the preceding sections the importance of the abiotic environment to the functioning of the system has been illustrated, especially of the physical and chemical nutrient aspects of it.

The importance of the abiotic environment has consequences for the use of the Ems model in predictive (management) applications. When the response of biological variables to physical or chemical variables has been correctly modelled, knowledge of the physico-chemical environment enables predictive applications of the model for these biological processes that are strongly forced by abiotic events. Where the major controls on biological variables are themselves biological in nature, however, predictive applications become much more uncertain. The question here is whether the reductionist systems postulate is entirely applicable to whole ecosystems (Mann, 1975). After all, when external conditions change, species composition and behaviour change. This may cause changes in the interaction between functional groups that are not included in the model, because in the concept of the functional group, its species composition is immaterial.

These uncertainties have not been resolved yet and until that moment, predictive applications of this and other models have to be viewed with skepticism.

On the other hand, estuarine ecosystem models such as this are the only available falsifiable syntheses of the different scientific disciplines concerned with estuaries. As such, incomplete and speculative as they are, they are the only comprehensive tool we have to explore the possible consequences of man-made stresses on estuarine systems.

References

Admiraal, W., 1977a. Influence of light and temperature on the growth rate of estuarine benthic diatoms in culture. Mar. Biol. **39**: 1–9.

Admiraal, W., 1977b. Salinity tolerance of benthic estuarine diatoms as tested with a rapid polarographic measurement of photosynthesis. Mar. Biol. **39**: 11–19.

Admiraal, W., 1977c. Tolerance of estuarine benthic diatoms to high concentrations of ammonia, nitrite ion, nitrate ion and orthophosphate. Mar. Biol. **43**: 307–315.

Admiraal, W., 1984. The ecology of estuarine sediment inhabiting diatoms. In: F.E. Round and D.J. Chapman (eds). Progress in Phycological Research **3**: 269–322. Biopress, Bristol.

Admiraal, W. and H. Peletier, 1979. Influence of organic compounds and light limitation on the growth rate of estuarine benthic diatoms. Br. Phycol. J. **14**: 197–206.

Admiraal, W. and H. Peletier, 1980a. Distribution of diatom species on an estuarine mudflat and experimental analysis of the selective effect of stress. J. Exp. Mar. Biol. Ecol. **46**: 157–175.

Admiraal, W. and H. Peletier, 1980b. Influence of seasonal variations of temperature and light on the growth rate of cultures and natural populations of intertidal diatoms. Mar. Ecol. Prog. Ser. **2**: 35–43.

Admiraal, W. and L.A.H. Venekamp, 1986. Significance of tintinnoid grazing during blooms of *Phaeocystis pouchetii* (Haptophyceae) in Dutch coastal waters. Neth. J. Sea Res. **20**: 61–66.

Admiraal, W., H. Peletier and H. Zomer, 1982. Observations and experiments on the population dynamics of epipelic diatoms from an estuarine mudflat. Estuarine Coastal Shelf Sci. **14**: 471–487.

Admiraal, W., L.A. Bouwman, L. Hoekstra and K. Romeyn, 1983. Qualitative and quantitative interactions between microphytobenthos and herbivorous meiofauna on a brackish intertidal mudflat. Int. Rev. Gesamten Hydrobiol. **68**: 175–191.

Admiraal, W., H. Peletier and T. Brouwer, 1984. Experimental analysis of the seasonal succession patterns of diatom species on an intertidal mudflat. Oikos **42**: 30–40.

Admiraal, W., J. Beukema and F.B. van Es, 1985. Seasonal fluctuations in the biomass and metabolic activity of bacterioplankton and phytoplankton in a well mixed estuary: the Ems-Dollard (Wadden Sea). J. Plankton Res. **7**: 877–890.

Afman, B.P., 1980. De voedselopname van de strandkrab *Carcinus maenas*. Interne Verslagen Nederlands Instituut voor Onderzoek der Zee, Texel, **1980–4**: 1–28.

Allan, D.J., T.G. Kinsey and C.J. Melody, 1976. Abundances and production of copepods in the Rhode River subestuary of Chesapeake Bay. Chesapeake Sci. **17**: 86–92.

Anonymus, 1964. Effects of polluting discharges on the Thames estuary. Dept. Sci. Ind. Research, Technical paper **11**, HMSO, London: 1–587.

Anonymus, 1979. Kwaliteitsonderzoek in de rijkswateren. Verslag van resultaten over 1e, 2e, 3e en 4e kwartaal 1979. Uitgave Rijkswaterstaat 's-Gravenhage.

Anonymus, 1984. WAQUA/DELWAQ berekeningen Eems-Dollard estuarium, Speurwerkverslag S296.02. Delft Hydraulics Laboratory: 36 pp.

Arkel, M.A. van and M. Mulder, 1979. Inventarisatie van de macrobenthische fauna van het Eems-Dollard estuarium. BOEDE Publ. en Versl. **1979–2**: 1–122.

Azam, F. and J.A. Fuhrman, 1984. Measurement of bacterioplankton growth in the sea and its regulation by environmental conditions. In: J.E. Hobbie and P.J.leB. Williams (eds). Heterotrophic activity in the sea. Plenum, New York: 179–196.

Baretta, J.W., 1980. Het zooplankton van het Eems-Dollard estuarium, soorten, aantallen, biomassa en seizoensfluctuaties. BOEDE Publ. en Versl. **1980–5**: 1–40.

Baretta, J.W., 1981. The zooplankton of the Ems estuary: quantitative data. In: N. Dankers, H. Kühl and W.J. Wolff. (eds). Invertebrates of the Wadden Sea. Rep. 4 of the Wadden Sea Working Group. F. Balkema, Rotterdam: 145–153.

Bause, K., 1982. Mass-scaled rates of respiration and intrinsic growth in very small invertebrates. Mar. Ecol. Prog. Ser. 9: 281–297.

Beek, F.A. van, 1976. Aantallen, groei, productie en voedselopname van de zandgrondel (P. minutus) en de wadgrondel (P. microps) op het Balgzand. Interne Verslagen Nederlands Instituut voor Onderzoek der Zee, Texel, 1976–9: 1–53.

Bennekom, A.J. van, E. Krijgman-van Hartingsveld, G.C.M. van der Veer and H.F.J. van Voorst, 1974. The seasonal cycles of reactive silicate and suspended diatoms in the Dutch Wadden Sea. Neth. J. Sea Res. 8: 174–207.

Bergman, M.J.N. and N. Dankers, 1978. De ecologische consequenties van het omleggen van de Eems door de Dollard. RIN rapport 79/1: 1–134.

Bergman, M., B. Kuipers, P. Spliethoff and H. van der Veer, 1976. Garnalen en krabben als mogelijke predatoren van 0-jarige schol op het Balgzand. Visserij 29: 432–438.

Berner, R.A., 1964. An idealized model of dissolved sulfate distribution in recent sediments. Geochim. Cosmochim. Acta 28: 1497–1503.

Beukema, J.J., 1974. Seasonal changes in the biomass of the macro-benthos of a tidal flat area in the Dutch Wadden Sea. Neth. J. Sea Res. 8: 94–107.

Beukema, J.J., 1976. Biomass and species richness of the macrobenthic animals living on the tidal flats of the Dutch Wadden Sea. Neth. J. Sea Res. 10: 236–261.

Beukema, J.J., 1981. The role of the larger invertebrates in the Wadden ecosystem. In: N. Dankers, H. Kühl and W.J. Wolff (eds). Invertebrates of the Wadden Sea. Rep. 4 of the Wadden Sea Working Group. Balkema, Rotterdam: 211–221.

Beukema, J.J. and J. de Vlas, 1979. Population parameters of the lugworm, Arenicola marina, living on tidal flats in the Dutch Wadden Sea. Neth. J. Sea Res. 13: 331–353.

Billen, G., 1982a. An idealized model of nitrogen recycling in marine sediments. Am. J. Sci. 282: 512–541.

Billen, G., 1982b. Modelling the processes of organic matter degradation and nutrients recycling in sedimentary systems. In: D.B. Nedwell and C.M. Brown (eds). Sediments microbiology, Academic Press, London.

Billen, G., 1984. Heterotrophic utilization and regeneration of nitrogen. In: J.E. Hobbie and P.J.leB. Williams (eds). Heterotrophic activity in the sea. Plenum, New York: 313–355.

Billen, G. and A. Fontigny, 1987. Dynamics of a Phaeocystis-dominated spring bloom in Belgian coastal waters. II. Bacterioplankton dynamics. Mar. Ecol. Prog. Ser. 37: 249–257.

BOEDE, 1983. Biologisch Onderzoek Eems-Dollard Estuarium. BOEDE Publ. en Versl. 1983–1: 1–267.

BOEDE, 1985. Biological research Ems-Dollard estuary. Rijkswaterstaat Communications no. 40/1985: 1–182.

Bogucki, M., 1953. The reproduction and development of Nereis diversicolor (O.F.M) in the Baltic. Polskie Arch. Hydrobiol.I: 251–270. (Polish with English summ.).

Bouwman, L.A., 1981. The meiofauna of the Ems estuary. In: N. Dankers, H. Kühl and W.J. Wolff (eds). Invertebrates of the Wadden Sea, Report 4 of the Wadden Sea Working Group. Balkema, Rotterdam: 153–158.

Bouwman, L.A., 1983a. A survey of nematodes from the Ems estuary: Part II: Ecology. Zool. Jahrb. Syst. 110: 345–376.

Bouwman, L.A., 1983b. Systematics, ecology and feeding biology of estuarine nematodes. Ph.D. Thesis Agricultural Academy Wageningen. BOEDE Publ. en Versl. 1983–3: 1–173.

Bouwman, L.A., K. Romeyn and W. Admiraal, 1984. On the ecology of meiofauna in an organically polluted estuarine mudflat. Estuarine Coastal Shelf Sci. 19: 633–653.

Broeker, W.S. and T.-H. Peng, 1974. Gas exchange rates between air and sea. Tellus 26: 21–35.

Butler, E.I., E.D.S. Corner and S.M. Marshall, 1970. On the nutrition and metabolism of zooplankton. VII Seasonal survey of nitrogen and phosphorus excretion by Calanus in the Clyde Sea area. J. Mar. Biol. Assoc. U.K. 50: 525–560.

Cadée, G.C., 1976. Sediment reworking by Arenicola marina on tidal flats in the Dutch Wadden Sea. Neth. J. Sea Res. 10: 440–460.

Cadée, G.C., 1979. Sediment reworking by the polychaete Heteromastus filiformis on a tidal flat in the Dutch Wadden Sea. Neth. J. Sea Res. 13: 441–456.

Cadée, G.C., 1982. Tidal and seasonal variation in particulate and dissolved organic carbon in the western Dutch Wadden Sea and Marsdiep tidal inlet. Neth. J. Sea Res. **15**: 228–249.

Cadée, G.C. and J. Hegeman, 1974. Primary production of phytoplankton in the Dutch Wadden Sea. Neth. J. Sea. Res. **8**: 240–259.

Cadée, G.C. and J. Hegeman, 1979. Phytoplankton primary production, chlorophyll and composition in an inlet of the western Wadden Sea (Marsdiep). Neth. J. Sea Res. **13**: 224–241.

Cadée, G.C. and R.W.P.M. Laane, 1983. Behaviour of POC, DOC and fluorescence in the freshwater tidal compartment of the River Ems. Mitt. Geol.-Paläont. Inst. Univ. Hamburg **55**: 331–342.

Cameron, W.M. and D.W. Pritchard, 1963. Estuaries. In: M.N. Hill (ed.) The Sea, vol.2, John Wiley, New York: 306–324.

Cauwet, G., 1981. Non-living particulate matter. In: E.K. Duursma and R. Dawson (eds). Marine Organic Chemistry, Elsevier, Amsterdam: 73–74.

Chambers, M.R. and H. Milne, 1975. Life cycle and production of *Nereis diversicolor*, O.F. Müller, in the Ythan estuary, Scotland. Estuarine Coastal Mar. Sci. **3**: 133–144.

Chervin, M.B., 1978. Assimilation of particulate organic carbon by estuarine and coastal copepods. Mar. Biol. **49**: 265–275

Chesson, P.L., 1986. Environmental variation and the coexistence of species. In: J. Diamond and T.J. Cose (eds). Community ecology, Harper and Row, New York: 240–256.

Cloern, J.F. and R.T. Cheng, 1981. Simulation model of *Skeletonema costatum* population dynamics in northern San Francisco Bay, California. Estuarine Coastal Shelf Sci. **12**: 83–100.

Colijn, F., 1982. Light absorption in the waters of the Ems-Dollard estuary and its consequences for the growth of phytoplankton and microphytobenthos. Neth. J. Sea Res. **15**: 196–216.

Colijn, F., 1983. Primary production in the Ems-Dollard estuary. Ph.D. Thesis, University of Groningen. BOEDE Publ. en Versl. **1983–2**: 1–123.

Colijn, F. and K. Dijkema, 1981. Species composition of benthic diatoms and distribution of chlorophyll a on an intertidal flat in the Dutch Wadden Sea. Mar. Ecol. Prog. Ser. **4**: 9–21.

Colijn, F. and V.N. de Jonge, 1984. Primary production of microphytobenthos in the Ems-Dollard estuary. Mar. Ecol. Prog. Ser. **14**: 185–196.

Conover, R.J., 1978. Transformation of organic matter. In: O. Kinne (ed.). Marine ecology Vol. **IV**: Dynamics. John Wiley, Chichester: 221–500.

Conover, R.J., 1981. Nutritional strategies for feeding on small suspension particles. In: A.R. Longhurst (ed.). Analysis of marine ecosystems. Academic Press, London: 363–396.

Corner, E.D.S., 1961. On the nutrition and metabolism of zooplankton. Preliminary observations in the feeding of the marine copepod, *Calanus helgolandicus* (Claus). J. Mar. Biol. Assoc. U.K. **41**: 5–16.

Corner, E.D.S., 1972. Laboratory studies related to zooplankton production in the sea. In: R.W. Edwards and J.D. Ganod (eds). Conservation and productivity of natural waters. Symp. zool. Soc. London (1972) no. **29**: 185–201.

Corner, E.D.S., R.N. Head and C.C. Kilvington, 1972. On the nutrition and metabolism of zooplankton VIII. The grazing of Biddulphia cells by *Calanus helgolandicus*. J. Mar. Biol. Assoc. U.K. **52**: 847–861.

Cuff, W.R. and M. Tomczak jr., 1983 (eds). Synthesis and modelling of intermittent estuaries. Springer, Berlin Heidelberg New York Tokyo: 302 pp.

Daan, N., 1978. Changes in cod stocks and cod fisheries in the North Sea. Rapp. P.-v. Réun. Cons. Perm. int. Explor. Mer. **172**: 39–57.

Daan, R., 1987. Impact of egg predation by *Noctiluca miliaris* on the summer development of copepod populations in the southern North Sea. Mar. Ecol. Prog. Ser. **37**: 9–17.

Dankers, N., M. Binsbergen, K. Zegers, R.W.P.M. Laane and M.M. Rutgers van der Loeff, 1984. Transportation of water, particulate and dissolved organic and inorganic matter between a salt marsh and the Ems-Dollard estuary, The Netherlands. Estuarine Coastal Shelf Sci. **19**: 143–165.

Donk, E. van and P.A.W.J. de Wilde, 1981. Oxygen consumption and motile activity of the brown shrimp *Crangon crangon* related to temperature and body size. Neth. J. Sea Res. **15**: 54–64.

Dorrestein, R., 1960. Einige klimatologische und hydrologische Daten für das Ems-Estuarium. In: J.H. van Voorthuysen and Ph.H. Kuenen (eds): Das Ems-Estuarium (Nordsee). Verh. K. Ned. Geol.-Mijnb.k. Gen. Geol. Ser. **19**: 39–42.

Dorrestein, R. and L. Otto, 1960. On the mixing and flushing of the water in the Ems-estuary. In: J.H. van Voorthuysen and Ph.H. Kuenen (eds): Das Ems-Estuarium (Nordsee). Verh. K. Ned. Geol.-Mijnb.k. Gen. Geol. Ser. **19**: 83–102.

Dronkers, J., 1986. Tidal asymmetry and estuarine morphology. Neth. J. Sea Res. **20**: 117–131.

Dücker, H.P., 1982. Suspensionsgehalte im Flachwassergebieten - Messungen in Watt von Scharhörn. Die Küste **37**: 85–184.

Durbin, E.G., 1977. Studies on the autecology of the marine diatom *Thalassiosira nordenskoldii*. II. The influence of cell size on growth rate, and carbon, nitrogen, chlorophyll a and silica content. J. Phycol. **13**: 150–155.

Ebenhöh, W., 1987. Coexistence of an unlimited number of algal species in a model system. Theor. Popul. Biol. **33**, in press.

Eggink, H.J., 1965. Het estuarium als ontvangend water van grote hoeveelheden afvalstoffen. RIZA, med. **2**: 1–144.

Eglington, G. and P.J. Barnes, 1978. Organic matter in aquatic sediments. In: W.E. Krumbein (ed.). Environmental biogeochemistry and geomicrobiology. Vol. 1: The Aquatic Environment. Ann Arbor Science, Ann Arbor, Michigan: 25–46.

Eisma, D., 1986. Flocculation and de-flocculation of suspended matter in estuaries. Neth. J. Sea Res. **20**: 183–199.

Eisma, D., P. Bernard, J.J. Boon, R. van Grieken, J. Kalf and W.G. Mook, 1985. Loss of particulate organic matter in estuaries as exemplified by the Ems and Gironde estuaries. Mitt. Geol.-Paläont. Inst. Univ. Hamburg SCOPE-UNEP Sonderband **58**: 397–412.

Eriksson, S., C. Sellei and K. Wallstrom, 1977. The structure of the plankton community of the Oregrundsgrepen (southwest Bothnian Sea). Helgol. Wiss. Meeresunters. **30**: 582–597.

Es, F.B. van, 1977. A preliminary carbon budget for a part of the Ems estuary: the Dollard. Helgoländer Wiss. Meeresunters. **30**: 283–294.

Es, F.B. van, 1982a. Community metabolism of intertidal flats in the Ems-Dollard estuary. Mar. Biol. **66**: 95–108.

Es, F.B. van, 1982b. Some aspects of the flow of oxygen and organic carbon in the Ems-Dollard estuary. Ph.D. Thesis, University of Groningen. BOEDE Publ. en Versl. **1982–5**: 1–121.

Es, F.B. van, 1984. Decomposition of organic matter in the Wadden Sea. In: R.W.P.M. Laane and W.J. Wolff (eds). Proceedings of the fourth international Wadden Sea symposium. The role of organic matter in the Wadden Sea. Neth. J. Sea Res. publication series **10**: 133–144.

Es, F.B. van and R.W.P.M. Laane, 1982. The utility of organic matter in the Ems-Dollard estuary. Neth. J. Sea Res. **16**: 300–314.

Es, F.B. van and L.-A. Meyer-Reil, 1982. Biomass and metabolic activity of heterotrophic marine bacteria. Adv. microb. Ecol. **6**: 111–170.

Es, F.B. van and P. Ruardij, 1982. The use of a model to assess factors affecting the oxygen balance in the water of the Dollard. Neth. J. Sea Res. **15**: 313–330.

Es, F.B. van, M.A. van Arkel, L.A. Bouwman and H.G.J. Schröder, 1980. Influence of organic pollution on bacterial, macrobenthic and meiobenthic populations in intertidal flats in the Dollard. Neth. J. Sea Res. **14**: 288–304.

Esselink, P. and J. van Belkum, 1986. De verspreiding van de zeeduizendpoot *Nereis diversicolor* en de kluut *Recurvirostra avosetta* in de Dollard in relatie tot een verminderde afvalwaterlozing. Rapport: GWAO-**86.155**: 1–50.

Eysink, W.A., 1979. Morfologie van de Waddenzee, gevolgen van zand- en schelpwinning. Verslag literatuur onderzoek. W.L. Rapport **R1336**, Delft.

Falkowski, P.G., 1980. Primary productivity in the sea. Environmental Sci. Research **19**. Plenum, New York, 531 pp.

Favejee, J.Ch.L., 1960. On the origin of the mud deposits in the Ems estuary. In: J.H. van Voorthuysen and Ph.H. Kuenen (eds). Das Ems-Estuarium (Nordsee). Verh. K. Ned. Geol.-Mijnb.k. Gen. Geol. Ser. **19**: 147–151.

Fenchel, T., 1982. Ecology of heterotrophic microflagellates. II. Bioenergetics and growth. Mar. Ecol. Prog. Ser. **8**: 225–231.

Fenchel, T. and T.H. Blackburn, 1979. Bacteria and mineral cycling. Academic Press, London: 1–225.

Fenchel, T.M. and B. Jørgensen, 1977. Detritus food chains of aquatic ecosystems: the role of bacteria. In: M. Alexander (ed.). Adv. Microb. Ecol. **1**. Plenum: 1–57.

Fonds, M, 1973. Sand gobies in the Dutch Wadden Sea (*Pomatoschistus*, Gobiidae, Pisces). Neth. J. Sea Res. **6**: 417–478.

Frankel, L. and D.J. Mead, 1973. Mucilaginous matrix of some estuarine sands in Connecticut. J. Sediment Petrol. **43**: 1090–1095.

Fransz, H.G. and W.W.C. Gieskes, 1984. The unbalance of phytoplankton and copepods in the North Sea. Rapp. P.-v. Réun. Cons. Perm. Int. Explor. Mer, **183**: 218–225

Frost, B.W., 1972a. Feeding behaviour of *Calanus pacificus* in mixtures of food particles. Limnol. Oceanogr. **17**: 805–815.

Frost, B.W., 1972b. Feeding processes at lower trophic levels in pelagic communities. In: C.B. Miller (ed.). The biology of the Oceanic Pacific. **33**rd Annual Biology Colloquium Corvallis. Oregon State University Press: 59–77.

Frost, B.W., 1975. A threshold feeding behavior in *Calanus pacificus*. Limnol. Oceanogr. **20**: 263–266.

Fuhrman, J.A. and F. Azam, 1980. Bacterioplankton secondary production estimates for coastal waters of British Columbia, Antarctica and California. Appl. environ. Microbiol. **39**: 1085–1095.

Fuhrman, J.A. and F. Azam, 1982. Thymidine incorporation as a measure of heterotrophic bacterioplankton production in marine surface waters: evaluation and field results. Mar. Biol. **66**: 109–120.

Gamble, J.C., 1978. Copepod grazing during a declining spring phytoplankton bloom in the northern North Sea. Mar. Biol. **49**: 303–315.

Gieskes, W.W.C. and G.W. Kraay, 1975. The phytoplankton spring bloom in Dutch coastal waters of the North Sea. Neth. J. Sea Res. **9**: 166–196.

Gocke, K., 1977. Heterotrophic activity. In: G. Rheinheimer (ed.). Microbial ecology of a brackish water environment. Springer, Berlin Heidelberg New York: 198–222.

Gocke, K. and H.G. Hoppe, 1977. Determination of organic substances and respiration potential. In: G. Rheinheimer (ed.). Microbial ecology of a brackish water environment. Springer, Berlin Heidelberg New York: 61–70.

Goes, E.R.F. van der, H. Rundberg and G.C. Visser, 1979. Erosie en sedimentatie in de westelijke Waddenzee. Rijkswaterstaat, Directie Waterhuishouding en Waterbeweging, District Kust en Zee, Studiedienst Hoorn. Nota WWKZ-79. H002 met 17 bijlagen: 1–29.

Goldhaber, M.B. and I.R. Kaplan, 1974. The sulfur cycle. In: E.D. Goldberg (ed.). The sea, vol. **5**, John Wiley, New York: 569–655.

Gordon, D.C. Jr., P.D. Keizer, G.R. Daborn, P. Schwinghamer and W.L. Silvert, 1986. Adventures in holistic ecosystem modelling: the Cumberland Basin ecosystem model. Neth. J. Sea Res. **20**: 239–251.

Gray, J.S., 1981. The ecology of marine sediments. An introduction to the structure and function of benthic communities. Cambr. Stud. in Modern Biol. **2**, University Press, Cambridge: 1–185.

Greve, W., 1971. Ökologische Untersuchungen an *Pleurobrachia pileus* 1: Freilanduntersuchungen, Helgol. Wiss. Meeresunters. **22**: 303–325.

Greve, W., 1972. Ökologische Untersuchungen an *Pleurobrachia pileus* 2: Laboratorium-Untersuchungen, Helgol. Wiss. Meeresunters. **23**: 141–164.

Groen, P., 1967. On the residual transport of suspended matter by an alternating tidal current. Neth. J. Sea Res. **3**(4): 564–574.

Hagström, A. and U. Larsson, 1984. Diel and seasonal variation in growth rates of pelagic bacteria. In: J.E. Hobbie and P.J.leB. Williams (eds). Heterotrophic activity in the sea. Plenum, New York: 249–262.

Hairston, N.G., F.E. Smith and L.B. Slobodkin, 1960. Community structure, population control and competition. Am. Nat. **94**: 421–425.

Hargrave, B.T., 1984. Sinking of particulate matter from the surface water of the ocean. In: J.E. Hobbie and P.J.LeB. Williams (eds). Heterotrophic activity in the sea. Plenum, New York: 155–178.

Hartmann-Schröder, G., 1971. Annelida, Borstenwürmer, Polychaeta. Tierwelt Deutschlands **58**: 1–594.

Heincke, F., 1913. Investigations on the plaice. General report. Rapp. P.-v. Réun. Cons. Perm. Int. Explor. Mer. **17**: 1–153.

Heip, C., N. Smol and V. Absillus, 1978. Influence of temperature on the reproductive potential of *Oncholaimus oxyuris* (Nematoda: Oncholaimidae). Mar. Biol. **45**: 255–260.

Helder, W., 1983. Aspects of the nitrogen cycle in Wadden Sea and Ems-Dollard estuary with emphasis on nitrification. Ph.D. Thesis, University of Groningen: 1–85.

Helder, W. and R.T.P. de Vries, 1983. Estuarine nitrite maxima and nitrifying bacteria (Ems-Dollard estuary). Neth. J. Sea Res. **17**: 1–18.

Helder, W. and P. Ruardij, 1982. A one-dimensional mixing and flushing model of the Ems-Dollart estuary: calculation of time scales at different river discharges. Neth. J. Sea Res. **15**: 293–312.

Helder, W., R.T.P. de Vries and M.M. Rutgers van der Loeff, 1983. Behaviour of nitrogen nutrients and dissolved silica in the Ems-Dollard estuary. Can. J. Fish. Aquat. Sci. **40**, suppl. 1: 188–200.

Hicks, G.R.F. and B.C. Coull, 1983. The ecology of marine meiobenthos harpacticoid copepods. Oceanogr. Mar. Biol. Annu. Rev. **21**: 67–175.

Hinrich, H., 1974. Schwebstoffgehalt, Gebietsniederschlag, Abfluß und Schwebstofffracht der Ems bei Rheine und Versen in den Jahren 1965 bis 1971. Dtsch. Gewässerk. Mitt. **18**: 85–95.

Hobbie, J.E., R. Daley and S. Jasper, 1977. Use of Nuclepore filters for counting bacteria by fluorescence microscopy. – Appl. Environ. Microbiol. **33**: 1225–1228.

Holling, C.S., 1975. Notes towards a science of ecological management. In: W.H. van Dobben and R.H. Lowe-McConnell (eds). Unifying concepts in ecology. Dr. W. Junk, The Hague, The Netherlands: 247–251.

Holling, C.S. (ed.), 1978. Adaptive environmental assessment and management. John Wiley, New York.

Hoppe, H.G., 1978. Relations between active bacteria and heterotrophic potential in the sea. Neth. J. Sea Res. **12**: 78–98.

Hughes, R.H., 1970a. Population dynamics of the bivalve *Scrobicularia plana* (Da Costa) on an intertidal mud-flat in North Wales. J. Anim. Ecol. **39**: 333–356.

Hughes, R.H., 1970b. An energy budget for a tidal flat population of the bivalve *Scrobicularia plana* (Da Costa). J. Anim. Ecol. **39**: 357–382.

Hulscher, J., 1982. The oyster-catcher *Haematopus ostralegus* as a predator of the bivalve *Macoma balthica* in the Dutch Wadden Sea. Ardea **70**: 89–152.

Höpner, Th., K. Worneberger and W. Ebenhöh, 1983. Die Nährstoffversorgung der Wattoberfläche. Wasser **61**: 77–89.

Iturriaga, R. and H.G. Hoppe, 1977. Observations of heterotrophic activity on photoassimilated organic matter. Mar. Biol. **40**: 101–108.

Jansen, G.M., 1980. De voedselopname van de garnaal, *Crangon crangon* (L). Interne Verslagen Nederlands Instituut voor Onderzoek der Zee, Texel, **1980–3**: 1–34.

Jelgersma, S., 1960. Die Palynologische und C14-Untersuchung einiger Torfprofile aus dem N.S.-Profil Meedhuizen-Farmsum. In: J.H. van Voorthuysen and Ph.H. Kuenen (eds). Das Ems-Estuarium (Nordsee). Verh. K. Ned. Geol.-Mijnb.k. Gen. Geol. Ser. **19**: 25–32.

Joint, I.R., J.M. Gee and R.M. Warwick, 1982. Determination of fine-scale vertical distribution of microbes and meiofauna in an intertidal sediment. Mar. Biol. **72**: 157–164.

Jonge, V.N. de, 1983. Relations between annual dredging activities, suspended matter concentrations and the development of the tidal regime in the Ems estuary. Can. J. Fish. Aquat. Sci. **40**(1): 289–300.

Jonge, V.N. de, 1985. The occurrence of 'epipsammic' diatom populations: a result of interaction between physical sorting of sediment and certain properties of diatom species. Estuarine Coastal Shelf Sci. **21**: 607–622.

Jonge, V.N. de and H. Postma, 1974. Phosphorus compounds in the Dutch Wadden Sea. Neth. J. Sea Res. **8**: 139–153.

Jonge, V.N. de and J. van den Bergs, 1987. Experiments on the resuspension of estuarine sediments containing benthic diatoms. Estuarine Coastal Shelf Sci. **24**: 725–740.

Jumars, P.A. and A.R.M. Nowell, 1984. Effects of benthos on sediment transport: difficulties with functional grouping. Continental Shelf Res. **3**: 115–130.

Jørgensen, B.B., 1983. The microbial sulphur cycle. In: W.E. Krumbein (ed.). Microbial geochemistry. Blackwell Sci., Oxford: 91–124.

Jørgensen, B.B. and N.P. Revsbeck, 1985. Diffusive boundary layers and the oxygen uptake of sediments and detritus. Limnol. Oceanogr. **30**, 111–122.

Jørgensen, B.B. and J. Sørensen, 1985. Seasonal cyclus of O_2, NO_3^- and SO_4^{--} reduction in estuarine sediments: the significance of an NO_3^- reduction maximum in spring. Mar. Ecol. Prog. Ser. **24**: 65–74.

Jørgensen, C.B., 1966. Biology of suspension feeding. Pergamon, New York: 1–375.

Kamps, L.F., 1963. Mud distribution and land reclamation in the eastern Wadden shallows. Internat. Instit. for land reclamation and improvement, Publ. **9**, Veenman, Wageningen.

Kinne, O., 1955. *Neomysis vulgaris* Thompson, eine Autökologisch-biologische Studie. Biol. Zentralb., **74**: Heft 3/4: 160–202.

Kinne, O., 1970. Marine Ecology, Vol. I, part 1. Wiley-Interscience, London: 1–681.

Kirchman, D. and R. Mitchell, 1982. Contribution of particle bound bacteria to total microheterotrophic activity in five ponds and two marshes. Appl. Environ. Microbiol. **43**: 200–209.

Kirchman, D., B. Peterson and D. Juers, 1984. Bacterial growth and tidal variation in bacterial abundance in the Great Sippewissett salt marsh. Mar. Ecol. Prog. Ser. **19**: 247–259.

Kjerfve, B., L.H. Stevenson, J.A. Proehl, T.H. Chrzanowski and W.M. Kitshens, 1981. Estimation of material fluxes in an estuarine cross-section: a critical analysis of spatial measurement density and errors. Limnol. Oceanogr. **26**: 325–335.

Kleiber, P. and T.H. Blackburn, 1978. Model of biological and diffusional processes involving hydrogen sulfide in a marine microcosm. Oikos **31**: 280–283.

KNMI, 1972. Klimaatatlas van Nederland, Staatsuitgeverij, Den Haag.

Kopylov, A.I., 1977. On the feeding of aquatic ciliates. Inform. Bull. Inst. Biol. Inlands Waters (Borok) **33**: 19–23 (In Russian).

Kremer, J.N. and S.W. Nixon, 1978. A coastal marine ecosystem: simulation and analysis. Springer, Berlin Heidelberg New York: 1–217.

Krouse, H.R. and R.G.L. McCready, 1979. Reductive reactions in the sulfur cycle. In: P.A. Trudinger and D.J. Swaine (eds). Biogeochemical cycling of mineral-forming elements. Elsevier, Amsterdam: 315–368.

Kuenen, J.G., 1975. Colourless sulfur bacteria and their role in the sulfur cycle. Plant Soil **43**: 49–76.

Kuipers, B.R. and R. Dapper, 1981. Production of *Crangon crangon* in the tidal zone of the Dutch Wadden Sea. Neth. J. Sea Res. **15**: 33–53.

Kuipers, B.R., P.A.W.J. de Wilde and F. Creutzberg, 1981. Energy flow in a tidal flat ecosystem. Mar. Ecol. Prog. Ser. **5**: 215–221.

Kühl, H. and H. Mann, 1968. Uber das Zooplankton der unteren Ems. Veröff. Inst. f. Meeresforschung, Bremerhaven, **XI**,1: 119–135.

Laanbroek, H.J. and H. Veldkamp, 1982. Microbial interactions in sediment communities. Phil. Trans. R. Soc. Lond. B. **297**: 533–550.

Laanbroek, H.J., J.C. Verplanke, P.R.M. de Visscher and R. de Vuyst, 1986. Distribution of phyto and bacterioplankton growth and biomass parameters, dissolved inorganic nutrients and free amino acids during a spring bloom in the Oosterschelde basin, The Netherlands. Mar. Ecol. Prog. Ser. **25**: 1–11.

Laane, R.W.P.M., 1980. Conservative behaviour of dissolved organic carbon in the Ems-Dollart estuary and the Western Wadden Sea. Neth. J. Sea Res. **14**: 192–199.

Laane, R.W.P.M., 1981. The composition and distribution of dissolved fluorescent substances in the Ems-Dollart estuary. Neth. J. Sea Res. **15**: 88–99.

Laane, R.W.P.M., 1982a. Sources of dissolved organic carbon in the Ems-Dollart estuary: the rivers and phytoplankton. Neth. J. Sea Res. **15**: 331–339.

Laane, R.W.P.M., 1982b. Chemical characteristics of the organic matter in the waterphase of the Ems-Dollart estuary. Ph.D. Thesis, University of Groningen. BOEDE Publ. en Versl. **1982-6**: 1–134.

Laane, R.W.P.M., 1984. Characteristics of organic matter in the Wadden Sea. Neth. J. Sea Res. publication series **1984-10**: 22–39.

Laane, R.W.P.M. and V. Ittekkot, 1983. Behaviour of dissolved organic waste in a part of the Ems-Dollart estuary: the Dollart. Mitt. Geol.-Paläont. Inst. Univ. Hamburg, **55**: 343–352.

Lancelot, C., 1983. Factors affecting extracellular release in the southern bight of the North Sea. Mar. Ecol. Prog. Ser. **12**: 343–347.

Lancelot, C., 1984. Metabolic changes in *Phaeocystis pouchetii* (Hariot) Lagerheim during the spring bloom in Belgian coastal waters. Estuarine Coastal Shelf Sci. **18**: 593–600.

Lancelot, C. and G. Billen, 1984. Activity of heterotrophic bacteria and its coupling to primary production during the spring phytoplankton bloom in the southern bight of the North Sea. Limnol. Oceanogr. **29**: 721–730.

Lee, J.J., J.H. Tietjen, N.M. Saks, G.G. Ross, H. Rubin and W.A. Mullder, 1975. Educing and modelling the functional relationships within sublittoral salt-marsh aufwuchs communities inside one of the black boxes. Estuarine Res. **1**: 710–734.

Lindeboom, H.J. and A.G.A. Merks, 1983. Annual changes in nutrient DOC and POC concentrations and their relationship with chemical and biological processes in a closed estuary. Mitt. Geol.-Paläont. Inst. Univ. Hamburg. SCOPE/UNEP Sonderband **55**: 315–329.

Lindeboom, H.J. and A.J.J. Sandee, 1983. The effect of coastal engineering projects on microgradients and mineralization reactions in sediment. Water Sci. Technol. **16**: 87–94.

Lindeboom, H.J., E.J. Wagenvoort and A.J.J. Sandee, 1984. Mineralisatie op en in de bodem van de Oosterschelde (BALANS VI). Nota BALANS VI, **1984-9**, Delta Dienst en Delta Instituut. 30 pp. + 1 Appendix.

Linley, E.A., R.C. Newell and M.I. Lucas, 1983. Quantitative relationships between phytoplankton, bacteria and heterotrophic microflagellates in shelf waters. Mar. Ecol. Prog. Ser. **12**: 77–89.

Lohmeyer, C., 1907. Übersicht der Fische der unteren Ems, Weser und Elbgebiets. Abh. Naturwiss. Ver., Bremen, **19**: 149–180.

Lopez, G., F. Riemann and M. Schrage, 1979. Feeding biology of the brackish-water oncholaimid nematode *Adoncholaimus thalassophygas*. Mar. Biol. **54**: 311–318.

Ludden, D.J.H., W. Admiraal and F. Colijn, 1985. The cycling of carbondioxide and oxygen in layers of marine microphytes, a simulation model and its ecological implications. Oecologia **66**: 50–59.

MacLaren, J.A., 1963. Predicting development rate of copepod eggs. Biol. Bull. **131**: 457–469.

Mague, G.W., E. Friberg, D.J. Hughes and I. Morris, 1980. Extracellular release of carbon by marine phytoplankton: a physiological approach. Limnol. Oceanogr. **25**: 262–279.

Mangum, Ch. and W. van Winkel, 1973. Responses of aquatic invertebrates to declining oxygen conditions. Am. Zool. **13**: 529–541.

Mann, K.H., 1975. Relationship between morphometry and biological functioning in three coastal inlets of Nova Scotia. In: E. Cronin (ed.). Estuarine res. Proc. 2nd Int. Estuarine Res. Conf., Myrtle Beach, S.C. Academic Press, New York: 1975, Vol. **I**.: 634–644.

Marchant, R. and W.L. Nicholas, 1974. An energy budget for the free-living nematode *Pelodera* (Rhabditidae). Oecologia **16**: 237–252.

Maschhaupt, J.G., 1948. Soil survey in the Dollard area. Verslagen van landbouwkundige onderzoekingen, no.**54.4**, 's Gravenhage: 1–222.

Meadows, P.W. and J.G. Anderson, 1968. Micro-organisms attached to marine sand grains. J. Mar. Biol. Assoc. U.K. **48**: 161–175.

Meyer-Reill, L.-A., 1977. Bacterial growth rates and biomass production. In: G. Rheinheimer (ed.). Microbial ecology of a brackish water environment. Springer, Berlin Heidelberg New York: 223–236.

Michaelis, H., 1981. Intertidal benthic animals communities of the estuaries of the rivers Ems and Weser. In: N. Dankers, H. Kühl and W.J. Wolff (eds). Invertebrates of the Wadden Sea. Rep. **4** of the Wadden Sea Working Group. Balkema, Rotterdam: 158–188.

Montagua, P.A., 1984. In situ measurement of meiobenthic grazing rates on sediment bacteria and edaphic diatoms. Mar. Ecol. Prog. Ser. **18**: 119–130.

Montagua, P.A., B.C. Coull, T.L. Herring and B.W. Dudley, 1983. The relationship between abundances of meiofauna and suspected microbial food (diatoms and bacteria). Estuarine Coastal Shelf Sci. **17**: 381–394.

Morris, I., 1980. The physiological ecology of phytoplankton. Studies in ecology, Vol. **7**. Blackwell Scientific, Oxford: 625 pp.

Morris, I. and H. Glover, 1981. Physiology of photosynthesis by marine coccoid cyanobacteria – some ecological implications. Limnol. Oceanogr. **26**: 957–961.

Nedwell, D.B., 1982. The cycling of sulphur in marine and freshwater sediments. In: D.B. Nedwell and C.M. Brown (eds). Sediment microbiology. Academic Press, London.

Newell, R.C., 1979. Biology of intertidal animals. Marine Ecological Survey Faversham, Kent: 1–781.

Newell, R.C., M.I. Lucas and E.A. Linley, 1981. Rate of degradation and efficiency of conversion of phytoplankton debris by marine microorganisms. Mar. Ecol. Prog. Ser. **6**: 123–136.

Newell, R.C., E.A. Linley and M.I. Lucas, 1983. Bacterial production and carbon conversion based on saltmarsh plant debris. Estuarine Coastal Shelf Sci. **17**: 405–420.

Nixon, S.W., J.R. Kelly, B.N. Furnas, C.A. Oviatt and S.S. Hale, 1980. Phosphorus regeneration and the metabolism of coastal marine bottom communities. In: K.R. Tenore and B.C. Coull (eds). Marine benthic dynamics. Univ. South Carolina: 219–242.

O'Kane, J.P., 1980. Estuarine water-quality management. Pitman, London: 1–155.

Ogura, N. and T. Gotoh, 1974. Decomposition of dissolved carbohydrates derived from diatoms of Lake Yuno-Ko. Int. Rev. Gesamten Hydrobiol. **59**: 39–47.

Ostlund, H.G. and J. Alexander, 1963. Oxidation rate of sulphide in seawater, a preliminary study. J. Geophys. Res. **68**: 3995–3997.

Paffenhöfer, G.A., 1971. Grazing and ingestion rates of nauplii, copepodids and adults of the marine planktonic copepod *Calanus helgolandicus*. Mar. Biol. **11**: 286–298.

Paffenhöfer, G.A. and S.C. Knowles, 1978. Feeding of marine planktonic copepods on mixed phytoplankton. Mar. Biol. **48**: 143–152.

Parsons, T.R., M. Takahashi and B. Hargrave, 1977. Biological oceanographic processes (2nd ed.). Pergamon, London: 1–332.

Peters, R.H., 1983. The ecological implications of body size. Cambridge studies in ecology: **2**. Cambridge University Press.

Petipa, T.S., Y.V. Pavlova and Y.I. Sorokin, 1971. Arucheniye pitaniya massorykh form planktona tropicheskoy oblasti Tikhogo Okeana radionglerodnym meiodom. In: M.Y. Vinogrador (ed.). Funktsionirovaniye pelagicheskikh Soobshchescv tropicheskikh Rayonov Okeana Izdatel'sivo Nauka, Moskow: 123–141. (A study of feeding by common planktonic forms of the tropical regions of the Pacific Ocean using the radiocarbon method).

Plagmann, J., 1939. Ernährungsbiologie der Garnele (*Crangon vulgaris* Fabr) Helgol. Wiss. Meeresunters. **2**: 113–162.

Platt, T. (ed.), 1981. Physiological bases of phytoplankton ecology. Can. Fish. Aquat. Sci. **210**, Ottawa: 1–346.

Platt, T. and W. Silvert, 1981. Ecology, physiology, allometry and dimensionality. J. Theor. Biol. **93**: 855–860.

Platt, T., K.L. Denman and A.D. Jassby, 1977. Modelling the productivity of phytoplankton. In: E.D. Goldberg (ed.). The sea: ideas and observations on progress in the study of the seas, Vol. **VI**. John Wiley, New York.

Pomeroy, L.R., 1984. Microbial processes in the sea: diversity in nature and science. In: J.E. Hobbie and P.J.leB. Williams (eds). Heterotrophic activity in the sea. Plenum, New York: 1–23.

Postma, H., 1954. Hydrography of the Dutch Wadden Sea. Arch. Néerl. Zool. **10**: 405–511.

Postma, H., 1961. Transport and accumulation of suspended matter in the Dutch Wadden Sea. Neth. J. Sea Res. **1**: 148–190.

Postma, H., 1967. Sediment transport and sedimentation in the estuarine environment. In: G.H. Lauff (ed.). Estuaries. Am. Assoc. Adv. Sci., Publication no 83, Washington D.C.: 158–179.

Postma, H., 1981. Exchange of materials between the North Sea and the Wadden Sea. Mar. Geol. **40**: 199–215.

Postma, H., 1982. Hydrography of the Wadden Sea: movements and properties of water and particulate matter. Wadden Sea Working Group, Report **2**: 1–75.

Preston, A., 1978. Input of pollutants to the Oslo Commission area. ICES, Cooperative Research Report **77**: 1–57.

Radford, P.J., 1981. Modelling the impact of a tidal power scheme on the Severn estuary ecosystem. In: W.J. Mitsch, R.W. Bosserman and J.M. Klopatek (eds). Energy and ecological modelling. International Energy Symposium, Louisville, 1981. Elsevier Scientific, Amsterdam: 235–248.

Radford, P.J. and I.R. Joint, 1980. The application of an ecosystem model to the Bristol Channel and Severn estuary. Institute of Water Pollution Control. Annual Conference. Water Pollution Control. Conference Paper **7**: 244–245.

Radford, P.J. and R.J. Uncles, 1980. Ecosystem models and the prediction of ecological effects. In: T.L. Shaw (ed). An environmental appraisal of tidal power stations with particular reference to the Severn Barrage. Pitman Publishers Ltd., London: 109–127.

Radford, P.J., I.R. Joint and A.R. Hilby, 1981. Simulation models of individual production processes. In: A.R Longhurst (ed.). Analysis of marine ecosystems. Academic Press, London: 677–700.

Redant, F., 1980. Population dynamics of brown shrimp (*Crangon crangon*) in the Belgian coastal waters. 1. Consumption-production model. ICES C.M. **1980/K32**.

Reenders, R. and D.H. van der Meulen, 1972. De ontwikkeling van de Dollard over de periode 1952–1969/'70. Rijkswaterstaat, directie Groningen, afd. Studiedienst, nota **72.1**: 23 pp. met 14 bijlagen.

Reeve, M.R., M.A. Walter and T. Ikeda, 1978. Laboratory studies of ingestion and food utilization in lobate and tentaculate ctenophores. Limnol. Oceanogr. **23**: 740–751.

Reijnders, P.J.H. and W.J. Wolff, 1981. Marine mammals of the Wadden Sea Report **7** of the Wadden Sea Working Group. Balkema, Rotterdam: 1–63.

Reise, K., 1985. Tidal flat ecology. An experimental approach to species interactions. Ecological Studies **54**. Springer, Berlin Heidelberg New York Tokyo: 1–191.

Revsbeck, N.P., J. Sørensen, T.H. Blackburn and J.P. Lomholt, 1980. Distribution of oxygen in marine sediments measured with microelectrodes. Limnol. Oceanogr. **25**: 403–411.

Riley, G.A., 1957. Phytoplankton of the North Central Sargasso Sea, 1950–52. Limnol. Oceanogr. **2**: 252–270.

Riley, G.A., H. Stommel and D.F. Bumpus, 1949. Quantitative ecology of the plankton of the western North Atlantic. Bull. Bingham Oceanogr. Collect. **12**: 1–169.

Roman, M.R. and P.A. Rublee, 1981. A method to determine in situ zooplankton grazing rates on natural particles assemblages. Mar. Biol. **65**: 303–309.

Romeyn, K. and L.A. Bouwman, 1983. Food selection and consumption by estuarine nematodes. Hydrobiol. Bull. **17**: 103–109.

Romeyn, K., L.A. Bouwman and W. Admiraal, 1983. Ecology and cultivation of the herbivorous brackish-water nematode *Eudiplogaster pararmatus*. Mar. Ecol. Prog. Ser. **12**: 145–153.

Round, F.E., 1979. A diatom assemblage living below the surface of intertidal sand flats. Mar. Biol. **54**: 219–223.

Ruardij, P., 1981. Een ééndimensionaal model van de transportprocessen in het Eems-Dollard oecosysteem. BOEDE Publ. and Versl. **1981–3**: 1–37 pp.

Rublee, P.A., S.M. Merkel and M.A. Faust, 1983. The transport of bacteria in sediments of a temperate marsh. Estuarine Coastal Shelf Sci. **16**: 501–509.

Rudert, M. and G. Müller, 1981. Mineralogy and provenance of suspended solids in estuarine and nearshore areas of the southern North Sea. Senckenb. Marit. **13**: 57–64.

Russell, R.J., 1967. Origins of estuaries. In: G.H. Lauff (ed.), Estuaries. Am. Assoc. Adv. Sci. publication no. **83**, Washington D.C., U.S.A.: 93–99.

Rutgers van der Loeff, M.M., F.B. van Es, W. Helder and R.T.P. de Vries, 1981. Sediment water exchanges of nutrients and oxygen on tidal flats in the Ems-Dollard estuary. Neth. J. Sea Res. **15**: 113–129.

Salomons, W., 1975. Chemical and isotopic composition of carbonates in recent sediments and soils from Western Europe. J. Sediment. Petrol. **45**: 440–449.

Samu, G., 1979. Die morphologische Entwicklung der Aussen-Ems von Dukegat bis zur See. Bundesanstalt für Wasserbau, Hamburg: 1–43.

Schiemer, F., 1982. Food dependence and energetics of free living nematodes II. Life history parameters of *Caenorhabditis briggsae* (nematode) at different levels of food supply. Oecologia **54**: 122–128.

Schiemer, F., A. Duncan and R.Z. Klekowski, 1980. A bioenergetic study of a benthic nematode, *Plectus palustris* de Man 1880, throughout its life cycle. II. Growth, fecundity considerations. Oecologia **44**: 205–212.

Schröder, H.G.J. and F.B. van Es, 1980. Distribution of bacteria in intertidal sediments of the Ems-Dollard estuary. Neth. J. Sea Res. **14**: 268–287.

Schwinghamer, P., B. Hargrave, D. Peer and C.M. Hawkins, 1986. Partitioning of production and respiration among size groups of organisms in an intertidal benthic community. Mar. Ecol. Prog. Ser. **31**: 131–142.

Seiderer, L.J., C.L. Davis, E.T. Robb and R.C. Newell, 1984. Utilization of bacteria as nitrogen resources by kelp-bed mussel *Choromyticus meridionalis*. Mar. Ecol. Prog. Ser. **15**: 109–116.

Sepers, A.B.J., 1979. De aerobe mineralisatie van aminozuren in natuurlijke aquatische milieus. Ph.D.Thesis, University of Groningen: 1–101.

Servais, P., G. Billen and J. Vives Rego, 1985. Rate of bacterial mortality in aquatic environments. Appl. Environ. Microbiol. **49**: 1448–1454.

Sibert, J., T.J. Brown, M.C. Healy, B.A. Kask and R.J. Naiman, 1977. Detritus-based food webs: exploitation by juvenile Chum-Salmon *Oncorhynchus keta*. Science **196**: 649–650.

Sieburth, J.McN., 1967. Seasonal selection of estuarine bacteria by water temperature. J. Exp. Mar. Biol. Ecol. **1**: 98–121.

Slagmolen, H.C., 1981. Voortgang slibonderzoek. RWS-nota **81.H206**.

Smet, L.A.H. de, 1960. Die holozäne Entwicklung des niederländischen Randgebietes des Dollarts und der Ems. In: J.H. van Voorthuysen and Ph.H Kuenen (eds). Das Ems-Estuarium (Nordsee). Verh. K. Ned. Geol.-Mijnb.k. Gen. Geol. Ser. **19**: 15–23.

Smet, L.A.H. de and A.J. Wiggers 1960. Einige Bemerkungen über die Herkunft und die Sedimentationsgeschwindigkeit der Dollart-Ablagerungen. In: J.H. van Voorthuysen and Ph.H. Kuenen (eds). Das Ems-Estuarium (Nordsee). Ver. K. Ned. Geol.-Mijnb.k. Gen. Geol. Ser. **19**: 129–135.

Smit, C.J., 1980a. The importance of the Wadden Sea for estuarine birds. In: C.J. Smit and W.J. Wolff (eds). Birds of the Wadden Sea. Balkema, Rotterdam: 280–289.

Smit, C.J., 1980b. Production of biomass by invertebrates and consumption by birds in the Dutch Wadden Sea. In: C.J. Smit and W.J. Wolff (eds). Birds of the Wadden Sea: 290–301.

Smith, J.C., T. Platt and W.G. Harrison, 1983. Photo adaptation of carboxylating enzymes and photosynthesis during a spring bloom. Prog. Oceanogr. **12**: 425–459.

Sorokin, I.Y., 1981. Microheterotrophic organisms in marine ecosystems. In: A.R. Longhurst (ed.): Analysis of marine ecosystems. Academic Press, London: 293–342.

Stam, A., 1979. De vissen, krabben en garnalen van het Eems-Dollard estuarium. I. Kwalitatieve inventarisatie. BOEDE Publ. en Versl. **1979–1**: 1–37.

Stam, A., 1982. De vissen, krabben en garnalen van het Eems-Dollard estuarium. II. Kwantitatieve inventarisatie van de garnalen. BOEDE Publ. en Versl. **1982–4**: 1–38.

Stam, A., 1984. De vissen, krabben en garnalen van het Eems-Dollard estuarium. III. Kwantitatieve inventarisatie van de platvissen. BOEDE Publ. and Versl. **1984–2**: 1–51.

Steele, J.H., 1974. The structure of marine ecosystems. Cambridge, Mass., Harvard University Press: 1–128.

Stommel, H. and H.G. Farmer, 1952. On the nature of estuarine circulation. Woods Hole Oceanogr. Inst. References no. 1952–51, 1952–63, 1952–88.

Straaten, L.M.J.U. van and Ph.H. Kuenen, 1958. Tidal action as a cause for clay accumulation. J. Sediment Petrol. **28**: 406–413.

Stratingh, G.A. and G.A. Venema, 1855. De Dollard of geschied-, aardrijks- en natuurkundige beschrijving van dezen boezem der Eems. J. Oomkens, J. Zoon en R.J. Schierbeek, Groningen (reprint 1979): 1–333.

Stroo, 1986. A method for validation. NIOZ, interne verslagen EON **1986–1**: 1–35.

Swennen, C., 1976. Wadden Seas are rare, hospitable and productive. In: M. Smart (ed.). Proc. Int. Conf. on the Conservation of Wetlands and Waterfowl, Heiligenhafen, 1974. JWRB, Slimbridge: 184–198.

Talling, J.F., 1971. The underwater light climate as a controlling factor in the production ecology of freshwater phytoplankton. Mitt. Int. Verh. Limnol. **19**: 214–243.

Tattersall, W.M. and O.S. Tattersall, 1951. The British Mysidacea. Royal Society, B. Quaritch, London: 1–460.

Theil, H., 1961. Economic forecasts and policy. North Holland, Amsterdam.

Tumantzev, N.I., 1979. Microzooplankton in pelagic areas of the Peruvian upwelling. In: M. Vinogradov (ed.). Ecosystems of pelagic areas of Peruvian upwelling. Nauka, Moscow.

Ustach, J.F., 1982. Algae, bacteria and detritus as food for the harpacticoid copepod, *Heteropsyllus pseudonunni* Coull and Palmer. J. Exp. Mar. Biol. Ecol. **64**: 203–214.

Vaeremans, M., 1977. De Copepoda – harpacticoidea van het Eems-Dollard estuarium. Licentiaatsverhandeling Rijksuniversiteit Gent: 1–78.

Valiela, I., 1984. Marine ecological processes. Springer, Berlin Heidelberg New York Tokyo: 1–546.

Veer, H.W. van der and M.J.N. Bergman, 1987. Predation by crustaceans on a newly settled 0-group plaice *Pleuronectes platessa* population in the western Wadden Sea. Mar. Ecol. Prog. Ser. **35**: 203–215.

Veer, H.W. van der and C.F.M. Sadée, 1984. The seasonal occurrence of the ctenophore *Pleurobrachia pileus* in the western Wadden Sea. Mar. Biol. **79**: 219–227.

Verwey, J., 1952. On the ecology of distribution of cockle and mussel in the Dutch Wadden Sea, their role on sedimentation and the source of their food supply. Arch. Néerl. Zool. **10**: 171–239.

Verwey, J., 1971. Die Folgen für das Milieu der Abfuhr organischer Abfallstoffe ins Mündungsgebiet der Ems. Contact-Commissie voor Natuur- en Landschapsbescherming, Amsterdam: 1–38.

Vlas, J. de, 1979. Annual food intake by plaice and flounder in a tidal flat area in the Dutch Wadden Sea, with special reference to consumption of regenerating parts of macrobenthic prey. Neth. J. Sea Res. **13**: 117–153.

Vlas, J. de, 1985. Weide van Vlees. The importance of the regeneration of browsed body parts of the benthic animals for the secondary production in a tidal flat area. Ph.D. Thesis, University of Groningen: 1–131.

Vollenweider, R.A., 1965. Calculation models of photosynthesis-depth curves and some implications regarding day rate estimates in primary production. Mem. Int. Ital. Idrobiol. **18**, Suppl.: 425–457.

Voorthuysen, J.H. van, 1960. Tektonische Vorgeschichte. In: J.H. van Voorthuysen and Ph.H. Kuenen (eds). Das Ems-Estuarium (Nordsee). Verh. K. Ned. Geol.- Mijnb.k. Gen. Geol. Ser. **19**: 11–13.

Voorthuysen, J.H. van and Ph.H. Kuenen (eds), 1960. Das Ems-Estuarium (Nordsee). Ein sedimentologisches Symposium. Verh. K. Ned. Geol.-Mijnb.k. Gen. Geol. Ser. **19**: 1–300.

Vooys, C.G.N. de, 1976. The influence of temperature and time of year on the oxygen uptake of the sea mussel, *Mytilus edulis*. Mar. Biol. **36**: 25–30.

Warren, Ch.E., 1971. Biology and water pollution control, Saunders, Philadelphia: 1–434.

Weisse, T., 1983. Feeding of calanoid copepods in relation to *Phaeocystis pouchetii* blooms in the German Wadden Sea area of Sylt. Mar. Biol. **74**: 87–94.

Widdows, J., P. Fieth and C.M. Worrall, 1979. Relationships between seston, available food and feeding activity in the common mussel *Mytilus edulis*. Mar. Biol. **50**: 195–207.

Wiebe, W.J., 1984. Physiological and biochemical aspects of marine bacteria. In: J.E. Hobbie and P.J.leB. Williams (eds). Heterotrophic activity in the sea. Plenum, New York: 55–82.

Wiggers, A.J., 1960. Die Korngrössenverteilung der Holozänen Sedimente im Dollart-Ems-Estuarium. In: J.H. van Voorthuysen and Ph.H. Kuenen (eds). Das Ems-Estuarium (Nordsee). Verh. K. Ned. Geol.-Mijnb.k. Gen. Geol. Ser. **19**: 111–133.

Wilde, P.A.W.J. de and E.M. Berghuis, 1978. Laboratory experiments on the spawning of *Macoma balthica* its implication for production research. In: D.S. McLusky and A.J. Berry (eds). Physiology and behaviour of marine organisms. Pergamon, Oxford: 375–384.

Wilde, P.A.W.J. de and J.J. Beukema, 1984. The role of zoobenthos in the consumption of organic matter in the Dutch Wadden Sea. In: R.W.P.M. Laane and W.J. Wolff (eds). Proceedings of the fourth international Wadden Sea Symposium: The role of organic matter in the Wadden Sea. Neth. J. Sea Res. publication series **10**: 145–158.

Williams, P., 1981. Detritus utilization by *Mytilus edulis*. Estuarine Coastal Shelf Sci. **12**: 739–746.

Williams, P.J. leB., 1984. A review of measurements of respiration rates of marine plankton populations. In: J.E. Hobbie and P.J. leB. Williams (eds). Heterotrophic activity in the sea. Plenum, New York: 357–389.

Wilson, G.T. and N. MacLeod, 1974. A critical appraisal of empirical equations and models for the prediction of the coefficient of reaeration of deoxygenated water. Water Res. **8**: 341–366.

Winter, D.F., K. Banse and G.C. Anderson, 1975. The dynamics of phytoplankton bloom in Puget Sound, a fjord in the north-western United States. Mar. Biol. **29**: 136–176.

Winter, J.E., 1976. Feeding experiments with *Mytilus edulis* L. at small laboratory scale. II. The influence of suspended silt in addition to algal suspensions on growth. In: G. Persoone and E. Jaspers (eds.). Proc. 10th Eur. Symp. Mar. Biol., Ostend, Belgium, Sept. 17–23, 1975. Universa, Wetteren: 583–600.

Winter, J.E., 1978. A review on the knowledge of suspension feeding in *lamellibranchiate bivalves*, with special references to artificial aquaculture systems. Aquaculture **13**: 1–33.

Wolter, K., 1982. Bacterial incorporation of organic substances released by natural phytoplankton populations. Mar. Ecol. Prog. Ser. **7**: 287–295.

Zaika, V.E. and T.V. Pavlovskaya, 1970. Feeding of marine protozoa with unicellular algae. Biology of the Sea vol. **19**: 80–90. Naukova Duma, Kiev (In Russian).

Zimmerman, J.T.F., 1976a. Mixing and flushing of tidal embayments in the western Dutch Wadden Sea. Part I: Distribution of salinity and calculation of mixing time scales. Neth. J. Sea Res. **10**: 149–191.

Zimmerman, J.T.F., 1976b. Mixing and flushing of tidal embayments in the western Dutch Wadden Sea, Part II: Analysis of mixing processes. Neth. J. Sea Res. **10**: 397–439.

Zimmerman, R., 1977. Estimation of bacterial number and biomass by epifluorescence microscopy and scanning electron microscopy. In: G. Rheinheimer (ed.). Microbial ecology of a brackish water environment. Springer, Berlin Heidelberg New York: 103–120.

Appendix A: Parameter Descriptions and Values

Appendix A-1: Parameters of the Physical Submodel

areaf(5)
Surface of tidal flats
u: m^2 See Table 4.1[1]

cCARASD
Conversion factor between particulate matter $(mg \cdot m^{-3})$ to particulate organic carbon $(mgC \cdot m^{-3})$.
u: mg/mgC – 2.2 –
MUD(I) = MUD(I) + C(I,J)*cCARASD

CONLIGHT
Control parameter for the LIGHT subroutine. In the LIGHT subroutine daily values of LIGHT for 5 years (1978–1982) are stored. The irradiance can be computed in different ways.(0 = field values)
u: – – 0 –
CALL SUBLIGHT(CONLIGHT,SUNQ,LIGHT)

ct0
Constant used in the computation of the saturation rate.
u: $Temp^{-1}$ – 0.023 –
SOX(I) = SOX(I) + REACON/depth(I)*EXP[ct0*TEMP(I)]*[OSAT-OX(I)]

depchFH(5)
Arithmetic mean depths in channels during the time the flats are flooded.
u: m See Table 4.2

depchLF(5)
Arithmetic mean depths in channels during the time the flats are dry.
u: m See Table 4.2

depfa(5)
Arithmetic mean depths above the tidal flats.
u: m See Table 4.2

depth
Harmonic mean depths above the tidal flats and in the channels.
u: m See Table 4.2

DIEMS(7)
Data series of exchange coefficients between compartment 3 and the Ems for different discharges of the Ems (see QE).
u: $m^3 \cdot d^{-1}$ –
RDIFEMS = SLP(QEMS,QE,DIEMS,7)

DIFF(35)
Data series of exchange coefficients between the compartments for 7 different discharges of the rivers (see QE and QS).
u: $m^3 \cdot d^{-1}$ –
RDIFF(I+2) = SLP[QWWA,QW,DIFF(1+I*7),7] (I=0,1)
           = SLP[QUPPER,QS,DIFF(1+I*7),7] (I=2,4)
The sequence in the array is:
D(1 → 2,Q1),D(2 → 3,Q1). 0. 0. 0. 0.D(5 → sea,Q1)
D(1 → 2,Q2), 0. 0. 0. 0. 0. 0. 0. 0. 0.D(5 → sea,Q2)
D(1 → 2,Q7). 0. 0. 0. 0. 0. 0. 0. 0. 0.D(5 → sea,Q7)

Emox(14)
Time series for OX on Ems. The monthly average values are derived from field measurements by RIZA.
u: $g\,O_2 \cdot m^{-3}$ See Fig. 4.10
EMSOX = TIMSER(tiox,Emox,14)

---

[1] u: The dimension (unit) of the parameter. This is follwed by a reference to the actual value(s) of the parameter of by the actual value itself. If known, the actual parameter value is preceded by the minimal value and followed by the maximal value, thus indicating the range of possible values.

Emsal(7)   Data series for SALT of the Ems. For each of the 7 different Ems discharges a value is given (see QE).
u: S
EMSSALT = SLP(QEMS,QE,Emsal,7)

MOVE(5)   Tidal volume at the seaward border of a compartment.
u: m^3 See Table 4.1
TSILT(I) = TSILT(I)-SILT(I)*MOVE(I)/VOL(I)

pDOWN0   Constant which control partition of redistributed mud over MUD and CMUD.
u: $-$ -0.8 $-$
pDOWN = pDOWN0*W50/(ACTWIND-6.+W50)

pOsW(6)   Time series of the oxygen saturation in the WWA. There are *no* field values available for OX on the WWA. We estimated that outside the waste discharge period there is a saturation of 80% and in the campaign a total lack of oxygen.
u: $-$ See Fig. 4.10
WWAOX = TIMSER(tpOsW,pOsW,6)*(475.-2.65*Wwsalt)/[33.5+TEMP(1)]

ptWET(5)   Fraction of a day that the flats are covered with water.
u: $-$ See Table 4.2

pWAFL(5)   Fraction of the total water volume over the tidal flats.
u: $-$ See Table 4.2
TSILT(I) = TSILT(I)-rTID*seddep/depfa(I)*pWAFL(I)*pDOWN*SILT(I)

QE(7)   Series of Ems discharges used as x-axis for the variable DIEMS in the linear interpolation function SLP.
u: $m^3 \cdot d^{-1}$
RDIFEMS = SLP(QEMS,QE,DIEMS,7)

QQEM(74)   Time series of discharges of the Ems. The discharges are mean monthly values for the years 1977–1982.
u: $m^3 \cdot s^{-1}$ See Fig. 2.5
QEMS = TIMCOS(TIMQ,QQEM,74)*86400.

QQWA(74)   Time series of discharges of the Westerwoldsche Aa. The discharges are mean monthly values for the years 1977–1982.
u: $m^3 \cdot s^{-1}$ See Fig. 2.5
QWWA = TIMCOS(TIMQ,QQWA,74)*86400.

QS(7)   Series of the sum of the river discharges used as x-axis for the variable DIFF and Sesal in the linear interpolation function SLP.
u: $m^3 \cdot d^{-1}$
RDIFF(I+2) = SLP[QUPPER,QS,DIFF(1+I*7),7]

QW(7)   Series of discharges of the Westerwoldsche Aa used as x-axis for the variable DIFF in the linear interpolation function SLP.
u: $m^{-3} \cdot d^{-1}$ $-$
RDIFF(I+2) = SLP[QWWA,QW,DIFF(1,I*7),7] where I = 0,1

REACON   Relative reaeration rate of oxygen per 1 m^2 surface.
u: $m \cdot d^{-1}$ 0.96 0.96 1.92
SOX(I) = SOX(I)+REACON/depth(I)*EXP[ct0*TEMP(I)]*[OSAT-OX(I)]

REDIS(5)   The asymmetry factors for mud. An element I stands for the asymmetry for compartment I compared to compartment I+1.
u: $-$ See Table 5.2

resdep   Resuspension height
u: m -0.14 $-$
TSILT(I) = TSILT(I) + rTID*totSILT(I)*pWAFL(I)*resdep/depfa(I)*(1.-pDOWN)

rTID   Number of tides per day
u: d^{-1} -1.95 $-$
TSILT(I) = TSILT(I) + rTID*totSILT(I)*pWAFL(I)*resdep/depfa(I)*(1.-pDOWN)

seddep   Settling water column
u: m -0.44 $-$
TSILT(I) = TSILT(I)-rTID*seddep/depfa(I)*pWAFL(I)*pDOWN*SILT(I)

Seox(14)   Time series of OX at sea. The monthly average values are derived from field
                  measurements by RIZA.
                  u: $g\,O_2 \cdot m^{-3}$
                  SEAOX = TIMSER(tiox,Seox,14)

Sesal(7)   Data series of salinity at sea for different river discharges (see QS).
                  u: S
                  SEASALT = SLP(QUPPER,QS,Sesal,7)

SIEM(13)   Time series of suspended matter concentration on the Ems at 0 salinity.
                  u: $mg \cdot m^{-3}$ See Fig. 4.8
                  EMSSUSM = TIMSER(TISI,SIEM,13)

SIWWA(6)   Time series of suspended matter concentration on the Westerwoldsche Aa.
                  u: $mg \cdot m^{-3}$ See Fig. 4.8
                  WWASUSM = TIMSER(TISIW,SIWWA,6)

TE3(131)   Time series of temperature for compartment 3 over the years 1978, 1979 and
                  1980 (RIZA-data).
                  u: °C
                  TEMP(3) = TIMSER(tTE,TE3,131) + TEMPEFFECT

TE4(131)   Time series of temperature for compartment 4 over the years 1978, 1979 and
                  1980 (RIZA-data).
                  u: °C   See Fig. 2.4a
                  TEMP(4) = TIMSER(tTE,TE4,131) + TEMPEFFECT

TE5(131)   Time series of temperature for compartment 5 over the years 1978, 1979 and
                  1980 (RIZA-data).
                  u: °C   See Fig. 4.13
                  TEMP(5) = TIMSER(tTE,TE5,131) + TEMPEFFECT

TED(131)   Time series of temperature for compartment 1 and 2 (Dollard) over the years
                  1978, 1979 and 1980 (RIZA data).
                  u: °C   See Fig. 4.13
                  TEMP(1) = TEMP(2) = TIMSER(tTE,TED,131) + TEMPEFFECT

TEex(4)   Times series of values with which the temperature can be increased. Used for
                  sensitivity analysis purposes.
                  u: °C
                  IF (TEex(2).ne.0.) TEMPEFFECT = TIMCOS(tTEex,TEex,4)

TIMQ(74)   Series of day numbers used as x-axis for the variables QQEM and QQWA in
                  the TIMCOS function.
                  u: −
                  QQWA = TIMCOS(TIMQ,QQWA,74)*86400

tiox(14)   Series of day numbers used as x-axis for the variables SeOX and WIND in re-
                  spectively TIMSER and TIMCOS functions.
                  u: −
                  SEAOX = TIMSER(tiox,Seox,14)

TISI(13)   Series of day numbers used as x-axis for the variable SIEM in TIMSER func-
                  tion.
                  u: −
                  EMSSUSM = TIMSER(TISI,SIEM,13)

TISIW(6)   Series of day numbers used as x-axis for the variable SIWWA in TIMSER func-
                  tion.
                  u: −
                  WWASUSM = TIMSER(TISIW,SIWWA,6)

tpOsW(6)   Series of day numbers used as x-axis for the variable pOsW in TIMSER func-
                  tion.
                  u: −
                  WWAOX = TIMSER(tpOsW,pOsW,6)*(475.-2.65*Wwsalt)/[33.5 + TEMP(1)]

TSILTSEA   The total silt concentration at sea: the sum of suspended silt and silt near the
                  bottom (based on RIZA data).
                  u: $mg \cdot m^{-3}$ $- 42 \cdot 10^3 -$
                  SEASILT = TSILTSEA*(1.-pDOWN)

tTE(131)	Series of day numbers used as x-axis for the variables TED, TE3, TE4 and TE5 in the TIMSER function. u: − TEMP(1) = TEMP(2) = TIMSER(tTE,TED,131) + TEMPEFFECT
tTEex(4)	Series of day numbers used as x-axis for the variable TEex in the TIMSER function. u: − IF (TEex(2).ne.0.) TEMPEFFECT = TIMCOS(tTEex,TEex,4)
VOL(5)	Volume of the 5 compartments. u: m^3 See Table 4.2 TC(I,J) = TC(I,J) + (CIM-CEM + CEXCH1 + CEXCH2)/VOL(I)
W50	Constant describing the influence of the wind on the partition variable pDOWN. u: − − 2 − pDOWN = pDOWN0*W50/(ACTWIND-6. + W50)
WIND(14)	Time series of monthly mean wind speed (measured on Borkum). u: $m \cdot s^{-1}$ See Fig. 2.4b ACTWIND = TIMCOS(tiox,WIND,14)
Wwsalt	Salinity of the water from the Westerwoldsche Aa. u: S − 0.3 − WWAOX = TIMSER(tpOsW,pOsW,6)*(475.-2.65*Wwsalt)/[33.5 + TEMP(1)]

Appendix A-2: Parameters of the Pelagic Submodel

BCOP(12)	Time series for mesozooplankton (PCOP) at sea. u: $mgC \cdot m^{-3}$ see Appendix C-3 SEAPCOP = TIMSER(TMAC,BCOP,12)
caFC	Constant defining the apportioning of excretion by flagellates (PFLAG) under nutrient stress to the detritus pool (PDET). u: − 0.2 0.5 0.7 flFC = caFC*LOCF
caGC	Constant defining the apportioning of excretion by diatoms (PDIA) under nutrient stress to the detritus pool (PDET). u: − 0.2 0.5 0.7 flFC = caGC*LOCF
CARMOR	Natural mortality of carnivores (CARN), independent of temperature, oxygen saturation or salinity. u: − − 0.03 − flIC = CARN(I)*[CARMOR + CARMOX*(1.-OXMIC)]
CARMOX	Extra mortality at low oxygen saturation of pelagic carnivores (CARN). u: d^{-1} − 0.2 − flIC = CARN(I)*[CARMOR + CARMOXR(1.-OXMIC)]
cBOC2	Conversion factor of biological oxygen consumption per 2 days to the one per 1 day. (BOC2 to BOC1). u: − − 0.6 − BOC2(I) = (rrsF + rrsG + rsE + rsM*corWAS + rsH)*corCOr/cBOC2

coFLeps	Correction of light attenuation (EPS) for high turbidity on on tidal flats.
	u: − − 0.25 −
	CORR = {MIN[1.,DEPFOT/depchFH(I)]*[1.-pwafl(I)]*OVERLAP
	      + MIN[1.,DEPFOT/depchLF(I)]*(SUNQ-OVERLAP)
	      + MIN[1.,coFLeps*DEPFOT/depfh(I)]*pWAFL(I)*OVERLAP}/24.

CONVCHL	Conversion factor chlorophyll-a to phytocarbon. u: − − 35 − CHLa = PHYT(I)/CONVCHL
corCOp	Conversion factor to calculate the oxygen produced by primary producers from the amount of carbon fixed. u: $gO_2 \cdot m^{-3} \cdot (mgC \cdot m^{-3})^{-1}$ 0.00267 0.00357 0.05 prodOX = (asF + asG)*corCOp

corCOr   Conversion factor to calculate the oxygen consumption from the amount of carbon respired.
u: $gO_2 \cdot m^{-3} \cdot (mgC \cdot m^{-3})^{-1}$ 0.0026   0.00333   0.00333
BOC2(I) = (rrsF + rrsG + rsE + rsM*corWAS + rsH)*corCOr/cBOC2

cVOLI   Fractional volume cleared by 1 mgC of carnivores (CARN).
u: $m^3 \cdot mgC^{-1}$ – 0.01 –
pVOLI = cVOLI*CARN(I)*zsI*ztI

dsedM   Pelagic bacteria concentration (PBAC) where the relative sedimentation rate doubles. Under circumstances with high pelagic bacterial biomass the percentage of attached and floc-forming bacteria increases and hence also the relative sedimentation of bacteria.
u: $mgC \cdot m^{-3}$ 20   250   500
M3 = sedm*[1. + PBAC(I)/dsedM]*PBAC(I)*[pWAFL(I)*ptWET(I)]

DTIM(14)   Series of day numbers used as x-axis for the variables EDP3(14) and EPOC(14) in the TIMSER function.
u: –   see Appendix C-3
EMSPDROC = TIMSER(DTIM,EDP3,14)

EDP3(14)   Time series for pelagic dissolved organic carbon (PDROC) on the Ems at 0 sal.
u: $mgC \cdot m^{-3}$ see Appendix C-3
EMSPDROC = TIMSER(DTIM,EDP3,14)

eff4H   Assimilation efficiency of mesozooplankton (PCOP) for suspended benthic diatoms (BDIA).
u: –   – 0.6 –
utilHf = ... + eff4H*SUSBDIA + ...
u4H = eff4H*fl4H*puw   exuH = (1.-effuH)*fluH

effCH   Assimilation efficiency of mesozooplankton (PCOP) for pelagic detritus.
u: –   – 0.005 –
utilHf = ... + effCH*frCH + ...
uCH = effCH*flCH   exCH = (1.-effCH)*flCH

effE   Assimilation efficiency of microzooplankton (PMIC).
u: –   – 0.3 –
rsE = rs12E*q10E*PMIC(I) + uE*(1.-effE)*(1.- pexE)
exE = uE*(1.-effE)* pexE

effEH   Assimilation efficiency of mesozooplankton (PCOP) for microzooplankton (PMIC).
u: –   – 0.8 –
utilHr = ... + effEH*PMIC(I) + ...
uEH = effEH*flEH*puw   exEH = (1.-effEH)*flEH*puw

effFH   Assimilation efficiency of mesozooplankton (PCOP) for flagellate phytoplankton (PFLAG).
u: –   – 0.4 –
utilHf = ... + effFH*pFH*PFLAG(I) + ...
uFH = effFH*flFH*puw   exFH = (1.-effFH)*flFH*puw

effGH   Assimilation efficiency of mesozooplankton (PCOP) for pelagic diatoms (PDIA).
u: –   – 0.6 –
utilHf = ... + effGH*PDIA(I) + ...
uGH = effGH*flGH*puw
exGH = (1.-effGH)*flGH*puw

effHH   Assimilation efficiency of mesozooplankton (PCOP) for mesozooplankton (PCOP)0.
u: –   – 0.85 –
utilHr = ... + effHH*pHH*PCOP(I) + ...
uHH = effHH*flHH*puw exHH = (1.-effHH)*flHH*puw

effI   Growth efficiency of pelagic carnivores (CARN).
u: –   – 0.1 –
uI = effI*upI   exI = (1-effI)*upI

effM   Assimilation efficiency of pelagic bacteria (PBAC) under saturated oxygen conditions.
                    u: –   0.1   0.3   0.7
                    rsM = [1.-effM*OXMIC-effMa*(1.-OXMIC)]*uM + rs12M*PBAC(I)*q10M

effMa   Assimilation efficiency of pelagic bacteria (PBAC) under low oxygen saturation conditions.
                    u: –   0.1   0.2   0.7
                    rsM = [1.-effM*OXMIC-effMa*(1.-OXMIC)]*uM + rs12M*PBAC(I)*q10M

EMSH(12)   Time series of the mesozooplankton (PCOP) biomass on the river Ems at 0–5 S. Boundary condition.
                    u: $mgC \cdot m^{-3}$ see Appendix C-3
                    EMSPCOP = TIMSER(TMAC,EMSH,12)

EPOC(12)   Time series of the concentration of particulate organic carbon (POC) at 0 S in the river Ems. Boundary condition.
                    u: $mgC \cdot m^{-3}$ see Fig. 4.15 and Appendix C-3
                    EMSPOC = TIMSER(DTIM,EPOC,12)

HTEMPE   The rise in temperature needed to double the temperature correction factor (q10E) of the microzooplankton (PMIC).
                    u: °C – 10 –
                    q10E = 2.**{[TEMP(I)-12.]/HTEMPE}

HTEMPM   The rise in temperature needed to double the temperature correction factor (q10M) for pelagic bacteria (PBAC).
                    u: °C 5 10 20
                    q10M = 2.**{[TEMP(I)-12.]/HTEMPM}

jTIM(26)   Series of day numbers used as x-axis for the variable SPD3(26) in the TIMSER function.
                    u: –   see Appendix C-3
                    SEAPDROC = TIMSER(jTIM,SPD3,26)

ksCO   Half-value constant in equation to calculate corWAS. Equation accomplishes that corWAS = 1 for OXMIC = 1 and corWAS = maxCO for OXMIC ≦ 0.001.
                    u: –   2   4   5
                    corWAS = maxCO*ksCO/(ksCO + OXMIC)

ksmfdE   Food concentration where microzooplankton (PMIC) has an uptake of half the maximal rate.
                    u: $mgC \cdot m^{-3}$ – 200 –
                    uE = rupE*foodE/(foodE + ksmfdE)*q10E*PMIC(I)

ksmfdHf   Food concentration where half the maximum food uptake of filter feeding mesozooplankton (PCOP) is reached.
                    u: $mgC \cdot m^{-3}$ – 300 –
                    FUPT1f = MAX(0.,[foodHf-minfdHf]/(ksmfdHf + foodHf-minfdHf)] *rupH* q10H

ksmfdHr   Food concentration where half the maximum food uptake of raptorial feeding mesozooplankton (PCOP) is reached.
                    u: $mgC \cdot m^{-3}$ – 20 –
                    FUPT1r = MAX(0.,[foodHr)/(ksmfdHr + foodHr)]*rupH*q10H

ksmfdI   The food concentration where predation by carnivores (CARN) is half the maximum.
                    u: $mgC \cdot m^{-3}$ – 30 –
                    uI = pVOLI*foodI*ksmfdI/(ksmfdI + foodI)

ksmoxE   Oxygen saturation where respiration rate of microzooplankton (PMIC) is reduced to one half of the potential respiration rate due to oxygen stress.
                    u: – – 0.25 –
                    miOX = MIN{1.,(1. + ksmoxE)*[OXMIC/(ksmoxE + OXMIC)]}

ksmoxH   Half-value constant, defining the fraction of oxygen saturation where half the population of mesozooplankton (PCOP) dies per day.
                    u: – – 1 –

	ZDQ = [ksmoxH/(OXMIC + ksmoxH)-ksmoxH/(1. + ksmoxH)] + q10H* morH
ksmoxM	Half-value constant for oxygen limitation equation for pelagic bacteria (PBAC).

u: – 0.001 0.01 0.2

miOXM = OXMIC/(OXMIC + ksmoxM)

KSPO4   Half-value constant for phosphate limitation of photo-synthesis of flagellates (PFLAG) and diatoms (PDIA).

u: μmol PO4·m^{-3} 20 100 200

miPO4 = PO4(I)/[PO4(I) + KSPO4]

KSSIL   Half-value constant for silicate limitation of photo-synthesis of diatoms (PDIA).

u: μmol Si·m^{-3} 500 1000 5000

miNU = MIN{miPO4,[SILIC(I)/(KSSIL + SILIC(I))]}

LIGHTMIN Minimum light energy at which optimal production is possible

u: J·cm^{-2}·h^{-1} – 8.1 –

DEPFOT = LOG[LIGHT/(LIGHTMIN*SUNQ)]/EPS(I)

maxCO   Constant in equation to calculate corWAS. Maximum value of CorWAS, reached at very low oxygen saturation values (OXMIC $\leq$ 0.001).

u: – 1 1.25 1.3

corWAS = maxCO*ksCO/(ksCO + OXMIC)

minfdHf Lower threshold concentration of food for filter-feeding mesozooplankton (PCOP).

u: mgC·m^{-3} 25 40 50

FUPT1f = MAX[0.,(foodHf-minfdHf)/(ksmfdHf + foodHf-minfdHf)] *rupH* q10H

morE   Relative microzooplankton (PMIC) mortality at 12 °C.

u: d^{-1} – 0.05 –

mortalE = [(1.-miOX)*morOXE + morE]*PMIC(I)

morH   Relative mesozooplankton (PCOP) mortality at 12 °C.

u: d^{-1} – 0.01 –

ZDQ = [ksmoxH/(OXMIC + ksmoxH)-ksmoxH/(1. + ksmoxH)] + q10H* morH

morM   Relative mortality of pelagic bacteria (PBAC) at 12 °C.

u: d^{-1} 0.01 0.2 0.3

flMD = morM*Q10M*PBAC(I)

morOXE   Relative microzooplankton (PMIC) mortality due to oxygen limitation.

u: d^{-1} – 0.25 –

mortalE = [(1.-zOX)*morOXE + morE]*PMIC(I)

p4G   The fraction of benthic diatoms (BDIA) in the upper 0.5 cm of the sediment which is resuspended.

u: d^{-1} – 0.15 –

TDIA(I) = PDIA(I) + CDIA(I)* p4G/depfa(I)*[pwafl(I)* ptWET(I)]

pAE   The fraction of refractory organic carbon (PROC) available for microzooplankton (PMIC).

u: – – 0.00004 –

foodE = ... + [pCE*PDET(I) + pAE*PROC(I)]*pSUSP(I) + ...

pAM   The fraction of particulate refractory organic carbon in the pelagic (PROC) available per day for aerobic bacterial (PBAC) mineralization.

u: d^{-1} 0.00005 0.0005 0.01

upAM = pAM*PROC(I)*pSUSP(I)

pBM   The fraction of dissolved refractory carbon in the pelagic (PDROC) available per day for aerobic bacterial (PBAC) mineralization.

u: d^{-1} 0.0001 0.0005 0.05

upBM = pBM*PDROC(I)

pCE   The fraction of pelagic detritus (PDET) available for microzooplankton (PMIC).
u: − − 0.008 −
foodE = ... + [pCE∗PDET(I) + pAE∗PROC(I)]∗pSUSP(I) + ...

pCH   The fraction of pelagic detritus (PDET) available for mesozooplankton (PCOP).
u: − − 0.5 −
foodHf = pFH∗PFLAG(I) + PDIA(I) + SUSBDIA
+ pSUSP(I)∗ pCH∗PDET(I)

pCM   The fraction of pelagic detritus (PDET) available per day for aerobic bacterial (PBAC) mineralization.
u: d^{-1} 0.001 0.01 0.1
upCM = pCM∗PDET(I)∗pSUSP(I)

pDE   The fraction of labile organic carbon (PLOC) available for microzooplankton (PMIC).
u: − − 0 −
foodE = pDE∗PLOC(I) + ...

pDM   The fraction of labile organic carbon (PLOC) available per day for aerobic bacterial (PBAC) mineralization.
u: d^{-1} 0.1 1 1
upDM = flFD + flGD + flED + PLOC(I)∗pDM

pDWB(8)   Time series of the fraction of dissolved refractory organic carbon (PDROC) in the total dissolved organic carbon in the Westerwoldsche Aa. Based on field data of 1979–1980.
u: − see Appendix C-3
frBWWA = TIMSER(tpDW,pDWB,8)

pED   The fraction of microzooplankton (PMIC) excretion going to labile organic carbon (PLOC).
u: − − 0.5 −
flED = (exE + mortalE)∗pED

pEE   Cannibalizable fraction available for microzooplankton (PMIC).
u: − − 0.2 −
foodE = ... + pEE∗PMIC(I) + ...

pexE   Excreted fraction of the microzooplankton (PMIC) uptake.
u: − − 0.5 −
exE = uE∗(1.-effE)∗ pexE

pFD   The fraction of assimilated carbon excreted by flagellates (PFLAG) into labile organic carbon (PLOC).
u: − − 0.1 −
flFD = LOCF-flFC + asF∗pFD∗miPO4

pFdet   Additional excretion by flagellates (PFLAG) as a maximum fraction of the assimilated carbon under nutrient limitation. Nutrient limitation gives rise to an enhanced excretion rate.
u: − 0.3 0.9 0.7
LOCF = asF∗pFdet∗(1.-miPO4)

pFE   The fraction of flagellate phytoplankton (PFLAG) available for the microzooplankton (PMIC).
u: d^{-1} − 0.4 −
foodE = ... + pFE∗PFLAG(I) + ...

pFH   The fraction of flagellates (PFLAG) available for mesozooplankton (PCOP). Not all flagellates can be eaten by mesozooplankton: they may be too small (nanno plankton) or too large (colonial forms of Phaeocystis) to be eaten or filtered.
u: − 0.2 0.5 0.8
utilHf = effFH∗pFH∗PFLAG(I) + effGH∗PDIA(I) + eff4H∗SUSBDIA

pfloc8   The fraction of sedimenting labile organic carbon (PLOC) not transformed into detritus (BDET).

pGD

u: − − 0.25 −
ADLOC(1) = pfloc8*FLOC
The fraction of assimilation which is excreted by diatoms (PDIA) into labile organic carbon (PLOC).
u: − 0.05 0.1 0.2
flGD = LOCG-flGC + pGD*asG*miNU

pGdet

Additional excretion by diatoms (PDIA) as a maximum fraction of the assimilated carbon under nutrient limitation. Nutrient limitation gives rise to an enhanced excretion rate.
u: − 0.7 0.5 0.9
LOCG = asG*pGdet*(1.-miNU)

pGE

The fraction of diatoms (PDIA) available for microzooplankton (PMIC).
u: − − 0.8 −
foodE = ... + pGE*PDIA(I) + pGE*SUSBDIA + ...

PH(8)

Time series of the flagellate biomass (PFLAG) at sea (boundary condition).
u: $mgC \cdot m^{-3}$ see Appendix C-3
SEAPFLAG = TIMSER(TMAC,PH,12)

pHC

The fraction of mesozooplankton (PCOP) faecal pellets going into PDET (remainder to PROC).
u: − − 0.5 −
flHA = exH*(1.-pHC)

pHH

Cannibalizable fraction of mesozooplankton (PCOP). Only nauplii are assumed to be small enough to be cannibalized. The fractional biomass of nauplii is taken to be a third of the total biomass.
u: − − 0.333 −
foodHr = PMIC(I) + pHH*PCOP(I)

pM8s

The fraction of the sedimentation of pelagic bacteria (PBAC) going to the labile organic carbon pool in the benthos (BLOC).
u: − 0.01 0.9 0.8
flM8 = M3*pM8s

PMAXPD

Relative primary productivity when RTPD and CORR both have a value of 1.0.
u: d^{-1} 0.5 3.77 2
asG = PDIA(I)*PMAXPD*RTPD*CORR

PMAXPH

Relative primary productivity when RTPH, miSAL and CORR all have a value of 1.0.
u: d^{-1} 0.5 3.77 4
asF = PFLAG(I)*PMAXPH*RTPH*miSAL*CORR

pME

The fraction of pelagic bacteria (PBAC) available for the microzooplankton (PMIC).
u: − − 1 −
foodE = ... + pME*PBAC(I)

PMS(20)

Time series for phosphate (PO4) on the Ems at 0 sal.
u: $\mu mol \cdot m^{-3}$ see Fig. 4.11 b and Appendix C-3
EMSPO4 = TIMSER(TNUA,PMS,20)

PO4CO

Constant linking the phosphate uptake to the net assimilation of flagellates (PFLAG) and diatoms (PDIA).
u: $\mu mol\ PO4 \cdot m^{-3} \cdot (mgC \cdot m^{-3})^{-1}$ 0.50 0.79 1
flKG = PO4CO*max[0.,(ppG-rsG)]

pPeA(10)

Time series for the fraction of particulate refractory organic carbon (PROC) in particulate organic carbon (POC) on the Ems.
u: −   see Appendix C-3
frAems = TIMSER(tPeA,pPeA,10)

pPOC

Particulate organic carbon (POC) as a fraction of total organic carbon (WWATOC) in the Westerwoldsche Aa.
u: − − 0.07 −
WWAPOC = pPOC*WWATOC

pPsA(6)   Time series of the fraction of particulate refractory organic carbon (PROC) in detrital particulate organic carbon at sea (boundary condition). Based on field data from 1979–1980.
u: −   see Fig. 4.17 and Appendix C-3
SEAPROC = SEAdeadet∗TIMSER(tPsA,pPsA,6)

pPWA(8)   Time series of the fraction of particulate refractory organic carbon (PROC) in detrital particulate organic carbon in the Westerwoldsche Aa (boundary condition). Based on field data from 1979–1980.
u: −   see Appendix C-3
frAWWA = TIMSER(tpPW,pPWA,8)

pPWC(8)   Time series of the fraction of particulate detritus (PDET) in detrital particulate organic carbon in the Westerwoldsche Aa (boundary condition). Based on field data from 1979–1980.
u: −   see Appendix C-3
frCWWA = TIMSER(tpPW,pPWC,8)

pRSF   The fraction of the assimilation used for activity respiration by flagellates (PFLAG).
u: −   0.1   0.1   0.3
arsF = pRSF∗asF

pRSFA   Maximum fraction of the assimilation respired under absolute nutrient stress by flagellates (PFLAG).
u: −   − 0.9 −
srsF = LOCF∗pRSFA

pRSG   The fraction of the assimilation used for activity respiration by pelagic diatoms (PDIA).
u: −   0.05   0.1   0.20
arsG = pRSG∗asG

pRSGA   Maximum fraction of the assimilation respired under absolute nutrient stress by pelagic diatoms (PDIA).
u: − − 0.9 −
srsG = LOCG∗pRSGA

PSE(13)   Time series for phosphate (PO4) at sea.
u: $\mu mol \cdot m^{-3}$ see Fig. 4.11 b and Appendix C-3
SEAPO4 = TIMSER(TNUC,PSE,13)

PWA(14)   Time series for phosphate (PO4) on the Westerwoldsche Aa.
u: $\mu mol \cdot m^{-3}$ see Fig. 4.11 b and Appendix C-3
WWAPO4 = TIMSER(TNUB,PWA,14)

q10PD   Constant which defines the change in the temperature correction function rtempG for diatoms (PDIA) with changes in temperature (rtempG = 1 at 12 °C).
u: $°C^{-1}$ 0.10   0.15   0.20
rtempG = 2.∗∗{[TEMP(I)-12.]∗q10PD}

q10PH   Constant which defines the change in the temperature correction function rtempF for flagellates (PFLAG) with changes in temperature (rtempF = 1 at 12 °C).
u: $°C^{-1}$ 0.1   0.2   0.25
rtempF = 2.∗∗{[TEMP(I)-12.]∗q10PH}

qrsHf   Activity respiration of filter-feeding mesozooplankton (PCOP). The activity respiration is expressed as carbon respired per m^3 filtered seawater.
u: $mgC \cdot m^{-3}$ − 40 −
arsHf = qrsHf∗RFVf

qrsHr   Activity respiration of raptorial-feeding mesozooplankton (PCOP). The activity respiration is expressed as carbon respired per m^3 seawater cleared of prey.
u: $mgC \cdot m^{-3}$ − 20 −
arsHr = qrsHr∗RFVr

rABC   The fraction of pelagic dissolved refractory organic carbon which goes over to
             particulate material due to processes which take place between 0 salinity and
             compartment 3.
             u: – – 0.5 –
             flABCems = rABC*EMSPOC

rs0F   Relative respiration rate of flagellates (PFLAG) at 12 °C.
             u: d^{-1} 0.01 0.03 0.05
             rrsF = PFLAG(I)*RTPH*rs0F

rs0G   Relative respiration rate of diatoms (PDIA) at 12 °C.
             u: d^{-1} 0.005 0.02 0.05
             rrsG = PDIA(I)*RTPD*rs0G

rs0I   Relative respiration rate of carnivores (CARN) at 0 °C.
             u: d^{-1} – 0.01 –
             RESCAR = rs0I*EXP[0.069*TEMP(I)]

rs12E   Relative respiration rate of microzooplankton (PMIC) at 12 °C.
             u: d^{-1} – 0.1 –
             rsE = rs12E*q10E*PMIC(I) + uE*(1.-effE)*(1.- pexE)

rs12H   Relative respiration rate of mesozooplankton (PCOP) at 12 °C.
             u: d^{-1} – 0.0125 –
             rrsH = rs12H*q10H

rs12M   Relative respiration rate of pelagic bacteria (PBAC) at 12 °C.
             u: d^{-1} 0.05 0.2 0.2
             rsM = ... + rs12M*PBAC(I)*q10M

RTMPD   Flattening coefficient in the equation to calculate temperature effect on rate
             constants of diatoms (PDIA).
             u: – 2 3.2 5
             RTPD = RTMPD*rtempG/(RTMPD + rtempG-1)

RTMPH   Flattening coefficient in the equation to calculate the temperature effect on rate
             constants of flagellates (PFLAG).
             u: – 5 25.9 15
             RTPH = RTMPH*rtempF/(RTMPH + rtempF-1)

rupE   Relative maximum food uptake of microzooplankton (PMIC) at 12 °C. This
             uptake is only realized under optimal food conditions.
             u: d^{-1} 0.46 2 5.88
             uE = rupE*foodE/(foodE + ksmfdE)*q10E*PMIC(I)

rupH   Relative maximum food uptake rate of mesozooplankton (PCOP) at 12 °C.
             u: d^{-1} – 0.8 –
             FUPT1f = MAX[0.,(foodHf-minfdHf)/(ksmfdHf + foodHf-minfdHf)] *rupH*
             q10H

rupM   Relative potential uptake of pelagic aerobic bacteria (PBAC) at 12 °C under op-
             timal conditions.
             u: d^{-1} 1 6.3 8
             upMmax = rupM*q10M*miOXM*PBAC(I)

SALCAR   Salinity which is limiting to ctenophores (CARN).
             u: S – 12 –
             zsI = MAX{0.,[SALT(I)-SALCAR]/[SALCAR + SALT(I)]}*3.

SALTM   Salinity where the salt limitation for growth of flagellates (PFLAG) reaches 0.5.
             No flagellates, especially Phaeocystis, are observed in the inner compartments.
             Hence a salinity-dependent growth limitation was introduced.
             u: S 5 10 15
             miSAL = SALT(I)/[SALT(I) + SALTM]

SDI(12)   Time series for pelagic diatoms (PDIA) at sea.
             u: mgC·m^{-3} see Appendix C-3
             SEAPDIA = TIMSER(TMAC,SDI,12)

sedM   Relative sedimentation of pelagic bacteria (PBAC) at normal PBAC concentra-
             tions. Bacteria attached to detritus and floc-forming bacteria sediment to the
             benthos as living (to benthic aerobic bacteria: BBAC) or dying bacteria (to
             benthic labile organic carbon: BLOC).

	u: d^{-1} 0.001 0.22 0.3
	M3 = sedm*[1. + PBAC(I)/dsedM]*PBAC(I)*[pWAFL(I)*ptWET(I)]
SEDPDQ	Net relative sedimentation rate of pelagic diatoms (PDIA) at 12 °C from a water column of 1 m. A sedimentation of diatoms from the water to the tidal flats is assumed, which is dependent on temperature and on the level of the tidal flats.
	u: m·d^{-1}·(m^{-1}) 0.01 0.05 0.25
	R = SEDPDQ*PDIA(I)*RTPD*[pWAFL(I)*ptWET(I)]
SILCO	Constant linking the silicate uptake to the net carbon assimilation of diatoms (PDIA). Silicate uptake is temperature-dependent.
	u: μmolSi·m^{-3}·(mgC·m^{-3})$^{-1}$ 15 30.8 40
	fILG = R*SILCO*[1.-0.033*TEMP(I)]
SLOP	Fraction of uptake by mesozooplankton (PCOP) lost due to breakup of particles.
	u: − − 0.1 −
	puW = (1.-SLOP)
	TOTSLOP = SLOP*(flFH + flGH + flEH + flHH + flfl4H)
SMS(20)	Time series for silicate (SILIC) on the Ems at 0 sal.
	u: μmol·m^{-3} see Fig. 4.11 a and Appendix C-3
	EMSSILIC = TIMSER(TNUA,SMS,20)
SPD3(26)	Time series pelagic dissolved refractory organic carbon (PDROC) at sea.
	u: mgC·m^{-3} see Appendix C-3
	SEAPDROC = TIMSER(jTIM,SPD3,26)
SPOC(12)	Time series of particulate organic carbon (POC) at sea.
	u: mgC·m^{-3} see Appendix C-3
	SEAPOC = TIMSER(TMAC,SPOC,12)-SEAPFLAG-SEAPDIA
SSE(13)	Time series for silicate (SILIC) at sea.
	u: μmol·m^{-3} see Fig. 4.11 a and Appendix C-3
	SEASILIC = TIMSER(TNUC,SSE,13)
SWA(14)	Time series for silicate (SILIC) on the Westerwoldsche Aa
	u: μmol·m^{-3} see Fig. 4.11 a and Appendix C-3
	WWASILIC = TIMSER(TNUB,SWA,14)
swit4pel	Switch (1 = on, 0 = off) for feeding on resuspended diatoms (SUSBDIA). Introduced for sensitivity analysis.
	u: − − 0 −
	SUSBDIA = swit4pel*p4G/depfa(I)*[pwafl(I)*ptWET(I)] *CDIA(I) + 1.E-20
TCAR(12)	Time series for carnivorous zooplankton (CARN) at sea.
	u: mgC·m^{-3} see Appendix C-3
	SEACARN = TIMSER(TMAC,TCAR,12)
TEMCAR	Temperature below which carnivore processes (CARN) slow down.
	u: °C − 6 −
	ztI = MAX[0.,TEMP(I)/TEMCAR]
TMAC(12)	Series of day numbers used as x-axis for the variables SBI(12), pH(12), BCOP(12), TCAR(12), EMSH(12) and SPOC(12) in the TIMSER function.
	u: − see Appendix C-3
	SEAPOC = TIMSER(TMAC,SPOC,12)-SEAPFLAG-SEAPDIA
TNUA(20)	Series of day numbers used as x-axis for the variables PMS(20) and SMS(20) in the TIMSER function.
	u: − see Appendix C-3
	EMSSILIC = TIMSER(TNUA,SMS,20)
TNUB(14)	Series of day numbers used as x-axis for the variables PWA(14) and SWA(14) in the TIMSER function.
	u: − see Appendix C-3
	WWAPO4 = TIMSER(TNUB,PWA,14)
TNUC(13)	Series of day numbers used as x-axis for the variables PSE(13) and SSE(13) in the TIMSER function.
	u: see Appendix C-3
	SEAPO4 = TIMSER(TNUC,PSE,13)

tpDW(8)   Series of day numbers used as x-axis for the variable pDW(8) in the TIMSER function.
u: –   see Appendix C-3
frBWWA = TIMSER(tpDW,pDWB,8)

tPeA(10)   Series of day numbers used as x-axis for the variable pPeA(10) in the TIMSER function.
u: –   see Appendix C-3
frAems = TIMSER(tPeA,pPeA,10)

tpPW(8)   Series of day numbers used as x-axis for the variables pPWA(8) and pPWC(8) in the TIMSER function.
u: –   see Appendix C-3
frCWWA = TIMSER(tpPW,pPWC,8)

TWAS(87)   Series of day numbers used as x-axis for the variable WTOC(87) in the TIMSER function.
u: –   see Appendix C-3
WWATOC = TIMSER(TWAS,WTOC,87)

WTOC(87)   Time series for total organic carbon (TOC) on the Westerwoldsche Aa.
u: $mgC \cdot m^{-3}$ see Fig.4.9 and Appendix C-3
WWATOC = TIMSER(TWAS,WTOC,87)

xfdI   Exponent regulating the apportioning between the different food sources for the pelagic invertebrate carnivores (CARN).
u: –   – 2 –
fdI = PBLAR(I)**xfdI + PCOP(I)**xfdI + [0.5*PMIC(I)]**xfdI   foodI = fdI** (1./xfdI)

ZZSILT   Relative attenuation coefficient of silt. A relative attenuation coefficient of silt is used for the calculation of the vertical attenuation coefficient (EPS), in combination with a constant background (0.4) and a minor contribution of the phytoplankton.
u: $(m \cdot mg \cdot m^{-3})^{-1}$ $0.03 \cdot 10^{-3}$ $0.04 \cdot 10^{-3}$ $0.05 \cdot 10^{-3}$
EPS(I) = 0.4 + ZZSILT*SILT(I) + 0.293E-3*PHYT(I)

Appendix A-3: Parameters of the Benthic Submodel

aex4   Activity excretion rate of benthic diatoms (BDIA). Excreted fraction of the actual primary production of benthic diatoms.
u: –   – 0.1 –
W48 = UDIA*rex4*zt4 + PP*aex4 + cPP*bex4*(1-p42)*PP

ALF0   Diffusion enhancement constant for the physical process of mixing of the sediment by wave action.
u: –   0.02   0.7   2
AMIX1 = ALF0*zMIX*zsm**2 + ALFt*TPUMP(I)

ALF5   Diffusion amplification constant per unit active biomass of the meiobenthos (BMEI).
u: $(mgC \cdot m^{-2})^{-1}$ $2 \cdot 10^{-5}$ $7 \cdot 10^{-3}$ –
act5 = BMEI(I)*zt5*zFOOD5*ALF5*zSM**2

ALF6   Diffusion amplification constant per unit active biomass of the deposit feeders (BBBM).
u: $(mgC \cdot m^{-2})^{-1}$ 0.0001   0.014   0.01
act6 = [DIFF6*ptWET(I) + BBBM(I)*cBUR6]*ALF6

ALF7   Diffusion amplification constant per active biomass of the filter feeders (BSF).
u: $(mgC \cdot m^{-2})^{-1}$ $2 \cdot 10^{-5}$ $1.4 \cdot 10^{-3}$ $2 \cdot 10^{-3}$
act7 = BSF(I)*zt7a*ptWET(I)*ALF7*CIND(I)*zOX7* zFOOD7*COR2

ALFt   Diffusion amplification constant for the process of tidal pumping.
u: $(cm \cdot d^{-1})^{-1}$ – 0.7 –
AMIX1 = ALF0*zMIX*zsm**2 + ALFt*TPUMP(I)

ALFY   Diffusion amplification constant per unit biomass of the mesoepibenthos (EMES).
              u: $(mgC \cdot m^{-2})^{-1}$ 0.0001 0.0085 0.01
              actY = EMES(I)*cBUR*ALFY

ardBAC   Fraction of benthic detritus (BDET) uptake respired by aerobic bacteria (BBAC).
              u: – 0.5 0.65 0.8
              W30 = W83*arlBAC + W23*ardBAC + W13*arrBAC + BBAC(I)*rrsBAC*zt3

ardBACA   Fraction of benthic detritus (ADET) uptake respired by anaerobic bacteria (ABAC).
              u: – 0.6 0.75 0.8
              W30A(I) = W83A*arlBACA + W23A*ardBACA + W13A*arrBACA + ABAC(I) *rrsBACA*zt3A

arlBAC   Fraction of labile organic carbon (BLOC) uptake respired by aerobic bacteria (BBAC).
              u: – 0.3 0.5 0.7
              W30 = W83*arlBAC + W23*ardBAC + W13*arrBAC + BBAC(I)*rrsBAC*zt3

arlBACA   Fraction of labile organic carbon (ALOC) uptake respired by anaerobic bacteria (ABAC).
              u: – 0.6 0.65 0.7
              W30A(I) = W83A*arlBACA + W23A*ardBACA + W13A*arrBACA + ABAC(I) *rrsBACA*zt3A

arrBAC   Fraction of benthic refractory organic carbon (BROC) uptake respired by aerobic bacteria (BBAC).
              u: – 0.7 0.9 0.9
              W30 = W83*arlBAC + W23*ardBAC + W13*arrBAC + BBAC(I)*rrsBAC*zt3

arrBACA   Fraction of benthic refractory organic carbon (BROC) uptake respired by anaerobic bacteria (ABAC).
              u: – 0.7 0.9 0.9
              W30A(I) = W83A*arlBACA + W23A*ardBACA + W13A*arrBACA + ABAC(I) *rrsBACA*zt3A

ars4   Activity respiration of the gross primary production.
              u: – 0.05 0.1 0.15
              W40a = PP*ars4 + cPP*brs4*PP

asWORK5   Relative respiration rate due to feeding activity by meiobenthos (BMEI) under standard conditions.
              u: d^{-1} – 0.02 –
              W50 = W50r + W5*prs5 + FACTOR5*asWORK5

asWORK6   Relative respiration rate due to feeding activity by deposit feeders (BBBM) under standard conditions.
              u: d^{-1} – 0.009 –
              W60 = W60r + W6*prs6*(1-PEX6A) + asWORK6*FACTOR6

AZMIX   Constant limiting the amplitude of the wind influence on sediment mixing (zMIX) and sedimentation (zSED).
              u: – 0.1 0.4 0.5
              zMIX = 1. + AZMIX*COS[PI/180.*(TIME + 30.)]
              zSED = 2.-zMIX

BAL0   Reference depth of the sulphide horizon, used for calculating the potential labile organic carbon (BLOC) uptake by the aerobic bacteria (BBAC).
              u: cm 1 3 5.
              zSM = BAL0/[BAL0 + BAL(I)]

bex4   Stress excretion and stress-mortality, associated with self-limitation of photosynthesis; a fraction of the difference between potential (PPR) and realized photosynthesis (PP).
              u: – – 0.2 –
              W48 = UDIA*rex4*zt4 + PP*aex4 + cPP$_{s}$-2bex4*(1-p42)*PP

BIOAM(5)   Compartment dependent enhancement of the apparent diffusion due to bioturbation.

u: – see Appendix Table A-3.1
BIOTUR = BTL*BIOAM(I)*DIFF6

brs4 Stress respiration, associated with self-limitation of photosynthesis; a fraction of the difference between potential and realized photosynthesis.
u: – -0.2 –
W40a = PP*ars4 + cPP*brs4*PP

BSL Relative feeding rate of deposit feeders (BBBM) above the sulphide horizon.
u: $cm \cdot d^{-1} \cdot (mgC \cdot m^{-2})^{-1}$ $-0.03 \cdot 10^{-3}$ –
FOOD6 = [BMEI(I) + UDIA + BBAC(I)]/BAL(I)*BSL + [LDIA + ABAC(I)]/
L BBBM*BTL

BTL Relative feeding rate of deposit feeders (BBBM) below the sulphide horizon.
u: $cm \cdot d^{-1} \cdot (mgC \cdot m^{-2})^{-1}$ $-0.01 \cdot 10^{-3}$ –
FOOD6 = [BMEI(I) + UDIA + BBAC(I)]/BAL(I)*BSL + [LDIA + ABAC(I)]/
LBBBM*BTL

CBOF(5) The depth in the sediment above which half of the total biomass of benthic diatoms (BDIA) occurs.
u: cm see Appendix Table A-3.1
CBO = CBOF(I)*BDIA(I)**2/[BDIA(I)**2 + kCBO**2]

cBUR6 Correction factor for diffusion enhancement due to the presence of deposit feeders (BBBM).
u: – -0.1 –
act6 = [DIFF6*ptWET(I) + BBBM(I)*cBUR6]*ALF6

cBURY Correction factor for diffusion enhancement due to the presence of mesoepibenthos (EMES).
u: – -1 –
cBUR = cBURY*zsm**   2 actY = EMES(I)*cBUR*ALFY

CIND(5) Factor to correct weight-specific rates for the differences in average size of suspension feeders (BSF) in the different compartments.
u: – see Appendix Table A-3.1
act7 = BSF(I)*zt7a*ptWET(I)*ALF7*CIND(I)*zOX7*zFOOD7*COR2

corCOp Conversion factor of produced carbon to oxygen
u: – -0.00357 –
PPox = PP*corCOr/corCOp-W40a

corCOr Conversion factor of respired carbon to oxygen
u: – -0.00333 –
PPox = PP*corCOr/corCOp-W40a

cSPAWN Parameter to obtain the correct shape of the spawning curve of macrobenthos.
u: – -5 –
zSPAWN = cSPAWN**4*
{0.75/[cSPAWN**2 + (TIME-100)**2]**2 +   0.25/[cSPAWN**2 + (TIME-250)**2]**2}/(PI2*cSPAWN)

ct0 Coefficient to calculate the temperature correction factor for molecular diffusion.
u: $(°C)^{-1}$ -0.0335 –
zt0 = EXP(ct0*TASED)

ct3m Coefficient giving the theoretical maximum for the temperature correction factors zt3 and zt3A in bacterial processes.
u: – -5 –
zt3 = ct3m*zt3/(ct3m-1. + zt3)

ct3s Slope coefficient used in calculating the temperature correction factors zt3 and zt3A. The value of the slope coefficient is also connected with the theoretical maximum coefficient (ct3m).
u: – -4 –
zt3 = ct3s**(TA/10.)

DAMP Maximum daily shift of the location of the sulphide horizon (BAL). DL is the calculated shift and is limited by DAMP to a maximum.
u: $cm \cdot d^{-1}$ – 0.1 0.2

DIFF

DL = [DIFFX*OXO-(BOXDIF+SOEQ)*BAL(I)]/ {BOXDIF
+SOEQ*[1.+BAL(I)/DAMP]}
Diffusion coefficient for molecular diffusion of oxygen and sulphide. The molecular diffusion coefficient is enhanced by bioturbation, storm mixing and tidal pumping (AMIX).
u: $m^2 \cdot d^{-1}$ 0.26 1.73 14.7
DIFFX = DIFF*zt0*AMIX

exdBAC

Fraction of benthic detritus (BDET) uptake excreted by aerobic bacteria (BBAC).
u: – 0.01 0.05 0.15
W38 = W83*exlBAC + W23*exdBAC + W13*exrBAC + mor3*(1.-pm32)

exdBACA

Fraction of benthic detritus (ADET) uptake excreted by anaerobic bacteria (ABAC).
u: – 0.01 0.05 0.12
W38A = W83A*exlBACA + W23A*exdBACA + W13A*exrBACA + mor3A*(1.-pm32)

exlBAC

Fraction of labile organic carbon (BLOC) uptake excreted by aerobic bacteria (BBAC).
u: – 0.01 0.1 0.2
W38 = W83*exlBAC + W23*exdBAC + W13*exrBAC + mor3*(1.-pm32)

exlBACA

Fraction of labile organic carbon (ALOC) uptake excreted by anaerobic bacteria (ABAC).
u: – 0.01 0.1 0.15
W38A = W83A*exlBACA + W23A*exdBACA + W13A*exrBACA + mor3A*(1.-pm32)

exrBAC

Fraction of benthic refractory organic carbon (BROC) uptake excreted by aerobic bacteria (BBAC).
u: – 0.001 0.05 0.15
W38 = W83*exlBAC + W23*exdBAC + W13*exrBAC + mor3*(1.-pm32)

exrBACA

Fraction of the benthic refractory organic carbon (BROC) uptake excreted by anaerobic bacteria (ABAC).
u: – 0.001 0.05 0.12
W38A = W83A*exlBACA + W23A*exdBACA + W13A*exrBACA + mor3A*(1.-pm32)

F5

Relative food uptake of meiobenthos (BMEI) at a defined food concentration (STANFO5).
u: d^{-1} – 0.075 –
zFOOD5 = 2*F5/(PupMEI + F5)

F6

Relative food uptake of deposit feeders (BBBM) at a defined food concentration.
u: d^{-1} – 0.03 –
zFOOD6 = 2*F6/(FOOD6 + F6)*zox6

FACT7

Relative respiration rate due to feeding activity by filter feeders (BSF) under standard conditions.
u: d^{-1} – 0.01 –
W70a = prs7*(1.-pex7)*FFIL + FACT7*FACTOR7*zFOOD7L*zFOOD7

FLSO

Fraction of acid-soluble sulphide (BSUL)) present as diffusible free sulphide.
u: – 0.001 0.05 0.1
SOEQ = FLSO*BSUL(I)*DIFFY*3./[TAL-BAL(I)]**2

FLSP

Relative pyrite formation rate.
u: d^{-1} 0.0001 0.0004 0.01
SP = FLSP*BSUL(I)

FRBD

Fraction of benthic detritus (BDET) available per day to aerobic bacteria (BBAC).
u: d^{-1} 0.0005 0.01 0.02
W23 = BDET(I)*FRBD*zt3

FRBDA Fraction of benthic detritus (ADET) available per day to anaerobic bacteria (ABAC).
u: d^{-1} 0.0005   0.01   0.02
$W23A = ADET(I)*FRBDA*zt3A$

FRBR Fraction of benthic refractory organic carbon (BROC) available per day to aerobic bacteria (BBAC).
u: d^{-1} $1 \cdot 10^{-6}$ $2 \cdot 10^{-4}$ $5 \cdot 10^{-4}$
$W13 = BROC(I)*FRBR*[BAL(I)/TAL]*zt3$

FRBRA Fraction of benthic refractory organic carbon (BROC) available per day to anaerobic bacteria (ABAC).
u: d^{-1} $1 \cdot 10^{-6}$ $2 \cdot 10^{-5}$ $5 \cdot 10^{-5}$
$W13A = BROC(I)*FRBRA*\{[TAL-BAL(I)]/TAL\}*zt3A$

KCBO The biomass at the depth CBO, where CBO is equal to 0.5*CBOF.
u: $mgC \cdot m^{-2} \cdot cm^{-1}$ – 1500 –
$CBO = CBOF*BDIA(I)/[BDIA(I)+kCBO]$

KD(5) Coefficient of light extinction in the top layer of sediment. Dependent mainly on the particle size distribution and organic carbon content of the sediment. Fixed values of KD, typical for compartments 1–5 are used.
u: cm^{-1} see Appendix Table A-3.1
$LALL = LOG[MAX(1.,PHOT/PHOTMIN)]/KD(I)$

ksBAC Concentration of aerobic bacteria (BBAC) where density-dependent rate limitation is 0.5
u: $mgC \cdot m^{-2} \cdot cm^{-1}$ 2000
$zBAC = ksBAC/[ksBAC + BBAC(I)/BAL(I)]$

ksBACA Concentration of anaerobic bacteria (ABAC) where density-dependent rate limitation is 0.5
u: $mgC \cdot m^{-2} \cdot cm^{-1}$ – 10000 –
$zBACA = ksBACA/[ksBACA + ABAC(I)/BAL(I)]$

LAD(5) Thickness of the sediment layer, just underneath the illuminated sediment layer (LALL) from which the diatom population is able to migrate into the illuminated layer (LALL).
u: $cm \cdot (h\ light)^{-1}$ see Appendix Table A-3.1
$LAL = LALL + LAD(I)*OVERLAP$

LBBBM Thickness of the layer, below the sulphide horizon exploited by deposit feeders (BBBM).
u: cm – 10 –
$FOOD6 = [BMEI(I) + UDIA + BBAC(I)]/BAL(I)*BSL + [LDIA + ABAC(I)]/LBBBM*BTL$

morBAC Relative mortality of aerobic bacteria (BBAC).
u: d^{-1} 0.110   0.035   0.03
$mor3 = morBAC*BBAC(I)*zt3*zmor3$

morBACA Relative mortality of anaerobic bacteria (ABAC).
u: d^{-1} 0.001   0.015   0.03
$mor3A = morBACA*ABAC(I)*zt3A*zmor3A$

OCP Optimal concentration of particles for suspension feeders (BSF). At particle concentrations above OCP the filtration rate decreases.
u: m^{-3} $1 \cdot 10^{10}$ $2 \cdot 10^{10}$ $3 \cdot 10^{10}$
$zPAR = 2.*OCP/(2.*OCP+PART)$

p42 Fraction of the excretion of benthic diatoms (BDIA) going to benthic detritus (BDET).
u: – – 0.7 –
$W52 = W45*pex5*p42$

PARPHY Conversion constant of biomass to the number of phytoplankton cells. It is assumed that an average phytoplankton cell has $120 \cdot 10^{-9}$ mgC biomass
u: number of cells $\cdot (mgC)^{-1}$ – $8 \cdot 10^6$ –
$PART = PARPHY*PHYTO + \{SILT(I) + cCARASD*pSUSP(I)*[PROC(I)+PDET(I)]\}*PARSIL$

PARSIL   Factor to convert silt mass into the number of silt particles relevant for suspension feeders (BSF). This constant is based on three assumptions: (1) Specific gravity is 1.6 (range: 1.6–2.0); (2) mean diameter of particles is 10 μm, particles smaller than 8 μm or larger than 32 μm are not taken up by suspension feeders; (3) 20% of silt consists of particles between 8 and 32 μm.
u: number of particles $\cdot$ (mg silt)$^{-1}$ $1.0 \cdot 10^5$ $2.38 \cdot 10^5$ $9.0 \cdot 10^5$
PART = PARPHY*PHYTO + {SILT(I) + cCARASD*
pSUSP(I)*[PROC(I) + PDET(I)]}*PARSIL

pex5   Fraction of uptake by meiobenthos (BMEI) excreted as detritus (BDET).
u: – – 0.2 –
W52 = W45*pex5*p42

pex6   Fraction of the food uptake that is not assimilated, but excreted as faeces by deposit feeders (BBBM).
u: – – 0.25 –
FAEC = {pex6*[1 + FOOD6/(FOOD6 + F6)]}*W6

pex7   Fraction of the food uptake that is not assimilated, but excreted as faeces by suspension feeders (BSF).
u: – 0.1 0.2 1
W72 = pex7*FFIL

PHOTMIN   Minimum daily irradiance, permitting benthic diatoms (BDIA) to achieve at least 50% of their maximum photosynthetic production. Diatoms in the sediment receiving a higher irradiance than PHOTMIN were assumed not to be light-limited. The photosynthetic production of buried cells, receiving an irradiance less than PHOTMIN, is usually small and was hence neglected.
u: $J \cdot cm^{-2} \cdot d^{-1}$ – 5 –
LALL = LOG[MAX(1.,PHOT/PHOTMIN)]/KD(I)

pm32   Fraction of dead bacteria (BBAC and ABAC) going into benthic organic detritus (BDET and ADET).
u: – 0.1 0.2 0.3
W32 = mor3*pm32   W32A = mor3A*pm32

POW5   Exponent in food-limiting function zFOOD5L of meiobenthos (BMEI). POW5 determines the steepness of the increase of function zFOOD5L at the switch point.
u: – – 2 –
zFOOD5L = FOOD5X**POW5/(FOOD5X**POW5 + rsex5r**POW5)

POW6   Exponent in food-limiting function zFOOD6L of deposit feeders (BBBM). High values of POW6 cause a steep change of zFOOD6L near the switch point.
u: – – 8 –
zFOOD6L = FOOD6X**POW6/(REST6**POW6 + FOOD6X**POW6)

POWM   Exponent in the calculation of the available amount of food for meiobenthos (BMEI). If POWM > 1, POWM effects a shift towards the more abundant food sources.
u: – – 2 –
PW35 = [BBAC(I)/BALA]**POWM

PPH   Maximum photosynthetic rate attainable by benthic diatoms (BDIA).
u: $mgC \cdot m^{-2} \cdot d^{-1}$ – 1500 –
PPX = (PPR + PPH + PPM)/2
PP = PPX-sqrt(PPX*PPX-PPH*PPR)

PPM0   Slope constant for the decrease in photosynthetic rate of benthic diatoms (BDIA) due to self-inhibition caused by the insufficient diffusion of substrates.
u: $mgC \cdot m^{-2} \cdot d^{-1}$ – 400 –
PPM = PPMO + (1.-zox4)*PPMox

PPMox   Addition to the slope constant (PPM) to account for the decrease in photosynthetic rate of benthic diatoms (BDIA) caused by exposure to oxygen-depleted waste water.
u: $mgC \cdot m^{-2} \cdot d^{-1}$ – 800 –
PPM = PPM0 + (1.-zox4)*PPMox

prs5   Relative activity respiration of meiobenthos (BMEI).
                u: $-$ -0.3 $-$
                $FOOD5X = zFOOD5*MAX\{0.,[PupMEI*(1.-pex5-prs5)-asWORK5]\}$
prs6   Relative activity respiration of deposit feeders (BBBM).
                u: $-$ -0.31 $-$
                $FOOD6X = (1-pex6A)*zFOOD6*FOOD6*(1-prs6)-zFOOD6*asWORK6$
prs7   Relative activity respiration of suspension feeders (BSF).
                u: $-$ -0.214 $-$
                $FILB = (1-pex7)*RRFIL*ptWET(I)*FOOD7*zPAR*zox7*(1.-prs7) -FACT7*$
                $zPAR*zox7$
pSO   Reduction factor indicating the fraction of the sulphide oxidation due to biotur-
                bation which is not measured during the determination of benthic oxygen con-
                sumption under bell jars. During the measurements the bioturbation, especially
                by macrobenthos, decreases.
                u: $-$ 0.3   0.5   1
                $MCONS = MBOX(I) + pSO*(BOXS-SO2)$
pup7   The maximum relative feeding rate of suspension feeders (BSF).
                u: d^{-1} 0.03   0.05   0.21
                $MAXBSF = pup7*zt7a*BSF(I)$
Q10BBM   Parameter in temperature function, governing the bioturbation activity of de-
                posit feeders (BBBM).
                u: $-$ -8 $-$
                $zt6 = Q10BBM**[(TA+TAM)/20.]$
Q10DIA   Constant in the temperature response function of benthic diatoms (BDIA).
                u: $-$ -4 $-$
                $zt4 = Q10DIA*(TA/10.)$
Q10MEI   Constant in the temperature response function of meiobenthos (BMEI).
                u: $-$ -8 $-$
                $zt5 = Q10MEI**(TA/10.)$
Q20BBM   Constant moderating the temperature response of deposit feeders (BBBM) at
                extreme temperatures.
                u: $-$ -1.5 $-$
                $zt6 = Q20BBM*zt6/(Q20BBM-1. + zt6)$
Q20DIA   Constant moderating the temperature response of benthic diatoms (BDIA) at
                extreme temperatures.
                u: $-$ -2 $-$
                $zt4 = Q20DIA*zt4/(Q20DIA-1. + zt4)$
Q20MEI   Constant moderating the temperature response of meiobenthos (BMEI) at ex-
                treme temperatures.
                u: $-$ -2 $-$
                $zt5 = Q20MEI*zt5/(Q20MEI-1. + zt5)$
QLOC   Fraction of labile organic carbon (BLOC or ALOC) available for uptake by
                bacteria (BBAC or ABAC).
                u: d^{-1} 0.4   1   1
                $RLOC = BLOC(I)*QLOC$ and $RLOCA = ALOC(I)*QLOC$
qox4   Threshold value for reaction of benthic diatoms (BDIA) to low oxygen satura-
                tion.
                u: $-$ -0.6 $-$
                $zox4 = MIN(1.,ROX/qox4)$
qox5   The relative oxygen saturation, below which mortality due to oxygen limitation
                occurs in meiobenthos (BMEI).
                u: $-$ -0.2 $-$
                $zox5 = MIN(1.,ROX/qox5)$
qox6   The relative oxygen saturation, below which mortality due to oxygen limitation
                occurs in deposit feeders (BBBM).
                u: $-$ -0.4 $-$
                $zox6 = MIN(1., ROX/qox6)$

qox7   The relative oxygen saturation, below which mortality due to oxygen limitation occurs in suspension feeders (BSF).
u: $-$ $-0.7-$
zox7 = MIN(1., ROX/qox7)

reacon   The relative reaeration rate of oxygen per m² surface. Reaeration is dependent on the difference between the saturation value and the oxygen concentration.
u: $m \cdot d^{-1}$ 0.96   1.44   1.92
COXP = COX/corCOr*.01 + (PPox-POX)/[reacon*zt0/ 0.01*
MAX(1.E-6,pOVERLAP)]

rex4   Relative rest excretion and rest mortality in benthic diatoms (BDIA) at 11 °C.
u: d^{-1} $-0.02-$
W48 = UDIA*rex4*zt4 + PP*aex4 + cPP*bex4*(1-p42)*PP

rex5   Relative rest excretion/mortality for meiobenthos (BMEI) at 11 °C.
u: d^{-1} $-0.005-$
W58r = BMEI(I)*[rex5 + rex5O*(1-zox5)]*zt5

rex5O   Relative mortality (treated as excretion) due to oxygen limitation for meiobenthos (BMEI).
u: d^{-1} $-0.02-$
W58r = BMEI(I)*[rex5 + rex5O*(1-zox5)]*zt5

rex6   Relative natural mortality rate of deposit feeders (BBBM).
u: d^{-1} $-0.0008-$
W68r = BBBM(I)*[rex6 + rex6O*(1-zox6)]*zt6c

rex6O   Relative mortality rate of deposit feeders (BBBM) due to low oxygen concentration.
u: d^{-1} $-0.06-$
W68r = BBBM(I)*[rex6 + rex6O*(1-zox6)]*zt6c

rmor7   Relative mortality rate of suspension feeders (BSF) due to low oxygen saturation.
u: d^{-1} $-0.01-$
W72S = rmor7*(1.-zox7)*BSF(I)

RPP   Relative photosynthesis of illuminated diatoms (FBDIA) per day per unit biomass at 11 °C.
u: d^{-1} $-1-$
PPR = RPP*FBDIA*zt4

RRFIL   Relative filtration rate of suspension feeders (BSF).
u: $m^3 \cdot mgC^{-1} \cdot d^{-1}$ $0.1 \cdot 10^{-3}$ $0.2 \cdot 10^{-3}$ $0.33 \cdot 10^{-3}$
FIL = FACTOR7*RRFIL*zFOOD7L*ptWET(I)*zPAR

rrs4   Relative rest respiration of benthic diatoms (BDIA).
u: d^{-1} $-0.01-$
WREST = BDIA(I)*zt4*(rex4 + rrs4)

rrs5   Relative rest respiration of meiobenthos (BMEI).
u: d^{-1} $-0.008-$
W50r = BMEI(I)*rrs5*zt5

rrs6   Relative rest respiration of deposit feeders (BBBM).
u: d^{-1} 0.003   0.003   0.011
W60r = BBBM(I)*rrs6*zt6c

rrs7   Relative rest respiration rate of suspension feeders (BSF).
u: d^{-1} 0.003   0.003   0.012
W70r = rrs7*BSF(I)*zt7c*CIND(I)

rrsBAC   Relative rest respiration of aerobic bacteria (BBAC).
u: d^{-1} 0.005   0.01   0.1
W30 = W83*arlBAC + W23*ardBAC + W13*arrBAC + BBAC(I)*rrsBAC*zt3

rrsBACA   Relative rest respiration of anaerobic bacteria (ABAC).
u: d^{-1} 0.01   0.01   0.1
W30A(I) = W83A*arlBACA + W23A*ardBACA + W13A*arrBACA + ABAC(I)*rrsBACA* zt3A

RSPAWN6   Relative annual spawning rate of deposit feeders (BBBM).
u: y^{-1} $-0.2-$

	SPAWN6 = BBBM(I)*RSPAWN6*zSPAWN
RSPAWN7	Relative annual spawning rate of suspension feeders (BSF).

u: y^{-1} $-0.2-$

SPAWN7 = BSF(I)*RSPAWN7*zSPAWN

SDL
Sedimentation rate under normal conditions. The sedimentation rate increases under calm weather conditions and decreases in windy periods. Annual sedimentation amounts to $0.80 \text{ cm} \cdot y^{-1}$.

u: $\text{cm} \cdot d^{-1}$ 0.001   0.0022   0.0033

SDLX = SDL*zSED

SML
Mixing rate of sediment under normal conditions. The mixing of sediment increases when storms occur and decreases when calm weather conditions prevail.

u: $\text{cm} \cdot d^{-1}$ 0.0001   0.01   0.05

SMLX = SML*zMIX*zsm

STANFO5
Reference food concentration for meiobenthos (BMEI). When the food concentration is equal to STANFO5, the reference uptake rate is realized.

u: $\text{mgC} \cdot m^{-2} \cdot cm^{-1}$ $-550-$

PupMEI = FOOD5/STANFO5*upMEI

susBDIA
Fraction of benthic diatoms (BDIA) that are suspended in the water over the flats at high tide.

u: $-$ $-0.16-$

PHYTO = PFLAG(I) + PDIA(I) + susBDIA*CDIA(I)/depfa(I)

T6T
Parameter in the temperature function for standard metabolism of deposit feeders (BBBM). Above this threshold standard metabolism is increased, representing heat stress.

u: °C $-19-$

zt6c = MAX[1.,(tem6-T6T)]

t7c
Parameter in the temperature function for standard metabolism of suspension feeders (BSF). Above this threshold standard metabolism is increased, representing heat stress.

u: °C $-18-$

zt7c = MAX[1.,(tem7-t7c)]

TAL
Thickness of the sediment layer with significant biological activity. Reworking and other signs of biological activity do not usually extend to greater depths than 30 cm. At this depth often a shell-fragment layer is found, indicating this to be the lower boundary of sediment reworking.

u: cm $-30-$

W13A = BROC(I)*FRBRA*{[TAL-BAL(I)]/TAL}*zt3a

tdrBAC
Fraction of benthic detritus (BDET) uptake by aerobic bacteria (BBAC) returned as refractory organic carbon (BROC).

u: $-$ 0.001   0.05   0.1

W31 = W23*tdrBAC

tdrBACA
Fraction of benthic detritus (ADET) uptake by anaerobic bacteria (ABAC) returned as refractory organic carbon (BROC).

u: $-$ 0.001   0.05   0.1

W31A = W23A*tdrBACA

TPUMP(5)
Tidal pump action, representing the percolating of water from tidal pools through the sediment and hence effectuating aeration of the sediment. Aeration by tidal pumping also enhances the molecular diffusion of oxygen.

u: $\text{cm} \cdot d^{-1}$ see Appendix Table A-3.1

AMIX1 = ALFO*zMIX*zsm**2 + ALFt*TPUMP(I)

TSA
Shift in day number for the time dependent temperature function of the sediment layer above the sulphide horizon (BAL). The minimum temperature in this sediment layer does not occur on January 1st, but 15 days later.

u: $-$ $-15-$

ACOS = -COS[(TIME-TSA)/180.*PI]

TSAN
Shift in day number for the time dependent temperature function of the sediment layer below the sulphide horizon (TAL-BAL). The minimum temperature in this sediment layer does not occur on January 1st, but 45 days later.

	u: $-$ -45 $-$
	ANCOS = -COS((TIME-TSAN)/180.*PI)
uM7	Fraction of pelagic bacteria (PBAC) available for suspension feeders (BSF). Free-living bacteria are too small to be retained by filter feeders.

uM7   Fraction of pelagic bacteria (PBAC) available for suspension feeders (BSF). Free-living bacteria are too small to be retained by filter feeders.
u: $-$ -0.1 $-$
FOOD7 = uM7*PBAC(I) + PBLAR(I) + PMIC(I) + PHYTO

upBAC   Relative uptake rate of aerobic bacteria (BBAC) at 11 °C
u: d^{-1} -0.75 $-$
FBAC = upBAC*BBAC(I)*zt3*zBAC

upBACA Relative uptake rate of anaerobic bacteria (ABAC) at 11 °C
u: d^{-1} -0.75 $-$
FBACA = upBACA*ABAC(I)*zt3A*zBACA

upMEI   Relative uptake rate of meiobenthos (BMEI) under standard food conditions.
u: d^{-1} -0.072 $-$
PupMEI = FOOD5/STANFO5*upMEI

Table A-3.1. Compartment-dependent sediment parameters in the benthic submodel

	BIOAM	CBOF	CIND	KD	LAD	TPUMP
Comp 1	2	0.2	1.51	320	0.004	0.1
Comp 2	1	0.6	1.51	150	0.008	5.0
Comp 3	3	1.0	1.00	90	0.014	3.0
Comp 4	4	1.0	1.00	70	0.017	0.8
Comp 5	10	1.0	0.92	55	0.012	1.0

Appendix A-4: Parameters of the Epibenthic Submodel

BIRD(12)   Monthly mean number of carnivorous birds in the Dollard. Numbers of birds in the Dollard (comp. 1 + 2) preying on macro- and mesoepibenthos, macrobenthic deposit and suspension feeders (EMAC, EMES, BBBM and BSF). With the preferences for the different compartments the number of birds in the whole estuary is calculated.
u: $-$ Appendix Table A-4.1
BIRDS = BIRD[1 + INT(MTIME/30)]*PFOODT/[PFOOD(1) + PFOOD(2)]

BNEC(12) The amount of food (monthly mean) that one bird tries to eat per day.
u: $mgC \cdot d^{-1} \cdot bird^{-1}$ Appendix Table A-4.1
BNEED = BNEC[1 + INT(MTIME/30.)]

cmorY   Relative mortality of mesoepibenthos (EMES) due to predation by pelagic fishes.
u: d^{-1} -0.08 $-$
rmorY = cmorY*{4.**4/[TEMP(I)**4 + 4.**4]}*EMES(I)/[10. + EMES(I)]

DCRIT   Potential start of emigration of macroepibenthos (EMAC). Emigration can start if [TIME.ge.DCRIT.and.TEMP(1).le.TEMI .and. FE.eq.1]. Daily emigration increases from 0 to routZ.
u: $-$ -190 $-$

DELT1   Time period of increasing immigration of macroepibenthos (EMAC).
u: d -40 $-$
TMAXIN = DSTART + DELT1

DELT2   Time period of decreasing immigration of macroepibenthos (EMAC).
u: d -20 $-$
TENDIN = TMAXIN + DELT2

kfdY   Monod half-saturation constant for the food uptake of mesoepibenthos (EMES).

	u: $mgC \cdot m^{-2}$ $-150-$
	$upY = ztY * rupYmax * ptWET(I) * fdY(I)/[fdY(I) + kfdY] * EMES(I)$
kfdZ	Monod half-saturation constant for the fooduptake of macroepibenthos (EMAC).
	u: $mgC \cdot m^{-2}$ $-260-$
	$upZ = ztZ * rupZmax * ptWET(I) * fdZ(I)/[fdZ(I) + kfdZ] * EMAC(I)$
koxY	The half-value constant of the oxygen limitation function for mesoepibenthos (EMES).
	u: $-$ $-0.2-$
	$okY(I) = [rOSAT(I)/koxY] ** xoxY$
koxZ	The half-value constant of the oxygen limitation function for macroepibenthos (EMAC).
	u: $-$ $-0.6-$
	$okoxZ = [rOSAT(I)/koxZ] ** xoxZ$
ksalZ	The half-value constant of the salinity limitation function for macroepibenthos (EMAC).
	u: S $-8-$
	$oksalZ = [SALT(I)/ksalZ] ** xsalZ$
ktZmax	The half-value constant of the upper temperature limitation function for macroepibenthos (EMAC).
	u: $^{\circ}C$ $-22-$
	$oktZmax = [TEMP(I)/ktZmax] ** xtZ$
ktZmin	The half-value constant of the lower temperature limitation function for macroepibenthos (EMAC). See koxZ
	u: $^{\circ}C$ $-2-$
	$oktZmin = [TEMP(I)/ktZmin] ** xtZ$
pexY2	The fraction of rest excretion and activity excretion of meso-epibenthos (EMES), going to benthic detritus (BDET). The excretion products of mesoepibenthos are partly going to benthic detritus, and partly to benthic labile organic carbon (BLOC).
	u: $-$ $-0.5-$
	$flY2 = pexY2 * (exYa + exYr)$
pexZ2	The fraction of rest excretion and activity excretion of macroepibenthos (EMAC), going to benthic detritus (BDET). see pexY2
	u: $-$ $-0.1-$
	$flZ2 = pexZ2 * (exZa + exZr) + exZm$
PREF(5)	The preference of the birds for the different compartments. Birds are distributed over the compartments according to food availability and preference. If there is a less than average amount of food in a certain compartment then the preference is lowered.
	u: $-$ $1/0.7/0.3/0.25/0.5$
	$PFOOD(I) = AMIN\{[BFOOD(I)/BFOODM * PREF(I)], PREF(I)\}$
pZ2	Fraction of macroepibenthos (EMAC) uptake going directly from macroepibenthos to benthic detritus. Part of the macroepibenthos species (shrimps) has an exo-skeleton which is lost at moulting. This moulting is related to the uptake.
	u: $-$ $-0.005-$
	$exZm = pZ2 * upZ$
Q10Y	Q10 value for the uptake of mesoepibenthos (EMES).
	u: $-$ $-2-$
	$ztY = 2.0 ** [TEMP(I)-12] * Q10Y/10$
Q10Z	Q10 value for the uptake of macroepibenthos (EMAC).
	u: $-$ $-2-$
	$ztZ = 2.0 ** [TEMP(I)-12] * Q10Z/10$
REDIST	Maximal fraction of biomass of macroepibenthos (EMAC) or meso-epibenthos (EMES) that is able to move to one of the neighbouring compartments. For EMAC this part of the biomass of the compartments I and I+1 is redistributed according to the preference factors over compartments I and I+1. For EMES

it is the part RDSTY, dependent on the comfort functions [okY(I)], which is redistributed in the same way as EMAC.
u: − − 0.25 −
traEMAC = traEMAC + EMAC(I)∗REDIST∗areaf(I)
RDSTY(I) = {0.05 + [1-okY(I)∗0.95]}∗REDIST

REFRAC | The respired fraction of uptake and biomass related losses for mesoepibenthos
(EMES) and macroepibenthos (EMAC).
u: − − 0.5 −
prsYa = REFRAC∗YUPLOS   rrsYr = REFRAC∗YBILOS

routZ | Maximal fraction of epmacrobenthos (EMAC), daily (e)migrating from com-
partment I to compartment I+1 in autumn. After DCRIT on the condition
TEMP(1)≤TEMI, the fraction: DEXPRT ∗routZ∗EMAC(I) starts moving
out. DEXPRT increases from 0 to 1 within a number of days.
u: d^{-1} − 0.067 −
EXPRT(I) = -DEXPRT∗routZ∗EMAC(I)

rupYmax | Maximal relative uptake rate per day of mesoepibenthos (EMES) at 12 °C.
u: d^{-1} − 0.7 −
upY = ztY∗rupYmax∗ptWET(I)∗fdY(I)/[fdY(I) + kfdY]∗EMES(I)

rupZmax | Maximal relative uptake rate per day of macroepibenthos (EMAC) at 12 °C.
u: d^{-1} − 0.35 −
upZ = ztZ∗rupZmax∗ptWET(I)∗fdZ(I)/[fdZ(I) + kfdZ]∗EMAC(I)

SETTLE | The total amount of macroepibenthos (EMAC) settling in the outer compart-
ment 5 during 1 year. The total immigrating biomass is divided by the size of
compartment 5 in m^2. The timing of the immigration is regulated by TIMM,
DELT1 and DELT2.
u: mgC·m^{-2} 170 225 330
MAXSETTLE = 2∗SETTLE/(DELT1 + DELT2)

TEMI | Water temperature in compartment 1, below which emigration can start. Emi-
gration of macroepibenthos (EMAC) can start if [MTIME.gt.DCRIT.and.
TEMP(1).le.TEMI.and.FE.eq.1.]
u: °C − 12 −

TIMM | The value of the sum of the daily water temperatures in compartment 5 (TOT-
TEMP) at the start of the immigration of macroepibenthos (EMAC). Immigration can start if [TOTTEMP.ge. TIMM.and.FI.eq.1.]
u: °C − 400 −

u2Y | Fraction of that part of the biomass of the benthic detritus (BDET) that is in
the upper 0.5 cm of the aerobic layer, accessible as food for mesoepibenthos
(EMES).
u: − − 0.001 −
fdY(I) = ... + [u2Y∗F(I)∗BDET(I)]∗∗xfd + ...

u2Z | Fraction of that part of the biomass of the benthic detritus (BDET) that is in
the upper 0.5 cm of the aerobic layer, accessible as food for macroepibenthos
(EMAC).
u: − − 0.005 −
fdZ(I) = ... + [u2Z∗F(I)∗BDET(I)]∗∗xfd + ...

u3Y | Fraction of that part of the biomass of the benthic bacteria (BBAC) that is in
the upper 0.5 cm of the aerobic layer, accessible as food for mesoepibenthos
(EMES). See also u2Y.
u: − − 0.2 −
fdY(I) = ... + [u3Y∗F(I)∗BBAC(I)]∗∗xfd + ...

u4Y | Fraction of the biomass of the diatoms in the upper 0.5 cm (CDIA) accessible
as food for mesoepibenthos (EMES). See also u2Y.
u: − − 0.2 −
fdY(I) = ... + [u4Y∗CDIA(I)]∗∗xfd + ...

u5Y | Fraction of that part of the biomass of the meiobenthos (BMEI) that is in the
upper 0.5 cm of the aerobic layer, accessible as food for mesoepibenthos
(EMES). See also u2Y.

u: − − 1 −

fdY(I) = ... + [u5Y∗F(I)∗BMEI(I)]∗∗xfd + ...

u5Z   Fraction of that part of the biomass of the meiobenthos (BMEI) that is in the upper 0.5 cm of the aerobic layer, accessible as food for macroepibenthos (EMAC). See also uZ2.

u: − − 0.5 −

fdZ(I) = ... + [u5Z∗F(I)∗BMEI(I)]∗∗xfd + ...

u6X   Fraction of the biomass of macrobenthic deposit feeders (BBBM), accessible as food for birds.

u: − − 0.4 −

BFOOD(I) = ... + [u6X∗BBBM(I)]∗∗xfdX + ...

u6Z   Fraction of the biomass of macrobenthic deposit feeders (BBBM), accessible as food for macroepibenthos (EMAC). See also uZ2.

u: − − 0.1 −

fdZ(I) = ... + [u6Z∗BBBM(I)]∗∗xfd + ...

u7X   Fraction of the biomass of macrobenthic suspension feeders (BSF), accessible as food for birds. See also u7X.

u: − − 0.4 −

BFOOD(I) = ... + [u7X∗BSF(I)]∗∗xfdX + ...

u7Z   Fraction of the biomass of macrobenthic suspension feeders (BSF), accessible as food for macroepibenthos (EMAC). See also uZ2.

u: − − 0.05 −

fdZ(I) = ... + [u7Z∗BSF(I)]∗∗xfd + ...

uHY   Fraction of the biomass of mesozooplankton (PCOP) in the whole water column, accessible as food for mesoepibenthos (EMES). See also u2Y.

u: − 1

fdY(I) = ... + [uHY∗PCOP(I)∗depfa(I)]∗∗xfd + ...

uYX   Fraction of the biomass of mesoepibenthos (EMES), accessible as food for birds. See also u7X.

u: − − 1 −

BFOOD(I) = ... + [uYX∗EMES(I)]∗∗xfdX + ...

uYY   Fraction of the biomass of mesoepibenthos (EMES), accessible as food for EMES. It is the fraction of EMES that is cannibalizable. The estimated value is based on the composition of EMES. See also u2Y.

u: − − 0.2 −

fdY(I) = ... + [uYY∗EMES(I)]∗∗xfd + ...

uYZ   Fraction of the biomass of mesoepibenthos (EMES), accessible as food for macroepibenthos (EMAC). See also uZ2.

u: − − 0.3 −

fdZ(I) = ... + [uYZ∗EMES(I)]∗∗xfd + ...

uZX   Fraction of the biomass of macroepibenthos (EMAC), accessible as food for birds, on the condition rOSAT(I) > 0.10. This fraction of EMAC is accessible as food for birds. If rOSAT(I) < 0.1 all the EMAC comes to the surface and will be accessible, so if rOSAT(I) ≦ 0.1 then uZX = 1. See also u7X.

u: − − 0.2 −

BFOOD(I) = ... + [uZX∗EMAC(I)]∗∗xfdX + ...

uZZ   Fraction of the biomass of macroepibenthos (EMAC), accessible as food for EMAC. The value of this cannibalizable fraction is strongly dependent on the composition of EMAC. See also uZ2.

u: − − 0.1 −

fdZ(I) = ...[uZZ∗EMAC(I)]∗∗xfd + ...

xfd   Exponent in calculation of the amount of food for macroepibenthos (EMAC) and for mesoepibenthos (EMES). If xfd > 1, xfd effects a shift towards the more abundant food sources.

u: − − 2 −

fdZ(I) = [u5Z∗BMEI(I)]∗∗xfd + [uZZ∗EMAC(I)]∗∗xfd + ... fdZ(I) = fdZ(I)∗∗1/xfd

xfdX   Exponent in calculation of the amount of food for BIRD. If xfdX = 1, the food
              sources are proportionally available.
              u: − −1 −
              BFOOD(I) = [uYZ*EMES(I)]**xfdX + [uZX*EMAC(I)]**xfdX + ...
              BFOOD(I) = BFOOD(I)**1/xfdX
xoxY   Exponent defining the steepness of the okY function. The okY function changes
              from a sigmoid function to a step function with increasing xoxY.
              u: − −8 −
              okY(I) = [rOSAT(I)/koxY]**xoxY
xoxZ   Exponent defining the steepness of the okoxZ function. See xoxY.
              u: − −8 −
              okoxZ = [rOSAT(I)/koxZ]**xoxZ
xsalZ   Exponent defining the steepness of the oksalZ function. See xoxY.
              u: − −10 −
              oksalZ = [SALT(I)/ksalZ]**xsalZ
xtZ   Exponent defining the steepness of the oktZmin en oktZmax functions. See
              xoxY.
              u: − −20 −
              oktZmin = [TEMP(I)/ktZmin]**xtZ
YBILOS   The daily relative biomass loss of mesoepibenthos (EMES) at 12 °C due to basal
              metabolism.
              u: d⁻¹ −0.07 −
              rrsYr = REFRAC*YBILOS
YUPLOS   The uptake loss of mesoepibenthos (EMES) at 12 °C, as a fraction of the total
              uptake of EMES.
              u: − −0.5 −
              pexYa = (1.-REFRAC)*YUPLOS
              prsYa = REFRAC*YUPLOS
ZBILOS   The daily relative biomass loss of macroepibenthos (EMAC) at 12 °C due to
              basal metabolism.
              u: d⁻¹ −0.01 −
              rrsZr = REFRAC*ZBILOS
              rexZr = (1.-REFRAC)*ZBILOS
ZUPLOS   The uptake loss of macroepibenthos (EMAC) at 12 °C, as a fraction of the total
              uptake of EMAC.
              u: − −0.7 −
              prsZa = REFRAC*ZUPLOS
              pexZa = (1.-REFRAC)*ZUPLOS

Table A-4.1. Compartment-dependent parameters in the
epibenthic submodel

MONTH	BNEC(12)	BIRD(12)
1	9 992	15 958
2	19 628	5 087
3	10 333	30 479
4	8 016	61 813
5	7 277	41 305
6	26 422	8 609
7	13 316	16 897
8	12 052	46 755
9	13 332	36 344
10	9 284	73 523
11	6 357	84 523
12	10 896	12 865

Appendix B: The FORTRAN Listing of the Submodels

Appendix B-1:

```
      SUBROUTINE BAHBOE
C ┌─────────────────────────────────────────────┐
C │ PHYSICAL SUBMODEL (main subroutine)         │
C └─────────────────────────────────────────────┘
$INCLUDE BAHBOE:COMM
      INTEGER NEL,NSTATE,ITR,L,M,IST
      PARAMETER NSTATE=33,NEL=5
      REAL RDIFF(6),totSILT(6),MUD(6)
      REAL C(5,NSTATE),SC(5,NSTATE),SEAC(NSTATE),WWAC(NSTATE)
     , , EMSC(NSTATE),TC(5,NSTATE),TS5C(NSTATE)
     , , TE3C(NSTATE),TW1C(NSTATE)
      EQUIVALENCE (C,OX),(SC,SOX),(TC,TOX),(SEAOX,SEAC),(WWAOX,WWAC)
     , , (EMSOX,EMSC),(TS5C,TS5OX),(TE3C,TE3OX),(TW1C,TW1OX)
      COMMON/TRANS/ITR(80)
$INCLUDE DOLFYS:COMM
      IF (IST.EQ.0) THEN
        IST=1
C---------------------------------------------------------------------
C    TYPE OF  TRANSPORT    STATE VARIABLE
C---------------------------------------------------------------------
         ITR( 1)=1          ;%     OX(5)
         ITR( 2)=3          ;%     SILT(5)
         ITR( 3)=2          ;%     PDET(5)
         ITR( 4)=1          ;%     SALT(5)
         ITR( 5)=1          ;%     PFLAG(5)
         ITR( 6)=1          ;%     PLOC(5)
         ITR( 7)=1          ;%     PCOP(5)
         ITR( 8)=1          ;%     PBAC(5)
         ITR( 9)=0          ;%     BBAC(5)
         ITR(10)=0          ;%     ABAC(5)
         ITR(11)=1          ;%     CARN(5)
         ITR(12)=1          ;%     PBLAR(5)
         ITR(13)=0          ;%     BDIA(5)
         ITR(14)=0          ;%     ALOC(5)
         ITR(15)=0          ;%     BMEI(5)
         ITR(16)=0          ;%     BBBM(5)
         ITR(17)=0          ;%     BLOC(5)
         ITR(18)=0          ;%     BDET(5)
         ITR(19)=1          ;%     PMIC(5)
         ITR(20)=2          ;%     PROC(5)
         ITR(21)=0          ;%     EMES(5)
```

```
          ITR(22)=0          ;%      EMAC(5)
          ITR(23)=0          ;%      BSUL(5)
          ITR(24)=0          ;%      BAL(5)
          ITR(25)=0          ;%      ADET(5)
          ITR(26)=0          ;%      BROC(5)
          ITR(27)=1          ;%      PDIA(5)
          ITR(28)=1          ;%      SILIC(5)
          ITR(29)=1          ;%      PO4(5)
          ITR(30)=0          ;%      BSF(5)
          ITR(31)=1          ;%      PDROC(5)
          ITR(32)=0          ;%      BPYR(5)
          ITR(33)=-1         ;%      CSILT(5)
       ENDIF
C
C    0=No transport
C    1=Advective/diffusive transport
C    2=Residual transport of particulate organic matter
C    3=SILT, -1=CCSILT
C
C Temperature--------------------------------------------------
C    Simulating of a warm or a cold winter with a time series
C       TEex(4):0.,X.,0.,0.  X>0 warm winter   X<0 cold winter
        TEMPEFFECT=0.;IF (TEex(2).ne.0.) TEMPEFFECT=TIMCOS(tTEex,TEex,4)
C    Normal temperature
        TEMP(1)=TEMP(2)=TIMSER(tTE,TED,131)+TEMPEFFECT
        TEMP(3)=TIMSER(tTE,TE3,131)+TEMPEFFECT
        TEMP(4)=TIMSER(tTE,TE4,131)+TEMPEFFECT
        TEMP(5)=TIMSER(tTE,TE5,131)+TEMPEFFECT
C Light-----------------------------------------------------------
C    Light is computed in a separate subroutine.
C    You can choose between 5 light functions
C CONLIGHT=-1: sinus function
C CONLIGHT= 0: daily values (1976–1980)
C CONLIGHT= 1: harmonic mean light curves over 5 years.
C CONLIGHT= 2: arithmetic mean light curves over 5 years
        CALL SUBLIGHT(CONLIGHT,SUNQ,LIGHT)
C
C Calculation of wind influence on turbidity --------------------------
        ACTWIND=TIMCOS(tiox,WIND,14)
        pDOWN=pDOWN0*W50/(ACTWIND-6.+W50)
C
C River discharges-------------------------------------------------
C Time series are recomputed from m3/s to m3/day
        QWWA=TIMCOS(TIMQ,QQWA,74)*86400.
        QEMS=TIMCOS(TIMQ,QQEM,74)*86400.
        QUPPER=QWWA+QEMS
C
C Calculation of the actual exchange coefficients-----------------------
        RDIFF(1)=0.
        DO FOR I=0,1
          RDIFF(I+2)=SLP(QWWA,QW,DIFF(1+I*7),7)
        ENDDO
        DO FOR I=2,4
          RDIFF(I+2)=SLP(QUPPER,QS,DIFF(1+I*7),7)
        ENDDO
        RDIFEMS=SLP(QEMS,QE,DIEMS,7)
C
C Calculation of default boundary conditions----------------------------
```

```
C If no boundary condition is given here or in the pelagic submodel then
C 1. Concentration on the WWA is zero
C 2. Concentration on the EMS is equal to the conc. in compartment 2
C 3. Concentration at SEA is equal to the conc. in compartment 5
      DO FOR J=1,NSTATE
        IF (ITR(J).NE.0) THEN
          WWAC(J)=0.
          EMSC(J)=C(2,J)
          SEAC(J)=C(5,J)
          TS5C(J)=TE3C(J)=TW1C(J)=0.
          DO FOR I=1,NEL
            TC(I,J)=0.
          ENDDO
        ENDIF
      ENDDO
C Boundary conditions-----------------------------------------------
C Boundary conditions for salt
      WWASALT=Wwsalt
      EMSSALT=SLP(QEMS,QE,Emsal,7)
      SEASALT=SLP(QUPPER,QS,Sesal,7)
C Boundary conditions for suspended matter (including C-compounds).
      WWASUSM=TIMSER(TISIW,SIWWA,6)
      EMSSUSM=TIMSER(TISI,SIEM,13)
C Boundary conditions for silt at silt at seaward boundary.
      SEASILT=TSILTSEA*(1.-pDOWN)
      SEACSILT=TSILTSEA-SEASILT
C Boundary conditions for oxygen
      WWAOX=TIMSER(tpOsW,pOsW,6)*(475.-2.65*Wwsalt)/(33.5+TEMP(1))
      EMSOX=TIMSER(tiox,Emox,14)
      SEAOX=TIMSER(tiox,Seox,14)
C
      DO FOR I=1,NEL
C Calculation of total suspended matter and total silt
      MUD(I)=totSILT(I)=CSILT(I)+SILT(I)
C Calculation of fraction of suspended matter in the water column
C (used in pelagic model)
      pSUSP(I)=SILT(I)/totSILT(I)
      DO FOR J=1,NSTATE
        IF (ITR(J).EQ.2) THEN
C Other substances in TSILT:
          MUD(I)=MUD(I)+C(I,J)*cCARASD
C Redistribution of these other susbstances
          TC(I,J)=TC(I,J)-C(I,J)*move(I)/VOL(I)
          IF (I.NE.1) TC(I,J)=TC(I,J)-C(I,J)*move(I-1)/VOL(I)
        ENDIF
      ENDDO
C Potential volume which is exchanged between two compartments
      TSILT(I) =TSILT(I) -SILT(I)*MOVE(I)/VOL(I)
      TCSILT(I)=TCSILT(I)-CSILT(I)*MOVE(I)/VOL(I)
      IF (I.NE.1) THEN
        TSILT(I) =TSILT(I) -SILT(I)*MOVE(I-1)/VOL(I)
        TCSILT(I)=TCSILT(I)-CSILT(I)*MOVE(I-1)/VOL(I)
      ENDIF
      TSILT(I)=TSILT(I)-rTID*seddep/depfa(I)*pWAFL(I)*pDOWN*SILT(I)
C Resuspension of silt
      TSILT(I)=TSILT(I)
     +        +rTID*totSILT(I)*pWAFL(I)*resdep/depfa(I)*(1.-pDOWN)
```

```
C Net sedimentation of particulate carbon: (sedimentation of PROC and
C PDET is in pelagic submodel)
      rRBEN(I)=rTID/depfa(I)*(seddep*pSUSP(I)*pDOWN-resdep*(1.-pDOWN))
      ENDDO
C
C---------------Benthic + pelagic + epibenthic submodels---------------
C
    CALL SUBBEN
    CALL SUBPEL
    CALL SUBEPI
C
C Boundary condition for silt-------------------------------------------
    EMSSILT=EMSSUSM-(EMSPDET+EMSPROC)*cCARASD
    WWASILT=WWASUSM-(WWAPDET+WWAPROC)*cCARASD
C
C Advective/diffusive transport-----------------------------------------
C
C QUNDER  = Inflow rate from compartment(I-1) in m**3/d
C QUPPER  = Flow rate to compartment(I) (outflow) in m**3/d
C CEL     = Concentration within the compartment in g/m**3
C CMIN    = Concentration in compartment(I-1) in g/m**3
C CPLUS   = Concentration in next compartment(I+1) in g/m**3
C
    DO FOR J=1,NSTATE
      IF (ITR(J).GT.0) THEN
C Start with WWA as compartment 0
        CUPPER=CEL=WWAC(J)
        CPLUS=C(1,J); IF (ITR(J).EQ.2) CPLUS=CPLUS*pSUSP(1)
        QUPPER=QWWA
        DO FOR I=1,NEL
C Give concentration of C(I-1) and C(I) to resp. CMIN and CEL by shifting
C
          CMIN=CEL
          CEL=CPLUS
C Define concentration and discharge at the lower border of the compartment
C
          CUNDER=CUPPER
          QUNDER=QUPPER
C Define concentration of CPLUS and correct discharge at upper border of
C the compartment for the Ems:
          IF (I.EQ.NEL) THEN
            CPLUS=SEAC(J)
          ELSE
            CPLUS=C(I+1,J); IF (ITR(J).EQ.2) CPLUS=CPLUS*pSUSP(I+1)
            IF (I.EQ.3) QUPPER=QUPPER+QEMS
          ENDIF
C Define concentration at the upper border of the compartment:
          CUPPER=(CEL+CPLUS)*.5
C River flow (Ems flow in compartment 3)
          CIM=QUNDER*CUNDER
          IF (I.EQ.3) THEN
            IF (ITR(J).EQ.1) THEN
              R=QEMS*(CEL+EMSC(J))*.5
            ELSE
              R=QEMS*EMSC(J)
            ENDIF
            CIM=CIM+R
            TE3C(J)=R
```

```
        ELSEIF (I.EQ.1) THEN
            TW1C(J)=CIM
        ENDIF
C Displacement
        CEM=QUPPER*CUPPER
C Diffusive exchange due to (Ems flow in compartment 3)
        CEXCH1=RDIFF(I)*(CMIN-CEL)
        IF(I.EQ.3.AND.ITR(J).EQ.1) THEN
            R=RDIFEMS*(EMSC(J)-CEL)
            CEXCH1=CEXCH1+R
            TE3C(J)=R+TE3C(J)
        ENDIF
        CEXCH2=RDIFF(I+1)*(CPLUS-CEL)
C Result of mixing processes passed to COMMON/SOURCE/SC(5,40)
TC(I,J)=TC(I,J)+(CIM-CEM+CEXCH1+CEXCH2)/VOL(I)
      ENDDO
      TS5C(J)=TS5C(J)-CEM+CEXCH2
    ENDIF
  ENDDO
C
C ------- Residual silt transport -------------------------------------
C
C Calculation of gross export of SILT + detrital POC from the esuary to
C the sea
C Mud and silt concentration at sea
    TS5SILT =TS5SILT -MOVE(5)*SILT(5)
    TS5CSILT=TS5CSILT -MOVE(5)*CSILT(5)
    MUD(6)=totSILT(6)=SEASILT
    DO FOR J=1,NSTATE
      IF (ITR(J).EQ.2) THEN
        MUD(6)=MUD(6)+SEAC(J)*cCARASD
        TS5C(J)=TS5C(J)- MOVE(5)*C(5,J)
      ENDIF
    ENDDO
    MUD(6)=MUD(6)/(1.-pDOWN)
    totSILT(6)=totSILT(6)/(1.-pDOWN)
C Redistribution process -----------------------------------------
    DO FOR I=1,NEL
C Total amount exchanged between two compartments
        TOT=(MUD(I)+MUD(I+1))
C Sum of two "attraction"-factors
        RTOT=REDIS(I)+1
        DO FOR M=I,I+1
C Redistribution (TOI = concentration suspended matter)
          IF (M.NE.6) THEN
            TOI=1.; IF (M.EQ.I) TOI=REDIS(I)
            TOI=TOI/RTOT*TOT*MOVE(I)/VOL(M)
C Sedimentation to channel bed
            TDOWN=TOI*pDOWN
C Calculate the redistribution of all substances
            DO FOR L=I,I+1
C Decompostion of TOI
              RTOI=totSILT(L)/TOT
              TSILT(M)=TSILT(M)+(TOI-TDOWN)*RTOI
              TCSILT(M)=TCSILT(M)+TDOWN*RTOI
              IF (I.EQ.5) THEN
                TS5SILT=TS5SILT+(TOI-TDOWN)*RTOI*VOL(5)
                TS5CSILT=TS5CSILT+TDOWN*RTOI*VOL(5)
              ENDIF
```

```
C Other substances redistributed in an analogous way
              DO FOR J=1,NSTATE
               IF (ITR(J).EQ.2) THEN
                 IF (L.NE.6) THEN
                   CH=C(L,J)
                 ELSE
                   CH=SEAC(J)/(1.-pDOWN)
                 ENDIF
                 RTOI=CH*cCARASD/TOT
                 R=TOI*RTOI/cCARASD
                 TC(M,J)=TC(M,J)+R
                 IF (I.EQ.5) TS5C(J)=TS5C(J)+R*VOL(5)
               ENDIF
             ENDDO
           ENDDO
         ENDIF
       ENDDO
     ENDDO
C
C Add total transport rates to /SOURCE/ ------------------------------
C
     DO FOR J=1,NSTATE
       DO FOR I=1,NEL
         IF (ITR(J).NE.0)SC(I,J)=SC(I,J)+TC(I,J)
       ENDDO
     ENDDO
C
C Calculation of reaeration of oxygen----------------------------------
C
     DO FOR I=1,NEL
       OSAT=(475.-2.65*SALT(I))/(33.5+TEMP(I))
       SOX(I)=SOX(I)+REACON/depth(I)*EXP(ct0*TEMP(I))*(OSAT-OX(I))
     ENDDO
     END
```

Appendix B-2:

SUBROUTINE SUBPEL

```
C
C   PELAGIC SUBMODEL
C
C   A=PROC    : Pelagic Refractory Organic Carbon (Particulate)
C   B=PDROC   : Pelagic Dissolved Refractory Organic Carbon
C   C=PDET    : Pelagic Detritus (Particulate)
C   D=PLOC    : Pelagic Labile Organic Carbon
C   E=PMIC    : Pelagic Microzooplankton (20–200 micron)
C   F=PFLAG   : Nondiatom Phytoplankton
C   G=PDIA    : Pelagic Diatoms
C   H=PCOP    : Pelagic Copepods
C   I=CARN    : Pelagic Carnivores (Invertebrates)
C   J=PBLAR   : Pelagic Larvae of Macrobenthos
C   K=PO4     : Phosphate (inorganic,dissolved)
C   L=SILIC   : Reactive Silicate (dissolved)
C   M=PBAC    : Pelagic Bacteria
C   0=CO2
C
```

```
$INCLUDE BAHBOE:COMM
    INTEGER ITR,ICOM(5),IST
    COMMON/TRANS/ITR(80)
    COMMON/DOLFYS/depfa(5),pWAFL(5),depth(5),VOL(5) ,depfh(5),
    & depca(5),ptWET(5),areaf(5),cCARASD ,depchLF(5), depchFH(5)
$INCLUDE DOLPEL:COMM
    DATA ICOM/1,2,3,4,5/
    IF (IST.EQ.0) THEN
      I = COMPN;ICOM(5) = I;ICOM(I) = 5;IST = 1
    ENDIF
C
C Boundary conditions for state variables -------------------
C Sea
C   Organisms
    SEAPFLAG = TIMSER(TMAC,PH,12)
    SEAPDIA = TIMSER(TMAC,SDI,12)
    SEAPCOP = TIMSER(TMAC,BCOP,12)
    SEACARN = TIMSER(TMAC,TCAR,12)
C   Detritus
    SEAPDROC = TIMSER(jTIM,SPD3,26)
    SEAPOC = TIMSER(TMAC,SPOC,12)
    SEATOC = SEAPDROC + SEAPOC
    SEAdeadet = SEAPOC-SEAPFLAG-SEAPDIA-SEAPMIC-SEAPBAC-SEAPCOP
    SEAPROC = SEAdeadet*TIMSER(tPsA,pPsA,6)
    SEAPDET = SEAdeadet-SEAPROC
C   Nutrients
    SEAPO4  = TIMSER(TNUC,PSE,13)
    SEASILIC  = TIMSER(TNUC,SSE,13)
C Ems
C   Organisms
    EMSPCOP = TIMSER(TMAC,EMSH,12)
    EMSPCOP = ((SALT(3)-EMSSALT)*EMSPCOP + EMSSALT*PCOP(3))/
    SALT(3)
    EMSPBAC = EMSSALT*PBAC(3)/SALT(3)
    EMSPFLAG = EMSSALT*PFLAG(3)/SALT(3)
    EMSPDIA = EMSSALT*PDIA(3)/SALT(3)
    EMSCARN = EMSSALT*CARN(3)/SALT(3)
    EMSPMIC = EMSSALT*PMIC(3)/SALT(3)
    EMSPBLAR = EMSSALT*PBLAR(3)/SALT(3)
C   Detritus
    EMSPLOC = EMSSALT*PLOC(3)/SALT(3)
    EMSPDROC = TIMSER(DTIM,EDP3,14)
    EMSPOC = TIMSER(DTIM,EPOC,14)
    EMSTOC = EMSPDROC + EMSPOC
C   Production from PDROC to POC
    flABCems = rABC*EMSPOC
    EMSPDROC = EMSPDROC + flABCems
    EMSPOC  = EMSPOC -flABCems
    EMSPDROC = ((SALT(3)-EMSSALT)*EMSPDROC + EMSSALT*PDROC(3))/
                   SALT(3)
    frAems = TIMSER(tPeA,pPeA,10)
    EMSPROC  = frAems*EMSPOC
    EMSPDET  = EMSPOC-EMSPROC
C   Nutrients
    EMSSILIC  = TIMSER(TNUA,SMS,20)
    EMSSILIC = ((SALT(3)-EMSSALT)*EMSSILIC + EMSSALT*SILIC(3))/SALT(3)
    EMSPO4  = TIMSER(TNUA,PMS,20)
    EMSPO4 = ((SALT(3)-EMSSALT)*EMSPO4 + EMSSALT*PO4(3))/SALT(3)
```

```
C Wwa
C   Detritus
    WWATOC=TIMSER(TWAS,WTOC,87)
    WWAPOC=pPOC*WWATOC ;WWADOC=WWATOC-WWAPOC
    frBWWA=TIMSER(tpDW,pDWB,8)
C   Dissolved speciation (two fractions: PDROC and PLOC)
    WWAPLOC=(1.-frBWWA)*WWADOC
    WWAPDROC=frBWWA*WWADOC
C   Particulate speciation (three fractions: PDET,PROC and BLOC(1) )
    frCWWA=TIMSER(tpPW,pPWC,8)
    frAWWA=TIMSER(tpPW,pPWA,8)
    WWAPDET=WWAPOC*frCWWA
    WWAPROC=WWAPOC*frAWWA
    DO FOR I=1,5
      ADLOC(I)=0.
      ADDET(I)=0. ;ADROC(I)=0.
    ENDDO
C   All particulate LOC to bottom... as floccules...
    FLOC=WWAPOC*(1.-frCWWA-fraWWA)*QWWA/areaf(1)
C   Some of it is transformed into detritus during sedimentation...
    ADLOC(1)=pfloc8*FLOC
    ADDET(1)=FLOC*(1.-pfloc8)
C   utrients
    WWAPO4 =TIMSER(TNUB,PWA,14)
    WWASILIC =TIMSER(TNUB,SWA,14)
C   Main loop over all compartments----------------------------------
    DO 1 J=1,5
    I=ICOM(J)
C
C Primary producers ( F and G)-----------------------------------
C
    TDIA(I)=PDIA(I)+CDIA(I)* p4G/depfa(I)*(pwafl(I)*ptWET(I))
    SUSBDIA=SWIT4PEL* p4G/depfa(I)*(pwafl(I)*ptWET(I))*CDIA(I)+1.E-20
C Abiotic conditions for primary production
C (SUNQ is day length as a fraction of the day)
    HDW=SUNQ/2.;HWW=ptWET(I)*6.3
    PW1=MOD(TIME*.8,24.)-12.
    PW2=MOD(TIME*.8+12.6,24.)-12.
    OVERLAP=MAX(0.,MIN(HDW,PW1+HWW)-MAX(-HDW,PW1-HWW))
    +    +MAX(0.,MIN(HDW,PW2+HWW)-MAX(-HDW,PW2-HWW))
    PHYT(I)=TDIA(I)+PFLAG(I)
    CHLA(I)=PHYT(I)/CONVCHL
    EPS(I)=.4+ZZSILT*SILT(I)+.293E-3*PHYT(I)
    DEPFOT=LOG(LIGHT/(LIGHTMIN*SUNQ))/EPS(I)
    CORR=(MIN(1.,DEPFOT/depchFH(I))*(1.-pwafl(I))*OVERLAP
    + +MIN(1.,DEPFOT/depchLF(I))*(SUNQ-OVERLAP)
    + +MIN(1.,coFLeps*DEPFOT/depfh(I))*pWAFL(I)*OVERLAP)/24.
C Flagellate phytoplankton dynamics (F)
    miPO4=PO4(I)/(PO4(I)+KSPO4)
    rtempF=2.**((TEMP(I)-12.)*q10PH)
    RTPH=RTMPH*rtempF/(RTMPH+rtempF-1.)
    miSAL=SALT(I)/(SALT(I)+SALTM)
C Potential primary production
    asF=PFLAG(I)*PMAXPH*RTPH*miSAL*CORR
C LOC excretion under nutrient stress
    LOCF=asF*pFDET*(1.-miPO4)
```

```
C Apportioning of LOC- and DET-fraction of excretion
      flFC = caFC*LOCF
      flFD = LOCF-flFC + asF* pFD *miPO4
C Potential primary production
      asF = asF*miPO4
C Gross primary production minus excretory losses
      ppF = asF-flFD-flFC
C Rest respiration
      rrsF = PFLAG(I)*RTPH*rs0F
C Activity respiration
      arsF = pRSF*asF
C Nutrient-stress respiration
      srsF = LOCF*pRSFA
C Total respiration
      rsF = rrsF + arsF + srsF
      flKF = MAX(0.,ppF-rsF)*PO4CO
C Pelagic diatom dynamics (G)
      miNU = MIN(miPO4,SILIC(I)/(KSSIL + SILIC(I)))
      rtempG = 2.**((TEMP(I)-12.)*q10PD)
      RTPD = RTMPD*rtempG/(RTMPD + rtempG-1.)
C Potential production (assimilation)
      asG = PDIA(I)*PMAXPD*RTPD*CORR
C LOC production under nutrient stress
      LOCG = asG* pGDET*(1.-miNU)
C Apportioning of LOC- and DET- fraction of excretion
      flGC = caGC*LOCG
      flGD = LOCG-flGC + pGD*asG*miNU
C Gross activity (assimilation)
      asG = asG*miNU
      ppG = asG-flGD-flGC
C Rest respiration
      rrsG = PDIA(I)*RTPD*rs0G
C Activity respiration
      arsG = pRSG*asG
C Nutrient-stress respiration
      srsG = LOCG*pRSGA
C Total respiration
      rsG = rrsG + arsG + srsG
      R = SEDPDQ*PDIA(I)*RTPD*(pwafl(I)*ptWET(I))
      flG8 = R*caGC
      flG2 = R-flG8
      R = MAX(0.,ppG-rsG)
      flKG = R*PO4CO
      flLG = R*SILCO*(1.-.033*TEMP(I))
C Primary production in the water column
      MPPP(I) = (ppF + ppG)*depca(I)
C
C Zooplankton (H)-------------------------------------------------
C
C The physiological temperature response
      q10H = 2.**((TEMP(I)-12.)/10.)
C Temperature effect on rest respiration rate
      rrsH = rs12H*q10H
C Amount of food available for filter feeding (mg/m3)
      foodHf = pFH*PFLAG(I) + PDIA(I) + SUSBDIA + pSUSP(I)* pCH*PDET(I)
C Amount of food for raptorial feeding
      foodHr = PMIC(I) + pHH*PCOP(I)
```

```
C Utilizable part of foodHf
    utilHf = effFH*pFH*PFLAG(I) + effGH*PDIA(I) + eff4H*SUSBDIA
   1  + effCH*pSUSP(I)*pCH*PDET(I)
C Utilizable part of foodHr
    utilHr = effEH*PMIC(I) + effHH*pHH*PCOP(I)
C Fraction of copepods filter feeding
    pHf = utilHf/(utilHf + utilHr)
C Fraction of copepods feeding raptorially
    pHr = 1.-pHf
C Effect of food concentration on uptake rate
C Relative rate of uptake dependent on amount of foodHf
    FUPT1f = MAX(0.,(foodHf-minfdHf)/(ksmfdHf + foodHf-minfdHf))*rupH*q10H
C Temporary solution for inedible food
    IF(utilHf .LE. minfdHf)FUPT1f = 0.
    FUPT1r = foodHr/(ksmfdHr + foodHr)*rupH*q10H
    IF(utilHr .LE. ksmfdHr)FUPT1r = 0.
    FUPTOTf = FUPT1f*PCOP(I)*pHf
    FUPTOTr = FUPT1r*PCOP(I)*pHr
C Cleared volume by filter feeding and by raptorial feeding (m3/m3)
    RFVf = (FUPTOTf/foodHf)
    RFVr = (FUPTOTr/foodHr)
C Filtration activity respiration
    arsHf = qrsHf*RFVf
C Raptorial activity respiration
    arsHr = qrsHr*RFVr
C Effective volume cleared due to double filtering
    TfVQ = 1.-EXP(-RFVf)
C Effective volume preyed due to competition
    TrVQ = 1.-EXP(-RFVr)
C Ml swept clear per mgC PCOP
    RfVH = TfVQ*1.E6/(PCOP(I)*pHf)
    RrVH = TrVQ*1.E6/(PCOP(I)*pHr)
C Grazing on flagellate phytoplankton, pelagic and benthic diatoms
    flFH = TfVQ*PFLAG(I)* pFH
    flGH = TfVQ*PDIA(I)
    fl4H = TfVQ*SUSBDIA
C Grazing on pelagic detritus
    flCH = TfVQ* pCH*PDET(I)*pSUSP(I)
C Grazing on juvenile stages of copepods
    flHH = TrVQ*pHH*PCOP(I)
C Grazing on microplankton
    flEH = TrVQ*PMIC(I)
C At low food concentration the zooplankton stop filtering and hence
C have no filtration activity respiration
    IF (TfVQ .GT. 0.) THEN
      rsH = arsHf + arsHr + rrsH*PCOP(I)
    ELSE
      rsH = arsHr + rrsH*PCOP(I)
    ENDIF
C Amount of food handled
    wH = flFH + flGH + flEH + flHH + fl4H
C Relative amount of food ingested
    puw = 1.-SLOP
C Total food lost to sloppy feeding
    TOTSLOP = SLOP*wH
C Amount of food assimilated
    uFH = effFH*flFH*puw
```

```
      uGH = effGH*flGH*puw
      u4H = eff4H*fl4H*puw
      uEH = effEH*flEH*puw
      uHH = effHH*flHH*puw
      uCH = effCH*flCH
      uH = uFH + uGH + uEH + uHH + uCH + u4H
C Amount of faecal pellets produced
      exFH = (1.-effFH)*flFH*puw
      exGH = (1.-effGH)*flGH*puw
      exEH = (1.-effEH)*flEH*puw
      exHH = (1.-effHH)*flHH*puw
      exCH = (1.-effCH)*flCH
      exH = exFH + exGH + exEH + exHH + exCH
      flHA = exH*(1.-pHC)
C Calculation of the actual oxygen saturation
      OSAT = (475.-2.65*SALT(I))/(33.5 + TEMP(I))
      OXMIC = MIN(1.,OX(I)/OSAT)
C Death rates (natural and at low oxygen-saturation)
      ZDQ = (ksmoxH/(OXMIC + ksmoxH)-ksmoxH/(1. + ksmoxH)) + q10H*morH
      flHC = ZDQ*PCOP(I)
C
C Microplankton (E)---------------------------------------------
C
C Temperature effect
      q10E = 2.**((TEMP(I)-12.)/HTEMPE)
      CORROX = 1. + ksmoxE
C Oxygen limitation
      miOX = MIN(1.,CORROX*(OXMIC/(ksmoxE + OXMIC)))
C Available food
      foodE = pDE*PLOC(I) + (pCE*PDET(I) + pAE*PROC(I))*pSUSP(I) +
     + pGE*PDIA(I) + pGE*SUSBDIA + pFE*PFLAG(I) + pEE*PMIC(I) + pME*PBAC(I)
C Uptake
      uE = rupE*foodE/(foodE + ksmfdE)*q10E*PMIC(I)
      pupE = uE/foodE
C Fluxes into microplankton
      flME =  pupE*pME*PBAC(I)
      flDE =  pupE*pDE*PLOC(I)
      flCE =  pupE*pCE*PDET(I)*pSUSP(I)
      flAE =  pupE*pAE*PROC(I)*pSUSP(I)
      flGE =  pupE*pGE*PDIA(I)
      fl4E =  pupE*pGE*SUSBDIA
      flFE =  pupE*pFE*PFLAG(I)
      flEE =  pupE*pEE*PMIC(I)
C Fluxes from microplankton
C Respiration, mortality and excretion
      rsE = rs12E*q10E*PMIC(I) + uE*(1.-effE)*(1.- pexE)
      mortalE = ((1.-miOX)*morOXE + morE)*PMIC(I)
      exE = uE*(1.-effE)* pexE
      flED = (exE + mortalE)*pED
      flEC = (exE + mortalE)*(1.-pED)
C
C Pelagic bacteria (M)---------------------------------------------
C
C Temperature effect
      q10M = 2.**((TEMP(I)-12.)/HTEMPM)
C Oxygen limitation
      miOXM = OXMIC/(OXMIC + ksmoxM)
C Available food
```

```
C Labile Organic Carbon
    upDM = flFD + flGD + flED + PLOC(I)* pDM
C Pelagic DETritus
    upCM =  pCM*PDET(I)*pSUSP(I)
C Pelagic Dissolved Refractory Organic Carbon
    upBM =  pBM*PDROC(I)
C Pelagic Refractory Organic Carbon
    upAM =  pAM*PROC(I)*pSUSP(I)
C Total substrate available
    upM = upCM + upDM + upAM + upBM
C Potential uptake
    upMmax = rupM*q10M*miOXM*PBAC(I)
C Actual uptake
    uM = MIN(upMmax,upM)
C Fluxes into bacteria
    flAM = uM*upAM/upM
    flBM = uM*upBM/upM
    flCM = uM*upCM/upM
    flDM = uM*upDM/upM
C Fluxes from bacteria
C Respiration
    rsM = (1.-effM*OXMIC-effMa*(1.-OXMIC))*uM + rs12M*PBAC(I)*q10M
C Mortality
    flMD = morM*q10M*PBAC(I)
C Sedimentation
    M3 = sedM*(1.+PBAC(I)/dsedM)*PBAC(I)*(pwafl(I)*ptWET(I))
    flM8 = M3* pM8S
    flM3 = M3-flM8
C Netto growth
    PRODM(I) = uM-rsM
C
C Carnivores (I) -------------------------------------------------------
C
C Available food
    fdI = PBLAR(I)**xfdI + PCOP(I)**xfdI + (.5*PMIC(I))**xfdI
    foodI = fdI**(1./xfdI)
C Salinity/temperature dependency
    zsI = MAX(0.,(SALT(I)-SALCAR)/(SALCAR + SALT(I)))*3.
    ztI = MAX(0.,TEMP(I)/TEMCAR)
C Salinity / temperature effect on volume cleared of prey
    pVOLI = cVOLI*CARN(I)*zsI*ztI
C Uptake, including effects of food limitation
    uI = pVOLI*foodI*ksmfdI/(ksmfdI + foodI)
C Fluxes into carnivores
    flHI = PCOP(I)**xfdI/fdI*uI
    flJI = PBLAR(I)**xfdI/fdI*uI
    flEI = (.5*PMIC(I))**xfdI/fdI*uI
C Gross growth
    GCARN = effI*uI
C Excretion
    exI = (1.-effI)*uI
C Respiration
    RESCAR = rs0I*EXP(.069*TEMP(I))
C Respiration
    rsI = RESCAR*CARN(I)
C Mortality is a constant, increased by low oxygen
    flIC = CARN(I)*(CARMOR + CARMOX*(1.-OXMIC))
```

```
C
C Detritus (A + B + C + D)----------------------------------------------------
C
C Sedimentation PROC & PDET equal to netto sedimentation of silt
      flA1 = PROC(I)*rRBEN(I)*pWAFL(I)
      flC2 = PDET(I)*rRBEN(I)*pWAFL(I)
C All particulate effluent from WWA sedimentates in compartment 1 at a
C rate that is at least equal to the natural sedimentation rate.
      IF (I.EQ.1) THEN
        flA1 = MAX(flA1,QWWA/VOL(1)*WWAPROC)
        flC2 = MAX(flC2,QWWA/VOL(1)*WWAPDET)
      ENDIF
      ADLOC(I) = (flM8 + flG8)*depfa(I)/pwafl(I) + ADLOC(I)
      ADDET(I) = (flC2 + flG2)*depfa(I)/pwafl(I) + ADDET(I)
      ADROC(I) = ADROC(I) + flA1*depfa(I)/pwafl(I)
      DOC(I) = PLOC(I) + PDROC(I)
      POC(I) = PBAC(I) + PMIC(I) + PHYT(I) + PCOP(I) + pSUSP(I)*(PDET(I) + PROC(I))
      TOC(I) = DOC(I) + POC(I)
      SUSM(I) = POC(I)*cCARASD + SILT(I)
C
C Source equations --------------------------------------------------------
C
      SPMIC(I) = SPMIC(I) + uE-rsE-flEC-flEH-flEI-flEE-flED
      SPLOC(I) = SPLOC(I) + flFD + flGD + TOTSLOP-flDE-flDM + flED + .5*exI + flMD
      SPDET(I) = SPDET(I) + flGC + flFC + pHC*exH + .5*exI-flC2-flCE-flCM-flCH
     +       + flHC + flIC + flEC
      SPDROC(I) = SPDROC(I)-flBM
      SPROC(I) = SPROC(I) + flHA-flA1-flAE-flAM
      SBROC(I) = SBROC(I) + ADROC(I)
      SBDIA(I) = SBDIA(I)-(fl4H + fl4E)/SUSBDIA*CDIA(I)*p4G*ptWET(I)
     +       *swit4pel
      SPFLAG(I) = SPFLAG(I) + ppF-flFH-rsF-flFE
      SPDIA(I) = SPDIA(I) + ppG-rsG-flGH-flGE-(flG8 + flG2)
      SPCOP(I) = SPCOP(I) + uH-rsH-flHI-flHC-flHH
      SBLOC(I) = SBLOC(I) + ADLOC(I)
      SBDET(I) = SBDET(I) + ADDET(I)
      SPBLAR(I) = SPBLAR(I)-flJI
      SCARN(I) = SCARN(I) + GCARN-rsI-flIC
      SSILIC(I) = SSILIC(I)-flLG
      SBBAC(I) = SBBAC(I) + flM3*depfa(I)/pwafl(I)
      SPO4(I) = SPO4(I)-flKG-flKF
      SPBAC(I) = SPBAC(I) + uM-rsM-flME-flM3-flMD-flM8
C Calculate the oxygen balance:
      corWAS = maxCO*ksCO/(ksCO + OXMIC)
      O2 = rsF + rsG + rsE + rsM*corWAS + rsH + rsI
      BOC2(I) = (rrsF + rrsG + rsE + rsM*corWAS + rsH )*corCOr/cBOC2
      prodOX = (asF + asG)*corCOp
      consOX = O2*corCOr
      SOX(I) = SOX(I) + prodOX-consOX
1     CONTINUE
      END
```

Appendix B-3:

```
      SUBROUTINE SUBBEN
C
C   BENTHIC SUBMODEL
C
C   1 = BROC                        :Benthic refractory organic carbon
C   2 = BDET/ADET                   :Benthic detritus in the aerobic/anaerobic layer
C   3 = BBAC/ABAC                   :Benthic bacteria in the aerobic/anaerobic layer
C   4 = BDIA                        :Benthic diatoms
C   5 = BMEI                        :Meiobenthos
C   6 = BBBM                        :Deposit feeders (macrobenthos)
C   7 = BSF                         :Suspension feeders (macrobenthos)
C   8 = BLOC/ALOC                   :Benthic labile organic carbon
C                                    in the aerobic/anaerobic layer
C   0 = CO2                         :Carbon dioxide
C
$INCLUDE BAHBOE:COMM
      INTEGER IFIRST,ICOM(5)
      COMMON/DOLFYS/depfa(5),pWAFL(5),depth(5),VOL(5) ,depfh(5),
     &     depca(5),ptWET(5),areaf(5),cCARASD
      PARAMETER PI = 3.14159,PI2 = PI/2.
$INCLUDE DOLBEN:COMM
      DATA IFIRST/0/,ICOM/1,2,3,4,5/
      IF(IFIRST.EQ.0)THEN
C select last compartment for computing
      IFIRST = 1;I = COMPN;ICOM(5) = I;ICOM(I) = 5
      ENDIF
C Main loop over all compartments
      DO 20 J = 1,5
      I = ICOM(J)
      BOX = 0.;BOXS = 0.
C
C Restriction factors----------------------------------------
C Temperature: ACOS = -1 at time TSA and +1 at time TSA + 180
C ANCOS = -1 at time TSAN and +1 at time TSAN + 180
C
      ACOS = -COS((TIME-TSA)/180.*PI)
      ANCOS = -COS((TIME-TSAN)/180.*PI)
      IF(I.EQ.1)THEN
      TA = 13.5*ACOS
      TAM = 11.5*ANCOS
      ELSE IF(I.EQ.2)THEN
      TA = 12.5*ACOS
      TAM = 10.5*ANCOS
      ELSE
      TA = 11.0*ACOS
      TAM = 9.0*ANCOS
      ENDIF
C Actual temperature in aerobic (TASED) and anaerobic (TANSED) sediment
      TASED  = 11.0 + TA
      TANSED = 11.0 + TAM
      tem6 = (TASED + TANSED)*.5
      tem7 = ptWET(I)*TEMP(I) + (1.-ptWET(I))*TASED
C zt. : Temperature functions-----------------------------------------
      zt6a = (tem6 + 3.)/13.
```

```
      IF (tem6.LT.4.0) zt6a = MAX(1.E-6,tem6/10.)
      zt6c = MAX(1.,(tem6-T6T))
      zt6 = Q10BBM**((TA+TAM)/20.)
      zt6 = Q20BBM*zt6/(Q20BBM-1.+zt6)
      zt7a = (tem7+3.0)/13.0
      IF (tem7.LT.4.0) zt7a = MAX(1.E-6,tem7/10.)
      zt7c = MAX(1.,(tem7-t7c))
      zt0 = EXP(ct0*TASED)
      zt5 = Q10MEI**(TA/10.)
      zt5 = Q20MEI*zt5/(Q20MEI-1.+zt5)
      zt4 = Q10DIA**(TA/10.)
      zt4 = Q20DIA*zt4/(Q20DIA-1.+zt4)
      zt3 = ct3s**(TA/10.)
      zt3 = ct3m*zt3/(ct3m-1.+zt3)
      zt3a = ct3s**(TAM/10.)
      zt3a = ct3m*zt3a/(ct3m-1.+zt3a)
C Oxygen functions------------------------------------------------
C Oxygen saturation
      COX = (475.-2.65*SALT(I))/(33.5+TEMP(I))
C Relative saturation
      ROX = OX(I)/COX
C zox. : Reduction factors due to (low) oxygen concentration
      zox4 = MIN(1., ROX/qox4)
      zox5 = MIN(1., ROX/qox5)
      zox6 = MIN(1., ROX/qox6)
      zox7 = MIN(1., ROX/qox7)
C Other RESTRICTION FACTORS--------------------------------------------
      zMIX = 1.+AZMIX*COS(PI/180.*(TIME+30.))
      zSED = 2.-zMIX
C Calculation of the OVERLAP between the daylight period and low tide
      HLW = SUNQ/2.
      HDRYW = (MIN(1.,1.-ptWET(I)))*6.3
      PW1 = MOD((TIME+7.5)*.8,24)-12.
      PW2 = MOD((TIME+7.5)*.8+12.6,24)-12.
      OVERLAP = MAX(0.,MIN(HLW,PW1+HDRYW)-MAX(-HLW,PW1-HDRYW))
     +    +MAX(0.,MIN(HLW,PW2+HDRYW)-MAX(-HLW,PW2-HDRYW))
      pOVERLAP = OVERLAP/24.
C Light for phytobenthos, LIGHT and SUNQ computed in transport model
      SUNDRY = OVERLAP/SUNQ
      PHOT = LIGHT*SUNDRY
      LALL = LOG(MAX(1.,PHOT/PHOTMIN))/KD(I)
      LADD = OVERLAP*LAD(I)
      LAL = LALL+LADD
C
C Calculation of the variations = = = = = = = = = = = = = = = = = = = = = = = = =
C
C Primary production----------------------------------------------
      CBO = CBOF(I)*BDIA(I)**2/(BDIA(I)**2+kCBO**2)
      CDIA(I) = BDIA(I)*ATAN(.5 /CBO)/PI2
      UDIA = BDIA(I)*ATAN(BAL(I)/CBO)/PI2
      LDIA = BDIA(I)-UDIA
      FBDIA = BDIA(I)*ATAN(LAL /CBO)/PI2
      PPR = RPP*FBDIA*zt4
      PPM = PPM0+(1.-zox4)*PPMox
      PPX = (PPR+PPH+PPM)/2.
      PP = PPX-SQRT(PPX*PPX-PPH*PPR)
      cPP = (PPR-PP)/MAX(1.E-6,PPR)
```

```
C Respiration + mortality
    W48 = UDIA*rex4 *zt4 + PP*aex4 + cPP*bex4*(1-p42)*PP
    W42 = cPP*bex4*p42*PP
    W48A = LDIA*rex4 *zt4
    W40a = PP*ars4 + cPP*brs4*PP
    W40r = BDIA(I)*rrs4 *zt4
    W40 = W40a + W40r
C Total changes
    MBPP(I) = + PP -W40a-W40r*CDIA(I)/BDIA(I)
    SBDIA(I) = SBDIA(I) + PP -W40 -W48-W48A-W42
    SBLOC(I) = SBLOC(I) + W48
    SALOC(I) = SALOC(I) + W48A
    SBDET(I) = SBDET(I) + W42
C
C Meiobenthos-----------------------------------------------------------
C
C The uptake of meiobenthos is expressed in cm sediment consumed.
C We have to calculate the concentrations of the food sources by
C dividing biomass (mgC/m**2) by the thickness of the inhabited layer
C (cm)-- > concentrations in mgC/10.000cc. The layers are:
C      Aerobic bacteria      BAL(I)
C      Benthic diatoms       CBO
C      Meiobenthos           BAL(I)
C      Anaerobic bacteria    (TAL-BAL(I))
C      Buried diatoms        (TAL-BAL(I))
C
C The food sources are selected if POWM > 1; in that case there is prefe-
C rence for the most abundant one. The coefficients Q35 etc. give the
C available fraction of each food source.
C
    PW35 = (BBAC(I)/BAL(I))**POWM
    PW45 = (UDIA/BAL(I))**POWM
    PW55 = (BMEI(I)/BAL(I))**POWM
    PW35A = (ABAC(I)/(TAL-BAL(I)))**POWM
    PW45A = (LDIA/(TAL-BAL(I)))**POWM
    PW5S = PW35 + PW45 + PW55 + PW35A + PW45A
    FOOD5 = PW5S**(1/POWM)
    PupMEI = FOOD5/STANFO5*upMEI
    zFOOD5 = 2*F5/(PupMEI + F5)
    FOOD5X = zFOOD5*MAX(0.,(PupMEI*(1.-pex5-prs5)-asWORK5))
    rsex5r = (rrs5 + rex5)
    zFOOD5L = FOOD5X**POW5/(FOOD5X**POW5 + rsex5r**POW5)
    zFOOD5L = MAX(0.05,ZFOOD5L)
    FACTOR5 = zFOOD5*zt5*zox5*BMEI(I)*ZFOOD5L
    W5    = FACTOR5*PupMEI
    W35   = PW35/PW5S*W5
    W45   = PW45/PW5S*W5
    W55   = PW55/PW5S*W5
    W35A = PW35A/PW5S*W5
    W45A = PW45A/PW5S*W5
C Mortality: natural + mortality due to low oxygen saturation
    W58r = BMEI(I)*(rex5 + rex5O*(1-zox5))*zt5
C Mortality + excretion to LOC
    W52 = W45*pex5*p42
    W58 = W58r + W5*pex5 -W52
C Total respiration: rest + activity + uptake
    W50r = BMEI(I)*rrs5*zt5
```

```
      W50a =  W5*prs5 + FACTOR5*asWORK5
      W50  = W50r + W50a
C Net changes
      SBMEI(I) = SBMEI(I)-W50-W55-W56-W58 + W5-W52
C
C Macrobenthos--------------------------------------------------------
C            Deposit feeders
C Function for spawning:
      zSPAWN = cSPAWN**4*(.75/(cSPAWN**2 + (TIME-100)**2)**2 +
     1     .25/(cSPAWN**2 + (TIME-250)**2)**2)/(PI2*cSPAWN)
C Amount consumed by 1 mgC BBBM under reference conditions
      FOOD6 = (BMEI(I) + UDIA + BBAC(I))/BAL(I)*BSL
     +     + (LDIA + ABAC(I) )/LBBBM*BTL
C Limitaton of deposit feeding in case of abundant food
      zFOOD6 = 2*F6/(FOOD6 + F6)*zox6
C Faeces production at reference conditions
      PEX6A = pex6*(1 + FOOD6/(FOOD6 + F6))
C Uptake needed to cover the standard metabolic loss
      FOOD6X = (1-pex6A)*zFOOD6*FOOD6*(1-prs6)-zFOOD6*asWORK6
      FOOD6X = MAX(0.,FOOD6X)
C Standard metabolic loss
      REST6 = (rrs6 + rex6)*zt6c/zt6a
C Limitation of uptake at low food concentrations
      zFOOD6L = FOOD6X**POW6/(REST6**POW6 + FOOD6X**POW6)
      zFOOD6L = MAX(0.05/ZFOOD6,ZFOOD6L)
C Calculation of the actual uptake
      DIFF6 = BBBM(I)*zFOOD6*zt6*ZFOOD6L
      FACTOR6 = DIFF6*zt6a/zt6
C
C BSLX: deposit feeding above the sulphide horizon (cm/day)
C BTLX: deposit feeding below the sulphide horizon (cm/day)
      BSLX = BSL*FACTOR6
      BTLX = BTL*FACTOR6
C Actual grazing fluxes
      W36  = BSLX*BBAC(I)/BAL(I)
      W46  = BSLX*UDIA /BAL(I)
      W56  = BSLX*BMEI(I)/BAL(I)
      W36A = BTLX*ABAC(I)/LBBBM
      W46A = BTLX*LDIA /LBBBM
      W6   = W36 + W46 + W56 + W36A + W46A
C Rest respiration
      W60r =  BBBM(I)*rrs6*zt6c
C Rest excretion
      W68r =  BBBM(I)*(rex6 + rex6O*(1-zox6))*zt6c
C Activity excretion/egestion
      FAEC = PEX6A*W6
      W62A = (W46 + W46A)/W6*FAEC*p42
      W68A = W68r +  FAEC        -W62A
C Total respiration
C activity respiration is expressed as a function of the actual uptake
      W60  = W60r + W6*prs6*(1-PEX6A) + asWORK6*FACTOR6
C Spawning
      SPAWN6 = BBBM(I)*RSPAWN6*zSPAWN
C Source equations
      SBBBM(I) = SBBBM(I) + W6-W60-W68A-W62A-SPAWN6
      SPBLAR(I) = SPBLAR(I) + SPAWN6/depfa(I)*pWAFL(I)
C
C Macrobenthos-------------------------------------------------------
```

```
C           suspension feeders
C
C Spawning
    SPAWN7 = RSPAWN7*BSF(I)*zSPAWN
    SPBLAR(I) = SPBLAR(I) + SPAWN7/depfa(I)*pWAFL(I)
C Calculation of available food
    PHYTO = PFLAG(I) + PDIA(I) + susBDIA*CDIA(I)/depfa(I)
    FOOD7 = uM7*PBAC(I) + PBLAR(I) + PMIC(I) + PHYTO
C Calculation of effect of particle concentration on filtration rate
    PART = PARPHY*PHYTO + (SILT(I)
   1          + cCARASD*pSUSP(I)*(PROC(I) + PDET(I)))*PARSIL
    zPAR = 2.*OCP/(2.*OCP + PART)
    FILB = (1-pex7)*RRFIL*ptWET(I)*FOOD7*zPAR*zox7*(1.-prs7)-
   1       FACT7*zPAR*zox7
    zFOOD7L = FILB**8/((rrs7*zt7c/zt7A)**8 + FILB**8)
    FACTOR7 = CIND(I)*BSF(I)*zt7A*zox7
    FIL = FACTOR7*RRFIL*zFOOD7L*ptWET(I)*zPAR
    COR0 = FIL/depfa(I)
    COR2 = 1.; IF(COR0 .GT. 0.001)COR2 = (1.-EXP(-COR0))/COR0
    FILA = FIL*FOOD7*COR2
    MAXBSF = pup7*zt7a*BSF(I)
    zFOOD7 = MIN(1.,2.*MAXBSF/(FILA + MAXBSF))
    FFIL = FILA*zFOOD7
    W72 = pex7*FFIL
    W70a = prs7*(1.-pex7)*FFIL + FACT7*FACTOR7*zFOOD7L*zFOOD7
    W70r = rrs7*BSF(I)*zt7c*CIND(I)
    PR7 = FFIL-W72-W70a-W70r-SPAWN7
    W72S = rmor7*(1.-zox7)*BSF(I)
    W47 = susBDIA*CDIA(I)*FFIL/depfa(I)/FOOD7
    pFIL = FFIL/depfa(I)/FOOD7*pWAFL(I)
C Source equations
    SBSF(I) = SBSF(I) + PR7-W72S
    SPFLAG(I) = SPFLAG(I)-PFLAG(I)*pFIL
    SPDIA(I) = SPDIA(I)-PDIA(I)*pFIL
    SPBAC(I) = SPBAC(I)-uM7*PBAC(I)*pFIL
    SPBLAR(I) = SPBLAR(I)-PBLAR(I)*pFIL
    SPMIC(I) = SPMIC(I)-PMIC(I)*pFIL
    SBDET(I) = SBDET(I) + W72 + W72S
    SBDIA(I) = SBDIA(I)-W47
C
C Bacteria-----------------------------------------------------------
C
C LOC uptake
    RLOC = BLOC(I)*QLOC
    zBAC  = ksBAC/(ksBAC + BBAC(I)/BAL(I))
    FBAC = UPBAC*BBAC(I)*zt3*zBAC
    W83   = RLOC*FBAC/(FBAC + RLOC)
    T8    = RLOC-W83
    RLOCA = ALOC(I)*QLOC
    zBACA = ksBACA/(ksBACA + ABAC(I)/BAL(I))
    FBACA = upBACA*ABAC(I)*zt3A*zBACA
    W83A = RLOCA*FBACA/(FBACA + RLOCA)
C DET uptake
    W23   = BDET(I)*FRBD *zt3
    W23A = ADET(I)*FRBDA*zt3a
C ROC uptake
    W13   = BROC(I)*FRBR *(BAL(I)/TAL)*zt3
    W13A = BROC(I)*FRBRA*((TAL-BAL(I)) /TAL)*zt3a
```

```
C
   zDup  = 2.*W83 /(W23 + W83 )
   zDupA = 2.*W83A/(W23A + W83A)
   zRup  = 2.*W83 /(W13 + W83 )
   zRupA = 2.*W83A/(W13A + W83A)
C
   W23  = W23 *zDup
   W23A = W23A*zDupA
   W13  = W13 *zRup
   W13A = W13A*zRupA
C Actual uptake
   W3   = W83 + W23 + W13
   W3A  = W83A + W23A + W13A
C Respiration
   W30  = W83 *arlBAC + W23 *ardBAC + W13 *arrBAC
   +     + BBAC(I)*rrsBAC *zt3
   W30A(I) = W83A*arlBACA + W23A*ardBACA + W13A*arrBACA
   +     + ABAC(I)*rrsBACA*zt3a
C Mortality + excretion
   zmor3 = 1.-MIN(1.,W83/W3)
   zmor3A = 1.-MIN(1.,W83A/W3A)
   mor3 = morBAC*BBAC(I)*zt3*zmor3
   W32  = mor3*pm32
   mor3A = morBACA*ABAC(I)*zt3A*zmor3A
   W32A = mor3A*pm32
   W38  = W83 *exlBAC + W23 *exdBAC + W13*exrBAC+mor3*(1.0-pm32)
   W38A = W83A*exlBACA + W23A*exdBACA + W13A*exrBACA + mor3a*(1.0-pm32)
   W31  = W23 *tdrBAC
   W31A = W23A*tdrBACA
C
C Active vertical transport---------------------------------------
C
C Bioturbation
C BIOAM(I) = compartment dependent bioturbation amplifier
   BIOTUR = BTL*BIOAM(I)*DIFF6
C SML : mixing by storms in cm/year
   zSM = BAL0/(BAL0 + BAL(I))
   SMLX = SML*zMIX*zSM
C SDL : sedimentation in cm/day
   SDLX = SDL*zSED
   pDISEXP = 1.-EXP(-(BIOTUR + SMLX)*.25)
C Carbon fluxes across the sulphide horizon
   T2   = BDET(I)/BAL(I)*(BIOTUR + SMLX + SDLX)-ADET(I)*pDISEXP
   T3   = BBAC(I)/BAL(I)*(BIOTUR + SMLX + SDLX)
   T3A  = ABAC(I)*pDISEXP
   T8   = T8 + BLOC(I)/BAL(I)*(BIOTUR + SMLX + SDLX)-ALOC(I)*pDISEXP
   TSUL = BSUL(I)/(TAL-BAL(I)) *(BIOTUR + SMLX )
   SBDET(I) = SBDET(I)-T2 + T3A*pM32
   SADET(I) = SADET(I) + T2 + T3*pM32
   SBLOC(I) = SBLOC(I)-T8 + T3A*(1.0-pM32)
   SALOC(I) = SALOC(I) + T8 + T3 *(1.0-pM32)
   SBBAC(I) = SBBAC(I)-T3
   SABAC(I) = SABAC(I)-T3A
   SBSUL(I) = SBSUL(I)-TSUL
   BOXS = BOXS + TSUL
C
C Source equations----------------------------------------------------
C
```

```
C Grazing by meiobenthos and macrobenthos
    SBBAC(I) = SBBAC(I)-W35 -W36
    SABAC(I) = SABAC(I)-W35A-W36A
    SBDIA(I) = SBDIA(I)-W45 -W46 -W45A -W46A
C Respiration and excretion by macrobenthos and meiobenthos
    BOX = BOX  + W50  + W60  + W40r + W70r
    SBLOC(I) = SBLOC(I)  + W58
    SBDET(I) = SBDET(I)  + W52
    SALOC(I) = SALOC(I)  + W68A
    SADET(I) = SADET(I)  + W62A
C Bacteria
    SBBAC(I) = SBBAC(I) + W3 -W30 -W38 -W31 -W32
    SABAC(I) = SABAC(I) + W3A-W30A(I)-W38A-W31A-W32A
    BOX = BOX  + W30
    SBSUL(I) = SBSUL(I)  + W30A(I)
    SBLOC(I) = SBLOC(I)-W83  + W38
    SALOC(I) = SALOC(I)-W83A  + W38A
    SBDET(I) = SBDET(I)-W23  + W32
    SADET(I) = SADET(I)-W23A + W32A
    SBROC(I) = SBROC(I)-W13 -W13A  + W31  + W31A
C
C Export ---------------------------------------------------------------
C
C Burial
    SEDROX = BROC(I)*SDLX/TAL
    SEDPYX = BPYR(I)*SDLX/(TAL-BAL(I))
    SEDSUL = BSUL(I)*SDLX/(TAL-BAL(I))
    SBROC(I) = SBROC(I)-SEDROX
    SBPYR(I) = SBPYR(I)-SEDPYX
    SBSUL(I) = SBSUL(I)-SEDSUL
C
C Phytobenthos contribution to benthic oxygen dynamics
    PPox = PP*corCOr/corCOp-W40a
    POX  = 0.5*LAL/BAL(I)*PPox
    COXC = COX/corCOr*0.01
    COXP = COXC + (PPox-POX)/(reacon*zt0/0.01*MAX(1.E-6,pOVERLAP))
C
C Location of the sulphide horizon (BAL)-------------------------------
C
    cBUR = cBURY*zsm**2
    act6  = (DIFF6*ptWET(I) + BBBM(I)*cBUR6)*ALF6
    act7  = BSF (I)*zt7a*ptWET(I)*ALF7*CIND(I)*zOX7*zFOOD7*COR2
    act5  = BMEI(I)*zt5*zFOOD5*ALF5*zSM**2
    actY  = EMES(I)*cBUR*ALFY
    ABIO = (act6 + act7 + act5 + actY + 1.0)
    AMIX1 = ALF0*zMIX*zsm**2 + ALFt*TPUMP(I)
    AMIX  = AMIX1 + ABIO
    DIFFX = DIFF*AMIX*zt0
    DIFFY = DIFF*zt0*(1.0 + actY + act5 + BBBM(I)*cBUR6*ALF6)
    BOXDIF = .5*(W50 + W60 + W70r + TSUL + W30 + W40R*CBO/BAL(I))-POX
C sulphide
    SOEQ = FLSO*BSUL(I)*DIFFY*3./(TAL-BAL(I))**2.
C Unit of OXO : mgC./m2/cm  = mgC/m3*.01
    OXO = (ptWET(I)*OX(I) + (1-ptWET(I)-pOVERLAP)*COX)/corCOr*
   + 0.01 + pOVERLAP*COXP
C Changes due to the shift DL of the sulphide horizon
    DL = (DIFFX*OXO-(BOXDIF + SOEQ)*BAL(I))/(BOXDIF + SOEQ*(1. + BAL(I)/
        DAMP))
```

```
      IF(DL.GT.DAMP) DL=DAMP
      IF(DL.LT.-DAMP)DL=-DAMP
      IF(DL.GT.0) THEN
        pDL=1.-EXP(-DL*.25)
        S2=ADET(I)*pDL
        S3=0
        S3A=ABAC(I)*pDL
      ELSE
        S2=BDET(I)/BAL(I)*DL
        S3=BBAC(I)/BAL(I)*DL
        S3A=0
      ENDIF
C Sulphide to pyrite and oxidation of sulphide-------------------------
      SO=SOEQ*(1.+DL/DAMP)
      SO2=SO*(AMIX1+ACT7)/AMIX
      SP=FLSP*BSUL(I)
      SBSUL(I)=SBSUL(I)-SO-SP
      BOXS =BOXS +SO
      SBPYR(I)=SBPYR(I) +SP
      SBDET(I)=SBDET(I)+S2+S3A*pM32
      SADET(I)=SADET(I)-S2-S3*pM32
      SBLOC(I)=SBLOC(I)+S3A*(1.0-pM32)
      SALOC(I)=SALOC(I)-S3*(1.0-pM32)
      SBBAC(I)=SBBAC(I)+S3
      SABAC(I)=SABAC(I)-S3A
      CONSOX=((W70a/ptWET(I)+BOX+BOXS)*corCOr)/depfa(I)
     1                                                    *pWAFL(I)*ptWET(I)
      SOX(I)=SOX(I)-CONSOX
C New location of the sulphide horizon
      SBAL(I)=SBAL(I)+DL
C Calculation of the community respiration
      MBOX(I) =BOX-W70r
      MSBOX(I)=BOXS+MBOX(I)
      MCONS(I)=MBOX(I)+PSO*(BOXS-SO2)
C Calculation of the total carbon concentration:
      BTOC(I)=BROC(I)+ADET(I)+BDET(I)+ALOC(I)+BLOC(I)+
     1        ABAC(I)+BBAC(I)+BDIA(I)+BMEI(I)+BBBM(I)+BSF(I)
   20 CONTINUE
      END
```

Appendix B-4:

```
      SUBROUTINE SUBEPI
C
C EPIBENTHIC SUBMODEL
C
C X=BIRDS    Birds (numbers/compartment)
C Y=EMES     Mesoepibenthos (mg C/m2)
C Z=EMAC     Macroepibenthos (mg C/m2)
C
$INCLUDE BAHBOE:COMM
      INTEGER ICOM(5),IFIRST
      COMMON/DOLFYS/depfa(5),pWAFL(5),depth(5),VOL(5) ,depfh(5),
     +    depca(5),ptWET(5),areaf(5)
      REAL PFOOD(5),PRMAC(5),PRMES(5),rOSAT(5),F(5),RDSTY(5)
```

```
      REAL BIRDN(5),fdY(5),fdZ(5),EXPRT(5),okZ(5),okY(5),BFOOD(5)
$INCLUDE DOLEPI:COMM
      DATA IFIRST/0/,ICOM/1,2,3,4,5/
      IF (IFIRST.EQ.0) THEN
C     Select last compartment for computing rates
      I=COMPN;ICOM(5)=I;ICOM(I)=5;IFIRST=1
      ENDIF
C
C   Redistribution of EMES and EMAC--------------------------------
C
      MTIME=AMOD(TIME,360.)
      IF (MTIME.LE.1.)THEN
       FI=1.; FE=1.
       TOTTEMP=0.
      ENDIF
      DO 10 J=1,5
       I=ICOM(J)
C
C   The comfort functions for oxygen, temperature and salinity influence
C   migration
      OSAT=(475.-2.65*SALT(I))/(33.5+TEMP(I))
      CTEMP=MAX(1.E-6,TEMP(I))
      rOSAT(I)=OX(I)/OSAT
      okoxZ=(rOSAT(I)/koxZ)**xoxZ
      okoxZ=okoxZ/(okoxZ+1.)
      okY(I)=(rOSAT(I)/koxY)**xoxY
      okY(I)=okY(I)/(okY(I)+1.)
      oksalZ=(SALT(I)/ksalZ)**xsalZ
      oksalZ=oksalZ/(oksalZ+1.)
      oktZmin=(CTEMP/ktZmin)**xtZ
      oktZmin=oktZmin/(oktZmin+1.)
      oktZmax=(CTEMP/ktZmax)**xtZ
      oktZmax=1.-oktZmax/(oktZmax+1.)
      okZ(I)=okoxZ*oksalZ*oktZmin*oktZmax
      okZ(I)=okZ(I)+I**2*1.E-6
C
C Calculation of the amount of food-----------------------------------
C
      F(I)=MIN(1.,.5/BAL(I))
      fdY(I)=(u3Y*F(I)*BBAC(I))**xfd+(u4Y*CDIA(I))**xfd
  +  +(u2Y*F(I)*BDET(I))**xfd+(u5Y*F(I)*BMEI(I))**xfd
  +  +(uYY*EMES(I))**xfd
  +  +(uHY*PCOP(I)*depfa(I))**xfd
      fdY(I)=fdY(I)**(1./xfd)
      fdZ(I)=(u5Z*F(I)*BMEI(I))**xfd  +(uZZ*EMAC(I))**xfd
  +  +(u6Z*BBBM(I))**xfd+(u7Z*BSF(I))**xfd+(uYZ*EMES(I))**xfd
      fdZ(I)=fdZ(I)**(1./xfd)
   10 CONTINUE
C
C Immigration/Emigration-----------------------------------------------
C
      DO FOR J=1,5
       EXPRT(J)=0.
      ENDDO
      TOTTEMP=TOTTEMP+TEMP(5)*DELT
      IF (TOTTEMP .GE.TIMM .AND. FI.EQ.1.) THEN
       DSTART=MTIME
       TMAXIN=MTIME+DELT1
```

```
        TENDIN = TMAXIN + DELT2
        FI = 0.
ENDIF
      MAXSETTLE = 2.*SETTLE/(DELT1 + DELT2)
      IMMI = 0.
      IF ((FI.EQ.0.).AND.(MTIME.LE.TMAXIN)) THEN
        IMMI = (MTIME-DSTART)*MAXSETTLE/DELT1
      ELSEIF((FI.EQ.0.).AND.
            (MTIME.GT.TMAXIN).AND.(MTIME.LE.TENDIN)) THEN
        IMMI = MAXSETTLE*(1.-(MTIME-TMAXIN)/DELT2)
      ENDIF
      TEMAC(5) = IMMI
      TS5EMAC = TS5EMAC + IMMI*areaf(5)
C Emigration of macroepibenthos: starts after day "DCRIT"
C "EXPRT" is defined as being < = 0
      IF(MTIME.GT.DCRIT.AND.TEMP(1).LE.TEMI.AND.FE.EQ.1.)THEN
        TCRIT = MTIME
        FE = 0.
      ENDIF
      IF (FE.EQ.0.) THEN
        DEXPRT = (MTIME/TCRIT)**(TCRIT-DCRIT)
        DEXPRT = DEXPRT/(DEXPRT + 40.)
        DO 15 J = 1,5
          EXPRT(J) = -DEXPRT*routZ*EMAC(J)
          TEMAC(J) = TEMAC(J) + EXPRT(J)
   15   CONTINUE
        DO 16 J = 2,5
          TEMAC(J) = TEMAC(J)-(EXPRT(J-1)*areaf(J-1))/areaf(J)
   16   CONTINUE
        TS5EMAC = TS5EMAC + EXPRT(5)*areaf(5)
      ENDIF
C
C Migration between compartments---------------------------------------
C
      DO 21 I = 1,5
        RDSTY(I) = (.05 + (1-OKY(I)*.95))*REDIST
   21 CONTINUE
      DO 20 J = 4,1,-1
        ENMES = ENMAC = 0. ;traEMES = traEMAC = 0.
        DO 201 I = J,J + 1
          traEMAC = traEMAC + EMAC(I)*REDIST*areaf(I)
          traEMES = traEMES + EMES(I)*RDSTY(I)*areaf(I)
          PRMAC(I) = okZ(I)*areaf(I)*(fdZ(I)*ptWET(I))
          ENMAC = ENMAC + PRMAC(I)
          PRMES(I) = okY(I)*areaf(I)*(fdY(I)*ptWET(I))
          ENMES = ENMES + PRMES(I)
  201   CONTINUE
        DO 202 I = J,J + 1
        IF (ENMAC.GT.0.) TEMAC(I) =
    1     (traEMAC*PRMAC(I)/ENMAC)/areaf(I) + TEMAC(I)-REDIST*EMAC(I)
          IF (ENMES.GT.0.) TEMES(I) =
    1     (traEMES*PRMES(I)/ENMES)/areaf(I) + TEMES(I)-RDSTY(I)*EMES(I)
  202   CONTINUE
   20 CONTINUE
C
C Birds----------------------------------------------------------------
C
        BNORM = 0.;BFOODT = 0.;areaft = 0.
```

```
      DO 50 I = 1,5
        uhZX = uZX;IF(rOSAT(I).LE..1) uhZX = 1
        BFOOD(I) = ((uhZX*EMAC(I))**xfdX + (uYX*EMES(I))**xfdX +
     +        (u6X*BBBM(I))**xfdX + (u7X*BSF(I))**xfdX)**(1./xfdX)
        BFOODT = BFOODT + BFOOD(I)*areaf(I)
        areaft = areaft + areaf(I)
   50 CONTINUE
      BFOODM = BFOODT/areaft
      PFOODT = 0
      DO 55 I = 1,5
        PFOOD(I) = MIN((BFOOD(I)/BFOODM*PREF(I)),PREF(I))
        BNORM = BNORM + (PFOOD(I))
        PFOODT = PFOODT + PFOOD(I)
   55 CONTINUE
      BIRDS = BIRD(1 + INT(MTIME/30.))*PFOODT/(PFOOD(1) + PFOOD(2))
      DO 60 I = 1,5
        BIRDN(I) = BIRDS*PFOOD(I)/BNORM
   60 CONTINUE
      DO 1 J = 1,5
        I = ICOM(J)
C Reduction-factors
      ztZ = 2.0**((TEMP(I)-12.0)*Q10Z/10.)
      ztY = 2.0**((TEMP(I)-12.0)*Q10Y/10.)
C The temperature/activity-curves are flattened at high temperature
      ztZ = 3.0*ztZ/(2.0 + ztZ)
      ztY = 3.0*ztY/(2.0 + ztY)
C
C Uptake------------------------------------------------------------
C
C Uptake by macro- and mesoepibenthos
      rupY = ztY*rupYmax*ptWET(I)*fdY(I)/(fdY(I) + kfdY)
      upY = rupY*EMES(I)
      rupZ = ztZ*rupZmax*ptWET(I)*fdZ(I)/(fdZ(I) + kfdZ)
      upZ = rupZ*EMAC(I)
C Uptake by birds: birds eat until they have enough:
      BNEED = BNEC(1 + INT(MTIME/30.))
      BIUP = MIN(BFOOD(I)*areaf(I),BNEED*BIRDN(I))
      BIUP = BIUP/areaf(I)
C Losses are in part proportional to activity ("UP"), in part to
C biomass ("BI"); of all losses, a fixed part is due to respiration
C ("REFRAC"), the rest to excretion/mortality.
      prsYa = REFRAC*YUPLOS
      pexYa = (1.-REFRAC)*YUPLOS
      rrsYr = REFRAC*YBILOS
      rexYr = (1.-REFRAC)*YBILOS
      prsZa = REFRAC*ZUPLOS
      pexZa = (1.-REFRAC)*ZUPLOS
      rrsZr = REFRAC*ZBILOS
      rexZr = (1.-REFRAC)*ZBILOS
C
C Respiration---------------------------------------------
C
C Activity respiration
      rsYa = prsYa*upY
      rsZa = prsZa*upZ
C Rest respiration
      rsYr = rrsYr*ztY*EMES(I)
      rsZr = rrsZr*ztZ*EMAC(I)
```

```
C Total respiration
      rsY = rsYa + rsYr
      rsZ = rsZa + rsZr
C
C Excretion,mortality---------------------------------------
C
C Activity excretion
      exYa = pexYa*upY
      exZa = pexZa*upZ
C Mortality due to pelagic fishes
      rmorY = cmorY*(4.**4/(TEMP(I)**4 + 4.**4))*EMES(I)/(10. + EMES(I))
C This mortality is concentrated on the migrating population
      morY = rmorY*EMES(I)
C Rest excretion/mortality
      exYr = rexYr*ztY*EMES(I)
      exZr = rexZr*ztZ*EMAC(I)
C Moulting
      exZm = pZ2*upZ
C Total
      flY2 = pexY2*(exYa + exYr)
      flY8 = (1-pexY2)*(exYa + exYr)
      flZ2 = pexZ2*(exZa + exZr) + exZm
      flZ8 = (1-pexZ2)*(exZa + exZr)
C Fluxes----------------------------------------------------
      fl4Y = upY*(u4Y*CDIA(I)/fdY(I))**xfd
      fl2Y = upY*(u2Y*F(I)*BDET(I)/fdY(I))**xfd
      fl5Y = upY*(u5Y*F(I)*BMEI(I)/fdY(I))**xfd
      fl5Z = upZ*(u5Z*F(I)*BMEI(I)/fdZ(I))**xfd
      fl3Y = upY*(u3Y*F(I)*BBAC(I)/fdY(I))**xfd
      fl6Z = upZ*(u6Z*BBBM(I)/fdZ(I))**xfd
      fl6X = BIUP*(u6X*BBBM(I)/BFOOD(I))**xfdX
      fl7Z = upZ*(u7Z*BSF(I)/fdZ(I))**xfd
      fl7X = BIUP*(u7X*BSF(I)/BFOOD(I))**xfdX
      flZZ = upZ*(uZZ*EMAC(I)/fdZ(I))**xfd
      flZX = BIUP*(uhZX*EMAC(I)/BFOOD(I))**xfdX
      flYZ = upZ*(uYZ*EMES(I)/fdZ(I))**xfd
      flYY = upY*(uYY*EMES(I)/fdY(I))**xfd
      flYX = BIUP*(uYX*EMES(I)/BFOOD(I))**xfdX
      flHY = upY*(uHY*PCOP(I)*depfa(I)/fdY(I))**xfd
      PRY = upY-rsY-FLY8-FLY2-FLYZ-FLYY-FLYX-morY + TEMES(I)
      PRZ = upZ-rsZ-FLZ8-FLZ2-FLZX-FLZZ + TEMAC(I)
C Oxygen equations --------------------------------------
      BOXD = (rsY + rsZ)
      SOX(I) = SOX(I)-BOXD/(depfa(I)*375.)*PWAFL(I)*ptWET(I)
C Carbon flows----------------------------------------------------
      SEMES(I) = SEMES(I) + PRY
      SEMAC(I) = SEMAC(I) + PRZ
      SBLOC(I) = SBLOC(I) + flY8 + flZ8
      SBBAC(I) = SBBAC(I)-fl3Y
      SBDET(I) = SBDET(I)-fl2Y + flZ2 + flY2
      SBBBM(I) = SBBBM(I)-fl6Z-fl6X
      SBSF(I) = SBSF(I)-fl7Z-fl7X
      SBDIA(I) = SBDIA(I)-fl4Y
      SBMEI(I) = SBMEI(I)-fl5Y-fl5Z
      SPCOP(I) = SPCOP(I)-FLHY/depfa(I)*pwafl(I)
1     CONTINUE
      END
```

Appendix C: Data Files

Appendix C-1:

Data of the Physical Model

* exchange depth benthic/pelagic:
depfa(5):.485,.634,1.139,1.027,.631
* average fraction of compartment volume above the tidal flats
pWAFL(5):.8522,.5969,.1457,.0917,.0792
* average water depth of the compartments
depth(5):.682,.797,1.374,1.611,1.213
* water volume of the 5 compartments:
VOL(5):6.26E06,70.1E06,172.E06,392.E06,797.E06
depfh(5):.249,.306,.454,.420,.300
depca(5):3.05,4.38,5.64,6.04,6.68
ptWET(5):.325,.390,.596,.586,.451
areaf(5):11.E06,66.E06,22.E06,35.E06,100.E06
* conversion factor from carbon to ash-free dry weight
cCARASD = 2.2
depchLF(5):2.17,3.66,4.86,4.03,5.99
depchFH(5):4.3,5.6,6.61,5.65,7.48
* Time series for water temperature
* tTE contains the observation dates (yymmdd)
tTE(131):780101,
    780109,780117,780124,780131,780207,780215,780302,780307,780314,
    780316,780320,780330,780411,780418,780424,780509,780517,780523,780606,
    780613,780620,780627,780703,780711,780720,780725,780731,780809,780822,
    780828,780905,780911,780919,780925,781003,781016,781023,781030,781106,
    781113,781120,781127,781205,781211,781220,
    790115,790207,790226,790306,790312,790320,790409,790417,790423,
    790501,790507,790515,790529,790606,790611,790620,790625,790702,790710,
    790716,790723,790730,790806,790903,790910,791001,791009,791023,791029,
    791106,791112,791120,791126,791203,791210,791217,
    800102,800107,800128,800204,800211,800218,800225,800303,800310,800317,
    800324,800401,800408,800415,800421,800506,800513,800519,800527,800602,
    800609,800616,800624,800630,800708,800714,800721,800729,800804,800811,
    800818,800825,800901,800909,800915,800922,800929,801008,801014,801020,
    801027,801103,801110,801117,801124,801201,801208,801215
* Water temperature in the Dollard (comp 1 and 2)
TED(131):   3.5, 4.0, 4.9, 2.4, 2.5, 3.0, 1.0, 5.5, 5.9, 7.9, 6.8, 6.2, 7.9,
            8.0, 8.6, 9.6,13.4,12.0,14.5,19.8,13.9,18.3,15.0,15.3,16.0,16.0,
            17.8,21.2,16.8,20.0,14.0,14.4,16.7,14.5,14.7,12.1,12.7,10.5,10.1,
            10.7, 7.7, 8.5, 7.8, 3.0, 2.9, 1.3,
            −0.4, 1.0, 0.5, 2.2, 4.1, 4.1, 6.5, 6.6,10.1, 9.2, 8.2, 9.5,15.9,
            17.4,21.4,17.6,17.9,17.4,17.2,16.7,17.2,15.9,18.9,18.4,18.2,18.6,
            14.1,12.3,10.5, 5.0, 6.8, 6.1, 4.4, 4.7, 3.8, 8.2, 5.8,

                 1.9, 2.6, –.2, 2.0, 6.0, 5.5, 3.6, 4.6, 4.8, 4.3, 2.6, 6.6, 7.3,
                 7.3, 8.2, 9.1,12.2,16.1,16.8,15.4,19.3,19.2,16.1,13.8,16.7,15.0,
                 16.0,19.5,21.3,18.2,20.0,14.9,15.5,17.4,15.4,17.2,15.3,15.4,11.5,
                 12.0, 8.9, 6.8, 5.4, 7.3,10.2, 4.5, 1.5, 3.2

* Water temperature in comp 3
TE3(131):   5.9, 4.1, 4.9, 3.1, 2.8, 2.9, 0.0, 5.2, 5.6, 7.0, 6.4, 5.1, 7.2,
            7.6, 8.3, 9.6,13.3,11.8,14.9,19.3,14.2,17.4,15.2,15.1,15.3,16.2,
            17.8,19.5,17.3,18.3,15.4,14.8,16.3,14.5,14.7,12.1,12.8,11.2,10.2,
            10.0, 9.3, 8.5, 7.7, 4.0, 3.2, 2.5,
          –.5, .4, 1.0, 1.5, 4.2, 4.0, 5.0, 5.6, 8.4, 8.6, 7.2, 8.2,13.1,
            15.8,19.3,18.0,16.8,17.8,16.5,16.3,18.0,15.8,18.2,18.3,17.8,18.3,
            13.7,12.1,10.0, 6.0, 7.1, 6.5, 4.4, 4.6, 3.0, 7.8, 6.0,
            1.9, 3.3, 0.2, 2.1, 3.8, 5.2, 3.2, 4.6, 3.6, 3.9, 2.6, 6.1, 7.4,
            7.1, 7.8, 9.3,12.5,14.2,13.8,14.8,17.9,19.0,16.0,15.2,15.8,15.0,
            16.0,19.0,20.3,17.8,20.0,14.8,15.8,17.2,14.8,16.5,15.3,16.0,11.4,
            11.0, 8.9, 6.2, 5.1, 7.2,10.2, 5.4, 2.0, 3.1

* Water temperature in comp 4
TE4(131):   5.9, 4.2, 4.9, 3.8, 3.0, 2.9, 0.5, 3.2, 5.6, 5.6, 6.0, 5.7, 6.6,
            7.0, 8.7, 9.3,11.8,11.6,14.8,18.3,14.8,16.4,15.3,15.1,15.3,16.3,
            18.2,20.3,19.0,18.5,15.7,15.2,15.6,15.2,14.5,12.9,12.4,11.4,10.8,
            10.1, 8.8, 8.5, 7.3, 4.0, 4.2, 1.9,
          –.3, 0.5, 0.0, 2.3, 6.0, 3.9, 7.0, 7.7, 8.3, 9.2, 7.4,11.0,14.2,
            15.7,18.1,17.9,17.0,18.0,17.0,18.0,18.0,15.5,17.8,19.2,17.8,18.3,
            13.4,13.2, 9.3, 7.5, 8.1, 7.0, 4.4, 4.3, 7.8, 8.3, 6.0,
            3.5, 3.3, 2.0, 1.9, 4.0, 4.2, 2.5, 4.0, 4.1, 3.5, 3.3, 6.3, 7.3,
            8.8, 8.5,10.5,11.7,14.8,14.8,15.2,18.0,18.8,15.9,14.9,16.2,17.0,
            17.0,19.5,19.9,18.1,19.6,14.3,15.9,16.0,16.0,16.9,15.6,15.1,11.9,
            10.7,10.8, 7.0, 7.0, 9.6,10.4, 4.2, 3.5, 4.0

* Water temperature in comp 5
TE5(131):   6.0, 4.1, 5.0, 4.0, 3.0, 2.9, 2.5, 3.2, 5.5, 5.3, 5.7, 5.5, 6.2,
            7.4, 8.4, 9.5,11.8,11.6,15.2,18.2,14.8,16.4,15.3,15.0,15.2,16.9,
            18.0,19.0,18.8,18.5,15.7,15.0,15.3,15.4,14.3,11.5,12.6,11.6,10.1,
            10.3, 9.0, 8.5, 6.8, 3.1, 3.8, 3.3,
          –.3, .8,–.2, 2.5, 4.1, 4.4, 6.7, 7.5, 8.1, 9.5,10.0,10.5,13.8,
            15.3,18.0,17.6,16.7,18.2,17.0,17.7,17.4,16.4,17.0,19.5,17.4,17.8,
            13.7,12.4, 9.2, 6.9, 7.6, 7.2, 4.4, 4.3, 8.0, 7.3, 6.0,
            3.0, 3.4, 1.8, 2.1, 3.7, 4.1, 2.2, 3.5, 3.7, 3.6, 3.0, 6.0, 6.7,
            8.2, 8.7,10.0,11.6,14.2,14.5,14.7,17.2,18.0,15.6,15.1,16.2,16.0,
            17.0,19.0,19.4,18.0,19.2,14.0,16.0,16.0,15.2,16.2,16.0,15.3,11.4,
            10.9,11.9, 6.7, 6.4, 9.5,10.6, 4.5, 3.5, 4.5

* Time series for the river discharges (yymmdd):
TIMQ(74):770101, 770115, 770215, 770315, 770415, 770515, 770615,
              770715, 770815, 770915, 771015, 771115, 771215,
              780115, 780215, 780315, 780415, 780515, 780615,
              780715, 780815, 780915, 781015, 781115, 781215,
              790115, 790215, 790315, 790415, 790515, 790615,
              790715, 790815, 790915, 791015, 791115, 791215,
              800115, 800215, 800315, 800415, 800515, 800615,
              800715, 800815, 800915, 801015, 801115, 801215,
              810115, 810215, 810315, 810415, 810515, 810615,
              810715, 810815, 810915, 811015, 811115, 811215,
              820115, 820215, 820315, 820415, 820515, 820615,
              820715, 820825, 820915, 821015, 821115, 821215,
              821231

* discharge of the WWA (m3/s)
QQWA(74):20.8, 20.8, 23.4, 19.7, 16.8, 14.8, 12.7,
             13.3, 18.2, 16.1, 17.6, 26.1, 21.3,
             22.2, 11.9, 18.6, 14.8,   8.9, 11.2,

```
                        13.9,   13.4,   14.4,   13.9,   11.7,   15.8,
                         9.8,    8.6,   26.5,   18.5,   17.7,   13.8,
                        12.8,   17.2,   11.9,   12.6,   19.2,   24.2,
                        17.8,   22.0,   12.7,   12.3,   09.7,   13.5,
                        24.8,   15.8,   13.7,   15.3,   18.6,   18.8,
                        25.6,   20.1,   25.8,   08.1,   11.3,   13.8,
                        16.5,   11.2,   11.9,   14.8,   19.0,   19.4,
                        19.4,   18.4,   19.3,    9.3,    8.0,   10.7,
                         5.9,    8.1,   10.8,   10.1,   10.1,   16.3,
                        16.3
```
* discharge of the EMS (m3/s)
```
QQEM(74):98.1,      8.1,  157.9,    91.1,  129.4,    69.8,    68.7,
                   42.3,   60.2,    39.9,   42.4,    36.9,    63.5,
                  167.8,  123.1,   172.0,  112.0,   097.9,    54.1,
                  102.9,   50.3,    60.6,   81.4,   117.9,   111.4,
                    134,    138,     307,    141,     151,      79,
                     52,     73,      56,     47,      51,     133,
                  129.4,    238,    98.6,  126.5,    73.9,    48.9,
                  142.7,   60.9,    58.9,   55.3,   114.4,   169.7,
                  298.7,  170.4,   363.4,  117.4,    69.6,    64.2,
                  190.9,   72.4,    59.9,  101.6,   167.7,   255.6,
                  198.6,  210.9,   193.0,   88.9,    86.6,    57.7,
                   42.4,   31.3,    27.6,   47.6,    70.3,   117.7,
                  117.7
```
* salinity of the WWA:
Wwsalt = 0.23
* salinity on the EMS boundary at different Q:
Emsal(7):22.7301,20.3079,16.986,14.381,13.5934,6.85997,4.96238
* salinity at the SEA boundary at different Q:
Sesal(7):32.3973,31.4676,30.0475,29.9323,29.5976,29.1096,27.5
= = = = = DIFFUSIVE EXCHANGE =
* DIFF in 7 series of 5 : exchange coefficients for different Q
DIFF(35): 1.68545E6,1.12401E6,1.8366E6,2.10708E6,3.3023E6,3.36134E6,4.43303E6,
            2.648E6,6.28038E6,5.61693E6,5.85777E6,6.38184E6,4.47407E6,6.94809E6,
            4.76864E7,3.70021E7,2.97896E7,2.81564E7,2.98004E7,3.45403E7,8.33724E7,
            3.70543E7,5.06964E7,4.09795E7,4.70473E7,4.71053E7,5.83032E7,1.38681E8,
            6.81834E7,5.54624E8,7.1488E8,1.23011E8,3.9926E8,8.27397E7,2.81439E8
* exchange between comp 3 and EMS for different Q (used together with QE)
DIEMS(7):1.67392E7,1.75071E7,1.62462E7,1.41283E7,2.34235E7,1.80126E7,2.22327E7
* Q-series of EMS for slp functions for SAEMS,DIEMS
QE(7):2.21184E6,3.3264E6,4.32E6,4.60512E6,6.58368E6,1.13357E7,2.7648E7
* Q-series WWA for slp functions of DIFF(1..14)
QW(7):440640,734400,777600,872640,1.27872E6,1.34784E6 ,2.6784E6
* Q-series of tot. discharge for slp functions of SASEA, DIFF(15..35)
QS(7):2.8512E6,4.06944E6,4.76064E6,5.36543E6,7.62048E6,1.22602E7,3.03264E7
= = = = = OXYGEN =
* OX-CONC. on WWA,EMS and SEA at time t (day number): (g O2/m3)
OXCONT = -1.
pOsW(6):0,0,.8,.8,0,0
tpOsW(6):-10,15,30,210,245,370
Emox(14):10,10.6,10.6,10,8.5,8.5,7,5.3,5.5,6.5,8.2,9.1,10,10.6
Seox(14):9.5,9.5,9.5,10.2,10.2,10.5,7.2,7.7,7.2,7.1,8.5,9,9.5,9.5
tiox(14):-15,15,45,75,105,135,165,195,225,255,285,315,345,375
REACON = .96   * relative rate of reaeration (1/m)
ct0 = .023
= = = = = SILT =
pDOWN0 = .800
W50 = 2.

* WIND IN M/S
WIND(14):8.6,9.2,7.2,7.0,6.6,6.6,6.4,6.8,6.4,6.8,7.1,7.4,8.6,9.2
MOVE(5):15.E6,113.E6,299.E6,549.E6,914.E6
rTID = 1.95   * number of tides per day
seddep = .4365
resdep = .1261
REDIS(5):1.69,2.37,1.82,2.62,1.83
= = = = = BOUNDARY CONDITIONS SILT = = = = = = = = = = = = = = =
TSILTSEA = 42.E3
* WWA:
SIWWA(6):140.E3,50.E3,50.E3,140.E3,140.E3,50.E3
TISIW(6):-10., 10., 220., 240., 350., 370.
* EMS:
SIEM(13): 30.E3,20.E3,10.E3,30.E3,50.E3,
            50.E3,50.E3,40.E3,40.E3,30.E3,30.E3,30.E3,30.E3
TISI(13):0,31,59,90,120,151,181,212,243,273,304,334,365
= = = = = LIGHT =
CONLIGHT = 1.   * light control (see program)
* used for sensitivity analysis
* tTEex(2): X > 0 WARM WINTER   X < 0 COLDWINTER
tTEex(4):0,37,75,365
TEex(4):0.,0.,0.,0.

Appendix C-2:

| Data of the Pelagic Model |

= =
= = = = = = = A = PROC   F = PFLAG   J = PBLAR   = = = = =
= = = = = = = B = PDROC   G = PDIA   K = PO4   = = = = =
= = = = = = = C = PDET   H = PCOP   L = SILIC   = = = = =
= = = = = = = D = PLOC   I = CARN   M = PBAC   = = = = =
= = = = = = = E = PMIC   = = = = =
= =
swit4pel = 0   * Switch (1 = on, 0 = off) for feeding on SUSBDIA
= = = = = = = = PHYTOPLANKTON PARAMETERS = = = = = = = = = = = = =
PMAXPH = 3.77   * maximal productivity of PFLAG(flagellates)
PMAXPD = 3.77   * maximal productivity of PDIA (diatoms)
q10PH = .2   * 1/Doubling temp. PFLAG
q10PD = .15   * 1/Doubling temp. PDIA
RTMPD = 3.2   * Flattening coefficient of temp.response PDIA
RTMPH = 25.9   * Flattening coefficient of temp.response PFLAG
ZZSILT = .04E-03   * Contribution of SILT to light extinction
coFLeps = .25   * Correction of EPS for high resusp. on flats
rs0F = .03   * Resp.rate of phytoplankton at 0 deg.C
rs0G = .02   * Resp.rate of pelagic diatoms at 0 deg.C
KSPO4 = 100.   * Half-value of PO4-limitation (mµmol/m3)
KSSIL = 1000.   * Half-value of SIO4-limitation (mµmol/m3)
pFD = .1   * Fraction of prim.prod. of PFLAG excreted as PLOC/PDET
pGD = .1   ,,   PDIA   ,,
pFdet = .9   * Fr. of act. which is excreted in case of nutr.lim.
pGdet = .9   * Fr. of act. which is excreted in case of nutr.lim.
SEDPDQ = .05   * Sedimentation rate of DIATOMS
SILCO = 30.76   * SILIC uptake per unit biomass at 0 deg.Celsius
PO4CO = .79   * PO4 uptake per unit biomass
p4G = .15   * Fraction of total diatoms being benthic diatoms

```
pRSF = .1              * Activity respiration of PFLAG
pRSG = .1              * Activity respiration of PDIA
pRSFA = .9             * Activity respiration of PFLAG under nutr.limit. circumst.
pRSGA = .9             * Activity respiration of PDIA under nutr.limit. circumst.
caFC = .5              * Fraction of activ.-excretion of PFLAG going to PDET
caGC = .5              * Fraction of activ.-excretion of PDIA going to PDET
SALTM = 10.            * Salinity limitation of PFLAG
* channel depth in the compartments
DEPG(5): 3.05,4.38,5.64,6.04,6.68
LIGHTMIN = 8.1         * min. light-intens. for optimal production
CONVCHL = 35.          * conversion factor chlorophyll-a to PHYT
* = = = = = = = = COPEPOD PARAMETERS = = = = = = = = = = = = = = = = = = = *
rs12H = .0125          * rest resp. rate at 12 deg. C of copepods
qrsHf = 40.            * mgC respired/m3 filtered
qrsHr = 20             * Activity respiration in rapt.feeding(mg/m3)
minfdHf = 40           * Min.amount of food still filterable
ksmfdHf = 300          * Ks of food concentration filter feeding
ksmfdHr = 20.          * Ks of food concentration rapt. feeding
cHH = .333             * Cannibalizable fraction of PCOP
pCH = .5               * Fraction PDET eatable for PCOP
pFH = .5               * Fraction PFLAG available for PCOP
SLOP = .1              * Fr. of food lost due to break up of particles
pHC = .5               * Fr. of faecal pellets going to PDET (rest to PROC)
ksmoxH = 1.            * Oxygen saturation at which 1/2 the pop. dies/day
morH = .01             * Oxygen independent mortality
rupH = .8              * Max.relative uptake
effFH = .4             * Assimilation efficiency of PCOP for FLAG
effGH = .6             *         ,,                      PDIA
eff4H = .6             *         ,,                      SUSBDIA
effEH = .8             *         ,,                      PMIC
effHH = .85            *         ,,                      PCOP
effCH = .005           *         ,,                      PDET
* = = = = = = = = PELAGIC CARNIVORE PARAMETERS = = = = = = = = = = = = *
xfdI = 2.              * Exponent regulating apportioning of food sources
PREDQ = 10             * Max.pred.rate of carnivores
cVOLI = .01            * fractional volume cleared by 1 mg CARN
rs0I = .01             * Respiration of carnivores TEMP = 0
TEMCAR = 6.            * Temp. below which there is an activity limitation
SALCAR = 12.           * limiting salinity for pelagic carnivores
effI = .1              * assimilation efficiency
ksmfdI = 30            * The food amount at which predation is half the max.
CARMOR = .03           * Carnivore mortality
CARMOX = .2            * Carnivore mortality at low oxygen
* = = = = = = = = MICROZOOPLANKTON PARAMETERS = = = = = = = = = = = = *
ksmfdE = 200.          * food concentration where relative uptake is .5
rupE = 2.              * max. relative daily uptake as a fraction of biomass
pDE = 0.               * Availability constants PMIC ~ PLOC
pCE = .008             *                            ~ PDET
pAE = .00004           *                            ~ PROC
pFE = .4               *                            ~ PFLAG
pGE = .8               *                            ~ PDIA
pEE = .2               *                            ~ PMIC
pME = 1.               *                            ~ PBAC
effE = .3              * PMIC assimilation efficiency
pED = .5               * Fraction of excretion by PMIC going to PLOC
HTEMPE = 10.           * Doubling TEMP of PMIC
rs12E = .1             * Respiration of Microplankton Temp = 12 oC
ksmoxE = .25           * Oxygen SATURATION where respiration is .5
```

```
pexE=.5              * Excreted fraction of uptake (activity excretion)
morOXE=.25           * Mortality due to oxygen limitation
morE=.05             * Temp. indep. mortality
*======== MICROBIAL PARAMETERS ==================*
HTEMPM=10.           * Doubling temperature PBAC
rupM=6.3             * Uptake PBAC (temp=12) as a fraction of their biomass
ksmoxM=.01           * miOXM constant
pDM=1.               * fraction of PLOC-pool available for PBAC
pCM=.01              * fraction of PDET-pool available for PBAC
pBM=.0005            * fraction of PDROC-pool available for PBAC
pAM=.0005            * fraction of PROC-pool available for PBAC
effMa=.2             * Bacterial efficiency under oxygen stress
effM=.3              * Bacterial Efficiency
rs12M=.2             * Rest respiration (in absence of external mort.)
morM=.2              * Mortality
pM8s=.9              * Sedimentation of PBAC into BLOC
sedM=.22             * sedimentation of bac at low PBAC conc.
dsedM=250.           * Doubling concentr. of PBAC for sedimentation
corCOp=.00357        * Conversion of C(produced) into OX
corCOr=.00333        * Conversion of C(respired) into OX
cBOC2=.6             * Conversion of BOC2 in BOC1
maxCO=1.25           * Maximum value of corWAS
ksCO=4               * Half-value constant of corWAS equation
pfloc8=.25           * Fr of sedimenting LOC not transformed into detritus
*Time series of day numbers used as x-axis in the following variables
DTIM(14): -16,15,46,74,105,135,166,196,227,258,288,319,349,380
* Time series for PDROC on the EMS. EDP3: The value at 0 sal.
EDP3(14): 8000,12000,11000,9000,9000,9000,9000,8000,8000,8000, 10000,9000,8000,12000
* Time series for POC on the EMS ** this is year 1978 **
EPOC(14): 6000,7500,7000,6500,6500,6000,6000,6000,6000,7000,
          6500,6000,7500
rABC=.5              * fraction of PROC to PDROC on Ems
* Time series for fraction PROC in POC on the EMS (values+times)
pPeA(10):.92,.92,.7,.7,.6,.6,.7,.70,.92,.92
tPeA(10):1,74,84,166,176,227,237,258,268,370
* Time series of POC at SEA
SPOC(12):400,500,1100,1500,1900,2300,2350,2000,1400,750,450,400
* Time series for fraction PROC in (POC-PHYT) on the SEA (values+times)
pPsA(6):.85,.85,.55,.55,.85,.85
tPsA(6):0,85,95,270,280,370
* Time series pelagic dissolved ROC (PDROC) at SEA
jTIM(26):780101,  780115,  780215,  780315,  780415,  780515,  780615,
         780715,  780815,  780915,  781015,  781115,  781215,
         790115,  790215,  790315,  790415,  790515,  790615,
         790715,  790815,  790915,  791015,  791115,  791215,
         791231
SPD3(26):  500,
         500,100,1200,3000,3900,4200,2400,2700,3200,2420,2200,2100,
         2000,1500,1400,1750,2700,3100,3400,2950,2500,2050,2100,2500, 2250
* Time series WWA: TOC, DOC etc.
TWAS(87):  780101,780108,780122,780215,780315,780415,780515,
           780615,780715,780815,780908,780922,781008,
           781022,781108,781122,781208,781222,
           790108,790122,790215,790315,790415,790515,
           790615,790715,790815,790908,790922,791008,
           791022,791108,791122,791208,791222,
           800108,800122,800215,800315,800415,800515,
           800615,800715,800815,800908,800922,801008,
```

```
                        801022,801108,801122,801208,801222,
                        810108,810122,810215,810315,810415,810515,
                        810615,810715,810815,810908,810922,811008,
                        811022,811108,811122,811208,811222,
                        820108,820122,820215,820315,820415,820515,
                        820615,820715,820815,820908,820922,821008,
                        821022,821108,821122,821208,821222,
                        821231
WTOC(87):70.0E3,  32.0E3,  26.0E3,  27.0E3,  28.5E3,  29.4E3,  34.4E3,
                  30.5E3,  27.6E3,  28.8E3,  43.6E3,  82.0E3,110.0E3,
                 161.0E3,171.7E3,142.5E3,187.6E3,118.0E3,
                  35.5E3,  27.7E3,  28.7E3,  27.8E3,  23.1E3,  25.6E3,
                  25.4E3,  23.6E3,  24.9E3,  55.0E3,  83.3E3,119.2E3,
                 124.4E3,142.4E3,111.0E3,115.1E3,  65.0E3,
                  26.3E3,  35.8E3,  24.5E3,  19.5E3,  22.2E3,  27.1E3,
                  24.5E3,  27.5E3,  37.3E3,  51.7E3,  90.5E3,104.9E3,
                 100.7E3,109.6E3,  83.2E3,  48.9E3,  23.0E3,
                  23.6E3,  25.4E3,  21.7E3,  24.6E3,  21.1E3,  24.4E3,
                  25.2E3,  26.5E3,  44.3E3,  56.9E3,  71.4E3,112.7E3,
                 125.5E3,  98.8E3,112.7E3,  78.5E3,  70.5E3,
                  28.8E3,  29.1E3,  21.8E3,  21.7E3,  22.3E3,  19.9E3,
                  27.3E3,  22.7E3,  32.1E3,  54.8E3,  62.3E3,  97.2E3,
                 121.9E3,190.1E3,164.8E3,107.7E3,  28.7E3,
                  28.7E3
```

pPOC = .07   * fraction POC of TOC in WWA
* Day numbers and the fractions PROC in POC and PDET in POC on WWA
tpPW(8):1,12,17,55,60,237,248,370
pPWA(8):.10,.10,.65,.65,.35,.35,.10,.10
pPWC(8):.30,.30,.25,.25,.50,.50,.30,.30
* Day numbers and the fraction PDROC in DOC on WWA
tpDW(8):1,12,17,55,60,237,248,370
pDWB(8):.1,.1,.8,.8,.7,.7,.1,.1
* Day numbers for the following time series
TMAC(12): -1,31,60,90,120,150,180,210,240,275,310,366
* Boundary concentrations of PDIA at SEA
SDI(12): 13,60,90,345,740,100,60,30,20,4,33,13
* Boundary concentrations of PFLAG at SEA
PH(12): 2,15,20,40,200,410,560,300,250,21,12,2
* Boundary concentrations of PCOP at SEA
BCOP(12): .31,1.82,4.02,8.54,14.7,5.58,6.51,2.71,3.51,6.95,6.50,.40
* Boundary concentrations of PCARN at SEA
TCAR(12): .01,.02,1,10,30,5,2,10,8,2,1,.1
* = = = = = = = = BOUNDARY CONCENTRATIONS ON THE RIVER EMS = = = *
* Boundary concentrations of PCOP on the Ems
EMSH(12): 10,76,7.7,7.4,132.7,31.8,36,4,6.3,8.5,.01,10
* Day numbers for series of PMS,SMS
TNUA(20): -10,23,54,70,109,115,154,175,
          194,203,225,243,252,262,295,300,329,347,350,383
* Time series for PO4 on EMS at 0 sal.
PMS(20): 8.33E + 3,2.73E + 3,1.59E + 3,3.35E + 3,3.49E + 3,2.95E + 3,3.80E + 3,
         3.30E + 3,4.31E + 3,4.37E + 3,5.47E + 3,4.98E + 3,7.15E + 3,5.24E + 3,
         8.54E + 3,8.21E + 3,11.21E + 3,8.391E + 3,8.33E + 3,2.73E + 3
*Time series for SILIC on EMS at 0 sal.
SMS(20): 115E + 3,214E + 3,232E + 3,180E + 3,170E + 3,153E + 3,37E + 3,33E + 3,111E + 3,
         93E + 3,93E + 3,63E + 3,203E + 3,100E + 3,175E + 3,182E + 3,239E + 3,140E + 3,
         115E + 3,214E + 3
* Day numbers for PSE, SSE
TNUC(13): -10,23,54,70,109,154,194,243,262,295,329,350,383

* Time series for PO4 at SEA
PSE(13): .85E3,1.78E3,.88E3,1E3,.5E3,1.45E3,1.07E3,1.2E3,1.2E3,1.78E3,.95E3,
       .85E3,1.78E+3
* Time series for SILIC at SEA
SSE(13): 13E3,24E3,33E3,30E3,9E3,4E3,5E3,1E3,1E3,6E3,13E3,13E3,24E3
* Day numbers for PWA, SWA
TNUB(14): -10,15,46,75,106,136,167,197,228,259,289,320,350,375
* Time series for PO4 on WWA
PWA(14): 11.1E+3,13.6E+3,5.40E+3,6.5E+3,6.9E+3,
      6.7E+3,8.5E+3,8.5E+3,7.6E+3,17.4E+3,24.9E+3,35.1E+3,11.1E+3,13.6E+3
* Time series for SI on WWA
SWA(14): 171.E3,127.E3,101.E3,128.E3,137.E3,98.E3,59.E3,74.E3,108.E3,100.E3,
      90.E3,100.E3,171.E3,127.E3

Appendix C-3:

+---------------------------------+
| Data of the Benthic Model |
+---------------------------------+

```
*=========================================*
*=======  1=BROC    4=BDIA   7=BSF    ==========*
*=======  2=BDET    5=BMEI   8=BLOC   ==========*
*=======  3=BBAC    6=BBBM   0=CO2    ==========*
*=========================================*
*======== MACROBENTHOS SEDIMENT FEEDERS =========*
cSPAWN=5.        *
RSPAWN6=.2       *
rrs6=.003        * rest respiration BBBM per day
rex6=.0008       * rest mortality   BBBM
rex6O=.06        *                                    OX
prs6=.31         * activity respiration, fraction of uptake
pex6=.25         *     excretion
F6=0.03          * constant in ZFOODB , /DAY
BSL=.03E-3       * CM/DAY sediment consumption in aerobic layer
                 * for 1 mg bbbm under normal conditions
BTL=.01E-3       *                                in anaerobic layer
LBBBM=10.        * cm of anaerobic sediment inhabited by the food
asWORK6=.009     * respiration prop. to reworked cm
POW6=8.
*======== MEIOBENTHOS =====================*
rrs5=.008        *   mg per day and per mg biomass
rex5=.005        *   mg per day and per mg biomass
rex5O=.02        *   Mortality due to ox-limitation
pex5=.2          * excreted part of uptake
prs5=.30 .05     * respired part of uptake
asWORK5=0.020    * amount needed for reworking upMEI cm sediment
F5=0.0750        * food concentration for opt. growth for 1 mg animal
POWM=2.0         * power of food preference
upMEI=.072       * cm per day which 1 mg BMEI reworks
POW5=2.0         *
STANFO5=550      *
pAE5=.9          *
*======== TRANSPORT THICKNESSES IN CM ===========*
SML=.010         *
SDL=.0 022       * cm/day transport by phys. processes
TAL=30.          * total layer
*======== OXYGEN AND DIFFUSION ==============*
```

```
BAL0 = 3.              *
REACON = 1.44          * rel. reaeration rate per m
*= = = = = = = = FURTHER PARAMETERS = = = = = = = = = = = = = = = = = = = *
ALF6 = 14.E-3          *  contribution to oxygen-diff by macrobenthos
ALF7 = 1.4E-3          *                           by filterfeeders
ALF5 = 7.0E-3          *                           by meiobenthos
ALFY = 8.5E-3          *                           by epimesobenthos
ALF0 = 0.7             *                           by physical proc
ALFt = 0.7             *                           by tidal pump
DIFF = 1.73            * Diffusion constant (cm**2/day)
DAMP = .1              * Max. daily shift of BAL in cm
AZMIX = 0.4
cBUR6 = 0.1            * amplification diffusion by BBBM burrows
cBURY = 1.0            * amplification diffusion by EMES burrows
BIOAM(5):2.0,1.0,3.0,4.0,10.0   *comp. dependent bioturbation amplifier
*= = = = = = = = SULPHIDE PARAMETERS = = = = = = = = = = = = = = = = = = *
FLSO = .05             *  Equilibrium of free sulphide/FeS
*                         DIFF*FLSO = eff. diff.constant for sulphide
FLSP = .0004           * /day sulphide into pyrite
*= = = = = = = = BENTHIC DIATOM PARAMETERS = = = = = = = = = = = = = *
* half value for phytobenthos distr
CBOF(5):0.2,0.6,1.0,1.0,1.0
kCBO = 1500.           * correction for CBO for low BDIA concentrations
                       * vertical extinction in bottom (CM-1)
KD(5):320., 150., 90., 70., 55.
* additional layer for reaction to light /hour light
LAD(5):.004,.008,.014,.017,.012
RPP = 1.0              * cell divisions per day under normal conditions
PPH = 1500             * maximal primary production in layer due to CO2 lack
PPM0 = 400.            * Michaelis value, much smaller then PPH
PPMox = 800.           * Michaelis value, under oxygen stress
PHOTMIN = 5.           * amount of light to which photosynthesis is possible
rrs4 = 0.01            * relative rest respiration (mgC/mgC/day)
ars4 = 0.10            * fraction of PP which is activity respiration
brs4 = 0.20            *            PPR-PP
rex4 = 0.020           * relative rest excretion (mgC/mgC/day)
aex4 = 0.10            * fraction of PP which is activity excretion
bex4 = 0.20            *            PPR-PP
p42 = .70              * fraction bdet in eaten diatoms
susBDIA = .16          * suspended diatoms to diatoms (1/m)
*= = = = = = = = BACTERIA = = = = = = = = = = = = = = = = = = = = = = = = = = *
arlBAC = .5            *
arlBACA = .65          *
exlBAC = .1            *
exlBACA = .1           *
FRBD = .01             *
FRBDA = .01            *
ardBAC = .65           *
ardBACA = .75          *
exdBAC = .05           *
exdBACA = .05          *
tdrBAC = .05           *
tdrBACA = .05          *
FRBR = 2.E-4           *
FRBRA = 2.E-5          *
arrBAC = .9            *
arrBACA = .9           *
exrBAC = .05           *
```

```
exrBACA = .05        *
rrsBAC = .010        *
rrsBACA = .010       *
upBAC = 0.75         *
upBACA = 0.75        *
ksBAC = 2000.        *
ksBACA = 10000.      *
morBAC = 0.035       *
pm32 = 0.2           *
morBACA = 0.015      *
QLOC = 1.            * Max. fraction of BLOC consumed/day (1/day)
*======== FILTERFEEDER PARAMETERS ===============*
PARPHY = 8.E06       * number of parts per mg of phytoplankton
PARSIL = 2.38E05     * number of parts per mg of silt
OCP = 2.E10          * Km for filtering
RRFIL = 0.200E-3     * (m3/mg) filtering rate of st.individuals
pex7 = 0.2           * excreted part of uptake
prs7 = 0.214         * respirated part of uptake
FACT7 = 0.01         * C/C/day needed for filtering
rrs7 = 0.003         * (1/day)rel.rest respiration
RSPAWN7 = 0.2        *
pup7 = .05           * max.food intake/day
uM7 = .1             * size-dependent availability of PBAC
t7c = 18.            *
rmor7 = .01          * relative mortality due to oxygen stress
* Metabolic correction for mean weight of individuals for 5 compartments
CIND(5):1.51,1.51,1.,1.,0.92
*TIDAL PUMP
TPUMP(5):0.1,5.0,3.0,0.8,1.0
*======== TEMPERATURE DEPENDENCES ==============*
ct0 = .0335          * constant for temp.dependent diffusion processes
Q10DIA = 4.          *
Q20DIA = 2.0         *
ct3s = 4.            *
ct3m = 5.            *
Q10BBM = 8.          *
Q20BBM = 1.5         *
Q10MEI = 8.          * parameter in temperature function (steepness)
Q20MEI = 2.          * parameter in temperature function (maximum)
TSA = 15.            * parameters in temperature function:
TSAN = 45.           * time of minima of acos and ancos
qox6 = .40           * parameter in oxygen function
qox7 = .7            * parameter in oxygen function
qox5 = .2            * oxygen-function coefficient
qox4 = .6            *
corCOp = .00357      * Conversion of C(produced) into OX
corCOr = .00333      * Conversion of C(respired) into OX
pSO = 0.50           * fraction sulphide-oxidation not measured in bell jars
T6T = 19.            *
```

Appendix C-4:

> Data of the Epibenthic Model

```
*= = = = = = = = = = = = = = = = = = = = = = = = = = = = = = = = = = = = =*
*= = = = = =    X = BIRDS     Y = EMES     Z = EMAC   = = = = = = = = = = =*
*= = = = = = = = = = = = = = = = = = = = = = = = = = = = = = = = = = = = =*
```

DELT1 = 40   * number of days between immigration start and maximum
DELT2 = 20   * number of days between maximum and end
kfdY = 150.   * half-saturation value for EMES food
kfdZ = 260.   *   ,,   EMAC food
SETTLE = 225   *
TIMM = 400   * temperature sum in comp 5, at start of immigration
DCRIT = 190   * potential start emigration EMAC
routZ = .067   * maximum percentage of EMAC emigrating out per day
TEMI = 12   * water temperature in comp 1, at start of emigration
rupZmax = .35   * max. fraction of body weight ingested/day by EMAC
rupYmax = .7   *   ,,   EMES
= = = = = = = AVAILIBILITY PARAMETERS = = = = = = = = = = =
uHY = 1.   * PCOP > EMES *fraction of copepods in the whole water column!
uYY = .2   * EMES > EMES
u3Y = .2   * BBAC > EMES
u2Y = .001   * BDET > EMES
u4Y = .2   * CDIA > EMES
u5Y = 1.   * BMEI > EMES
u7Z = .05   * BSF > EMAC
u6Z = .1   * BBBM > EMAC
uYZ = .3   * EMES > EMAC
u5Z = .5   * BMEI > EMAC fraction available in upper .5 cm
u2Z = .005   * BDET > EMAC
uZZ = .1   * EMAC > EMAC
xfd = 2.   * proportionality parameter
YUPLOS = .5   * uptake-losses EMES
ZUPLOS = .7   * uptake-losses EMAC
YBILOS = .07   * biomass-related loss EMES (12 oC)
ZBILOS = .01   * biomass-related loss EMAC (12 oC)
REFRAC = .5   * fraction of total loss due to respiration
Q10Z = 2.   * Q10 EMAC
Q10Y = 2.   * Q10 EMES
* The mean number of birds in the DOLLARD each month
BIRD(12):15958,5087,30479,61813,41305,8609,16897,46755,36344,73523,84523, 12865
* Preference the birds show for the different compartments
PREF(5): 1,.7,.3,.25,.5
* mgC needed per bird per day
BNEC(12): 9992,19628,10333,8016,7277,26422,13316,12052,13332,9284,6357, 10896
rbmin = 1.   * if there is less food than rbmin*BNEC limitation starts
u6X = .4   * BBBM > BIRDS
u7X = .4   * BSF > BIRDS
uZX = .2   * EMAC > BIRDS
uYX = 1.   * EMES > BIRDS
koxZ = .6   * relative oxygen saturation which no longer limits EMAC
koxY = .2   *   ,,   EMES
xoxZ = 8   * steepness of oxygen-OK function EMAC
xoxY = 8   *   ,,   EMES
ksalZ = 8.   * minimal salinity for EMAC
xsalZ = 10   * steepness of salinity OK function
ktZmax = 22   * maximum temperature
ktZmin = 2.   * minimum temperature

xtZ = 20   * steepness of ok function
REDIST = .25   * fraction of biomass, able to move to neighbouring comp
pexY2 = .5   * fraction of excretion of EMES to BDET
pexZ2 = .1   * fraction of excretion of EMAC to BDET
pZ2 = .005   * fraction of MAUP, going directly from EMAC to BDET
xfdX = 1   * selectivity exponent
cmorY = .08   * Relative mortality due to fish predation

Appendix C-5:

The Initial Values of the State Variables

OX(5):7.23772,9.27828,11.5209,11.6227,10.9471
SILT(5):351746.,252273.,117430.,80630.5,37951.8
PDET(5):1511.67,1021.25,455.873,288.183,125.423
SALT(5):8.48982,12.5545,15.5038,22.1915,27.1272
PFLAG(5):1.73340,2.67940,4.40730,7.04519,9.46819
PLOC(5):24908.8,1284.03,52.6224,7.57983,1.76986
PCOP(5):2.42178,3.88352,2.92021,3.80156,4.83751
PBAC(5):823.712,727.324,104.068,35.0991,12.7100
BBAC(5):891.934,3895.59,1713.35,1125.72,1209.36
ABAC(5):12575,4440.,1414.35,1221.68,687.051
CARN(5):.0381600,.0727063,.205311,.381457,.632791
PBLAR(5):.002542,.003845,.005149,.007605,.009656
BDIA(5):7471.68,1755.90,1417.13,1682.83,2156.76
ALOC(5):1551.81,226.933,37.1003,27.2179,16.1208
BMEI(5):634.544,148.092,96.8076,104.241,147.120
BBBM(5):1,807.300,206.477,281.899,302.081
BLOC(5):290.290,215.520,50.8671,27.9312,25.5084
BDET(5):23313.3,50000,28463.7,9599.64,4474.16
PMIC(5):80.1148,104.667,20.5246,9.34366,4.99537
PROC(5):8995.45,6209.77,2892.53,1774.62,743.332
EMES(5):26.7962,25.1892,37.8697,46.0138,34.1447
EMAC(5):1.39066,.586496,38.0788,5.86595,15.2697
BSUL(5):601758.,180000.,150000.,110000.,47310.7
BAL(5):.309153,1.39409,1.76037,1.64648,2.80985
ADET(5):101000.,131969.,47508,17279,4617
BROC(5):8010000,5600000,5000000,3500000,1100000
PDIA(5):3.18717,5.13487,10.7679,18.6470,26.4458
SILIC(5):122386.,97855.0,72425.3,46194.6,23589.3
PO4(5):8832.05,7476.46,5131.78,3176.33,1577.41
BSF(5):1,573.911,549.234,892.573,1630.53
PDROC(5):8603.47,8025.18,6563.05,4401.48,2767.86
BPYR(5):1500000,870000.,820000.,700000.,165000.
CSILT(5):231618.,141932.,65558.5,43940.6,19617.8

Subject Index

Acartia bifilosa 89
Acartia clausi 89
Acartia discaudata 89
Acartia tonsa 89, 91, 93
Aerobic layer, see Sulphide horizon
aex4 112
Algal exudates 194–196
Algal mats 111
Ammonia 27, 112
Amphipoda 33
Amphora 32
Anaerobic layer 10, 53, 211
Anas crecca 35
Anas platyrhynchos 35
Annual cycle 174
Annual production 217
ardBAC 151
ardBACA 151
areaf 45
Arenicola marina 33, 40, 136
arlBAC 151
arlBACA 151
arrBAC 151
arrBACA 151
ars4 112
Aselective grazing 204
Assimilation efficiency, see Efficiency,
  assimilation
Asterionella glacialis 30
Asterionella kariana 30
asWORK5 118
asWORK6 126
Atrochromadora microlaima 33
Autumn bloom 221
AZMIX 135

Bacteria
  aerobic 117, 147, 205
  anaerobic (ABAC) 6, 115, 117, 130, 137,
    147–151, 205, 211
  attached 30, 100
  benthic (ABAC, BBAC) 6–7, 32, 101,
    137, 146–151, 162, 217, 224
    biomass (TBAC) 205–206

biomass determination 97
colonization 60
fermentative 147
free-living 30
pelagic (PBAC) 6, 52, 89, 96–98, 100,
  147, 191–192, 198
  assimilation 197
  biomass 182
  biovolume 97
  production 187–188
sulphate-reducing 130, 134, 140, 147
sulphur 97
BAHBOE simulation software 12
BAL 211
Balanus 35
BAL0 136
Benthic diatoms (BDIA) 6–7, 32, 105–111,
  126, 137, 162, 206, 208, 217, 221, 224–225,
  227
  suspended 81, 113, 129, 198–199
  validation 205
Benthic flora 8
bex4 113
Biddulphia aurita 30
BIOAM 136
Biological oxygen consumption
  (BOC2) 103, 186, 189, 204
  validation 179, 188
Biological oxygen demand 8
Bioturbation 10, 33, 70, 130, 135, 136, 139,
  141–142
BIRD 167
Birds 35, 153, 165, 237
  distribution 166
  uptake 167, 232–233
BNEC 167
Boundaries 42
  hydraulic 16
Boundary conditions 12, 49–51, 73–74, 142,
  171, 189, 204, 240, 245, 261–263, 266
  mesozooplankton 181
  suspended matter 49, 254
brs4 112
BSL 123
BTL 123

caFC   84
caGC   84
Calanoids   31, 89
Calanus helgolandicus   94
Cannibalism   34, 193, 214–215
   meiobenthos   115, 117
Carbon
   budget   38, 248, 250–251
      benthic   256
      dissolved (DOC)   248–249
      particulate (POC)   247
   flux   10, 38, 105, 175, 183, 186, 189, 204,
      212, 215, 240, 250, 260
      benthic   256
      grazing   192–193
   sink   220
Carcinus maenas   54, 232
Cardium edule, see *Cerastoderma*
CARMOR   96
CARMOX   96
cBOC2   102
CBOF   109
Centropages hamatus   89
Cerastoderma edule   33, 264
Ciliates   87–89
CIND   122, 129
Clay content   16, 18, 246
cmorY   165
Cometabolism   147, 149, 220
Comfort functions   155–158, 261
Compartmentalization   42–43
Copepoda harpacticoida   32, 34
corCOp   101, 138
corCOr   101, 138
Corophium   33, 136, 154
Crangon crangon   34, 155, 229, 239
Ctenophores   77, 94–95, 181
cVOLI   95
Cyanobacteria   32

DAMP   145
Data sets   174
Daylight period   55
DCRIT   161
Deaeration   102
Decomposers   77
DELT1   160
DELT2   160
depchFH   45
depchLF   45
depfa   45
depth   45
Desiccation   113
Detritus   6
   algal   194–195
   benthic   137, 162, 220
   burial   255

carbon   6, 11, 57, 77, 186, 218, 256
   compounds   186
   consumption   255
   food web   3, 6, 196
   particulate   79, 255
   pathway   11
   pelagic   94, 98
   production   255
   sedimentation   79, 220
   state variables   4–5
   uptake of   149
Detrivory   62
Diatoms
   blooms   109, 180, 201
   centric   30
   pelagic   52, 175, 191
   pennate   30, 32
Dichromadora geophila   33
DIFF   141
Diffusion   65–67, 130, 137, 139, 141
   apparent   139, 141
   bicarbonate   111
   molecular   139, 141–142
   vertical   221
Diploneis   32
Double filtering   128
   correction   92
   suspension feeders   123
dsedM   101
Dynamic equilibrium   172, 261

eff4H   90
effCH   90
effE   88
effEH   90
effFH   90
effGH   90
effHH   90
effI   96
Efficiency
   assimilation   39, 191
      deposit feeders   124
      mesozooplankton   93
      pelagic bacteria   197–198
      suspension feeders   129
   bacterial   220
   ecological   190–191, 216
   growth
      benthic bacteria   151
      macrobenthos   120
      pelagic bacteria   197–198
effM   99, 197
effMa   99
Egestion   39, 94
   carbon fluxes   190
   pelagic   191

Emigration
    macroepibenthos   156, 161
    mesoepibenthos   264
Epifauna   33, 224–225
    macroepibenthos (EMAC)   153–165
    mesoepibenthos (EMES)   153–165
Epifluorescence microscopy   32, 97, 147
Epipelon   107
Epipsammon   107
Escherichia coli   115
Eudiplogaster pararmatus   33, 115
Eurytemora affinis   89, 91
Exchange coefficient   8, 66–67
Exchange volumes   73
Excretion
    activity   37
    annual fluxes   190
    benthic bacteria   150, 220
    macroepibenthos   164, 230
    meiobenthos   116–117
    microzooplankton   197
    pelagic   191
    phytobenthos   112, 113
    phytoplankton   195, 196
    rest   37
exdBAC   150
exdBACA   150
exlBAC   150
exlBACA   150
exrBAC   150
exrBACA   150
Extinction coefficient (EPS)   70, 81, 110
    validation   178

F5   116
F6   124
FACT7   127
Faecal pellets   24, 193
Feedback mechanisms   81, 128, 223
    macrobenthos   122
Feeding modes of meiobenthos   114
Feeding relationships   79
Feeding threshold   42
Ferrous sulphide   131
Filterfeeders   6
    benthic, see suspension-feeding
        macrobenthos
Fish fauna   8, 34
Flagellate blooms   180
Flatfish   34, 230
Flow profiles   16
FLSO   143
FLSP   145
Fluid mud (CSILT)   68–71, 243, 246
Food
    available   40–41, 224
        birds   166–167

    macroepibenthos   163
    meiobenthos   114–115
    mesoepibenthos   162, 231
    limitation in meiobenthos   116
    particles   127
    preference   40
    saturation in meiobenthos   116
    sources   37, 40, 192, 215, 234, 260, 261
        deposit feeders   123–124
        macroplankton   95
        meiobenthos   114
        mesoepibenthos   162, 232
    web   6, 79, 215, 218, 260
Forcing functions   12, 38, 171
FRBD   149
FRBDA   149
FRBR   149
FRBRA   149
Free sulphide   130, 134–135, 140–144, 211
Fresh water discharge   8, 50, 65–66, 171, 247
Functional groups   36, 261, 268

Gammarus   33
Gobius   232
Gradient
    depth   3
    DOC   241
    POC   245, 254
    salinity   22, 31, 240
    silt   66, 246
    sulphide   144
    suspended matter   254
Grazers   77
Grazing   41, 215
Grazing pressure   40, 224, 263, 267

Haematops ostralegus   35
Harpacticoid copepods   33, 114–115
Heteromastus   33
Heteromastus filiformis   *136*
Holoplankton   *31*
HTEMPE   88
HTEMPM   99
Humic compounds   81, 186
Hydrography   8
    parameters   45

Immigration of macroepibenthos   156,
    159–160
Ingestion   39, 91, 126
Initial conditions   171
Irradiance   53, 81, 180, 221, 267
    benthic   110

KCBO   109
kfdY   163
kfdZ   163

koxY 157
koxZ 157
ksalZ 158
ksBAC 148
ksBACA 148
ksCO 101
ksmfdE 88
ksmfdHf 91
ksmfdHr 91
ksmfdI 95
ksmoxE 89
ksmoxH 94
ksmoxM 99
KSPO4 82
KSSIL 82
ktZmax 157
ktZmin 157

LAD 110
Land reclamation   15
LBBBM 123
LIGHTMIN 81

Macoma balthica 33, 124, 136, 264
Macrobenthos (BBBM   BSF), 6, 136, 207,
    209, 215, 217, 224, 233, 236–237
    biomass   118
    deposit–feeding   33, 119, 121–124, 126,
        163, 218, 227, 238–239, 267
    energy budget   120
    species   133
    suspension–feeding   40, 119, 121–123,
        127–129, 225, 256–258, 263, 265
Macroplankton (CARN)   52, 86, 94, 191
    excretion   96
    mortality   96
    respiration   96
Marine ecosystem   3
maxCO 101
MCONS 208, 212
Mean square prediction error (MSE)   173
    mean component (MC)   173
    random component (RC)   173
    slope component (SC)   173
Median grain size   16
Meiobenthos (BMEI)   6–7, 32, 115–117, 162,
    207–209, 214–215, 224–225, 227, 233,
    236–237
Meroplankton (PBLAR)   31
Mesoepibenthos (EMES)   154, 162
    uptake   163
Mesopodopsis slabberi 34
Mesozooplankton (PCOP)   31, 52, 86–87,
    89–91, 93–94, 162, 181, 191, 257
    assimilation   92
    biomass   178

excretion   94
    faecal pellets   94
    feeding mode   90
    feeding strategy   90
    filter feeding   91
    functional response   91
    grazing   198
    grazing fluxes   193
    ingestion   92
    mortality   94
    production   199
    productivity   192
    raptorial feeding   91
    respiration   93
    uptake   91
    validation   178
Methanogenesis   148
Microbial loop   87, 249
Microphytobenthos, see Benthic diatoms
Microzooplankton (PMIC)   30–31, 52,
    86–88, 181, 191, 257
    assimilation   88
    efficiency   88
    excretion   88–89
    food sources   87
    grazing fluxes   193
    mortality   89
    production   199
    productivity   192
    respiration   88
    seasonal distribution   203
    sensitivity analysis   200
    uptake   88
Migration   234, 239
    macroepibenthos   231
Mineralization   27, 29–30, 59, 149, 217, 227,
    248, 251, 263
    aerobic   32, 130, 146–147, 151, 211
    anaerobic   32, 130, 146–147, 151, 208, 212
    benthic   256
    pelagic   182
    validation   208, 212
minfdHf 91
Model
    carbon flow   10
    hydraulic   8
    hydrodynamic   8, 262
    mathematical   7
    performance   77
    simulation   7, 11
Monopostia mirabilis 33
morBAC 150
morBACA 150
morE 89
morH 94
morM 100
morOXE 89

Morphology   8, 15–16
  parameters   43, 45–46, 262
Mortality   37–38
  benthic bacteria   150, 220
  epibenthos   164–165
  macrobenthos   123, 126, 129
  meiobenthos   117
  phytobenthos   113
Mya arenaria   33, 264
Mysidaceae   34

Navicula   32
Nematoda   32–33, 114–116, 207
Neochromadora poecilosoma   33
Neochromadora trichophora   33
Neomysis integer   34
Nereis diversicolor   30, 33–34, 136, 154, 229, 266
Net growth   38
Nitrate reduction   147, 211–212
Nitrogen   25, 27, 52, 183
Nitzschia   32
Numerical response   33
Nursery function   231, 239
Nutrients (PO4, SILIC)
concentration   27, 80, 82–84, 175, 267
  limitation   52, 79–80, 84–86, 183, 202–204
  regeneration   83, 183, 204
  uptake   39
  validation   178

OCP   127
Odontophora rectangula   33
Oligochaeta   32–33, 114
Organic carbon
  allochthonous   59, 258
  detrital (PDET, BDET, ADET)   59, 60, 117, 178
  dissolved   58, 60–61, 186, 194, 241–242, 248–250, 258
    export   250
    production   249
    transport   249
  import   250
  labile (PLOC, BLOC, ALOC)   52, 60, 62, 89, 100, 113, 117, 148, 150, 182, 195, 197–198, 209, 249–250, 262
  particulate   58, 61, 186, 194, 240, 242, 245, 247, 250, 252, 254–255, 258
  refractory (PDROC, PROC, BROC)   94, 135, 149, 209, 217, 240, 250, 252–253
  riverine   60
  sediments   57
  sources   61
  total   50, 208, 209
    benthic (BTOC)   208–210
    pelagic   50, 185–186

transport   250
validation   178
Oscillatoria   30, 32
Ostracoda   32–33
Oxygen (OX)
  concentration   51, 101–103, 137–139, 142, 157, 182, 262
  consumption   102–103, 129, 137, 143, 182, 188–189, 204, 212, 220, 263
  demand   104, 130, 137, 188, 265
  depletion   32
  diffusion   138
  dynamics   10, 101
    sediments   132
  exchange coefficient   102
  production   103, 138, 143
  saturation   103, 112, 157–159, 166, 229, 263
  stress, in meiobenthos   116
  validation   178

P/B ratios   216
p42   113
pAE   88
pAM   98
Paracalanus parvus   89
Paracanthonchus caecus   33
Parameters
  dimensional   261
  physiological   260
  structural   260
PARPHY   127
PARSIL   127
Particulate material   68, 70, 255, 258
pBM   98
pCE   88
pCH   90
pCM   98
pDE   88
pDM   98
pDOWN0   70
Pediastrum   30
pEE   88
Pelagic bacteria (PBAC)
  assimilation efficiency   99–100
  excretion   100
  flocculation   101
  growth efficiency   100
  growth rate   98
  mortality   100–101
  oxygen dependence   99
  production (PRODM)   179, 187–188
  respiration   99–100
  seasonal succession   99
  sedimentation   100–101
  temperature dependence   99
  uptake   99
  validation   178–179

pex5 116
pex6 124
pex7 127
pexE 88
pFD 84
pFdet 84
pFE 88, 202
pFH 90, 202
pGD 84
pGdet 84
pGE 88, 202
pGH 202
Phaeocystis pouchetii 30, 79–80, 90, 201
pHH 90
Phoca vitulina 31
Phosphate (PO4) 27, 52, 80, 183, 202, 204
Phosphorus 25, 27
Photic layer 81–82, 107, 221
PHOTMIN 110
Photoperiod 32, 55
Photosynthesis 81, 107
  benthic 110, 221
  gross 82
  validation 179, 187, 208, 212
Phytobenthos, see Benthic diatoms
Phytocarbon 82, 175
Phytoplankton (PDIA, PFLAG) 6–7, 59,
  79–85, 179, 194, 245, 258
  bloom 30, 267
  density 79
  diatoms 80, 175, 257
  excretion 84, 195–196
  exudates 182, 188
  flagellates 80, 175, 189, 191, 257
  fresh water 79
  grazing fluxes 193
  productivity 192
  respiration 84, 85
  seasonal succession 180, 201–204
  species composition 79
  state variables 80
  validation 178
Pleurobrachia pileus 94
Pleuronectes flesus 232
Pleuronectes platessa 34, 155, 232, 239
pm32 150
pM8s 101
PMAXPD 83
PMAXPH 83
pME 88
PO4CO 83
Pore water 141
POW5 116
POW6 124
POWM 115
PPH 111, 221
PPM0 112

PPMox 112
Praunus flexuosus 34
Precipitation 20
Predation 38, 40, 215
  by birds 6, 237
  flux 234
  internal 234
  on mesoepibenthos 165, 231
  pressure 234–235, 237
PREF 166
Primary production 29, 248, 252, 258, 263,
  267
  benthic (MBPP) 9, 56, 107–112, 138, 208,
    212, 221–223, 225, 256
  gross 84
  pelagic 56, 82–85, 186–187, 257
    validation (MPPP) 179, 187
Production of epibenthos 232
Productivity
  benthic 216
  macroepibenthos 230–231
  pelagic 190–192
prs5 116
prs6 124
prs7 127
pRSF 85
pRSFA 85
pRSG 85
pRSGA 85
Pseudocalanus minutus 89
ptWET 45
pup7 128
pWAFL 45
Pyrite (BPYR) 135, 208–209, 211
  formation 146, 210, 212
pZ2 164

q10PD 83
q10PH 83
Q10Y 165
Q10Z 165
QLOC 148
qrsHf 93
qrsHr 93

Rates
  accretion 25, 145
  assimilation 82–84
  decomposition 59
  diffusion 111
  filtration 93, 128, 237, 258
  growth 32, 40, 201, 231, 237
  reaeration 138–139
  respiration 54, 205
  sediment mixing 136
  sedimentation 146, 210–211, 246, 253,
    257

sulphate reduction   32, 212, 218
  sulphide production   142
  temperature dependence of   54
  uptake   38, 229, 234
Ratio C:ChlorA   83, 175, 179
REACON   102, 139
Reaeration   102
REASIL   252
Recycling algal material   197
REDIS   75
Redistribution
  epibenthos   155–156
  suspended matter   73–75
REFRAC   164
resdep   71
Residual sediment transport   70
Respiration   10, 54, 103
  activity   37, 116
  anaerobic   212, 218, 220
  annual fluxes   190
  benthic bacteria   151
  community (MCONS)   208, 212, 214
  deposit feeding macrobenthos   126
  macroepibenthos   164, 230
  meiobenthos   116–117
  pelagic   191
  phytobenthos   112–113
  phytoplankton   186
  rest   37
  suspension feeding macrobenthos   129
Resuspension   24–25, 57, 68, 70–71, 73, 243,
  246, 252–253
  microphytobenthos   113
rex4   113
rex5   116
rex6   124
rex5O   117
routZ   161
RPP   110
RRFIL   127
rrs4   112
rrs5   118
rrs6   124
rrs7   129
rrsBAC   151
rrsBACA   151
rs0F   85
rs0G   85
rs12E   88
rs12H   91
rs12M   99
RSPAWN6   126
RSPAWN7   126
rTID   71
RTMPD   83
RTMPH   83
rupE   88

rupH   91
rupM   98
rupYmax   163
rupZmax   163

Sabatiera pulchra   33
SALCAR   95
Salinity (SALT)   20, 65–66, 80, 157, 159, 180,
  240, 242
  distribution   8, 50, 175
  measurement   8, 66
Salt marsh   29
SALTM   80
SDL   145
SDO   246
Seasonal
  cycle   27
  succession   31, 79
seddep   71
Sediment   131, 246
  aggregates   18
  characteristics   105
  composition   16
  consolidation   24
  origin   25
  surface   107
  suspended   8, 81, 243, 245–247, 253
Sedimentation   32, 49, 57, 68, 70–71, 73, 135,
  145–146, 217, 248–249, 255–257
  detritus   79, 220
  diatoms   85
sedM   101
SEDPDQ   85
Self-shading of phytoplankton   81
Sensitivity analysis   260–261, 268
  exudate production   194
  POC transport   252
SETTLE   160, 234
Severn estuary   266
SILCO   83, 202
Silicate (SILIC)   27, 52, 80, 183, 202
Silt (SILT)   65, 68, 70, 257
  budget   247
Skeletonema costatum   30, 79
SLOP   92
Sloppy feeding   39
SML   136
Spatial compartments   65
Spawning   225
  deposit feeding macrobenthos   126
  suspension feeding macrobenthos   129
Species composition   40, 215, 233, 261
  birds   167
  macrobenthos   118
  meiobenthos   114
  microphytobenthos   107
Spring bloom   79–80, 201, 221, 225

STANFO5 115
State variables 4–6, 9–10, 49, 77
  annual cycle 172
  benthic bacteria 147
  detrital 184
  DOC 185
  initial concentrations 171
  macrobenthos 118–119
  macroepibenthos 153
  meiobenthos 114
  mesoepibenthos 153
  POC 185
  sulphide horizon 137
Steady state 66, 77, 172
Storm-induced mixing 135
Stratification 22
Sulphate reduction 32, 144, 147, 212, 218
Sulphide (BSUL) 208–209
  diffusion 140, 143
  horizon (BAL) 53, 105, 131, 135, 136,
    138–140, 142–145, 208, 211, 218, 265
  production 130, 131, 134, 140, 143, 211
Sulphur cycle 131
Suspended matter (SUSM) 50, 70–71, 77,
  127, 178, 242, 252
  concentrations 74
  validation 178
Suspension feeders, see macrobenthos

tdrBAC 150
tdrBACA 150
TEMCAR 95
TEMI 161
Temora longicornis 89
Temperature
  adaptation 189, 225
  air 18
  correction 54, 165
  dependence
    macrobenthos metabolism 122
    macroplankton 95
    phytoplankton 201
  sediment 53–54
  water 53–54, 157–161
Temperature-dependent migration 34
Thalassionema nitzschioides 30
Thalassosira 30
Thymidine 97
  incorporation 32
Tidal
  amplitude 20
  asymmetry 70
  excursion 21–22, 240
  flat
    estuary 3
    exposure 9, 54
    surface 6

  prism 21
  pumping 134, 139, 141–142, 144
TIMM 160
Tintinnidae 87
Topography 15
Transport
  advective 9, 65–66, 70–71, 73–75
  diffusive 65–66, 70–71, 73–75, 130, 137, 143
    macroplankton 95, 181–182
  mass 66
  model 9, 240, 266
  nutrient 27
  process 9, 65, 240
  suspended matter 9, 24, 57, 67, 70–71, 73,
    75, 77, 252
  tidal effects 66
  vertical 33, 70, 105, 130, 135–137, 146
TS5PLIV 248
TS5POC 248
TSILTSEA 74
Turbellaria 32, 207
Turbidity 68, 81, 180, 237, 267
  maximum 42, 50
Turbulence 67
Tychoplankton 107

u2Y 162
u2Z 163
u3Y 162
u4Y 162
u5Y 162
u5Z 163
u6X 167
u6Z 167
u7X 163
u7Z 167
uHY 162
upBAC 148
upBACA 148
upMEI 115
Uptake 38–39, 215, 256
  benthic fauna 215
    bacteria 148
    meiobenthos 115–117
    suspension feeding macrobenthos 128
  birds 167, 232–233
  of detritus 196
  epibenthos 163, 215
    macroepibenthos 163, 230, 233
    mesoepibenthos 163, 230, 233
  macroplankton 95
  mesozooplankton 90
  microzooplankton 200
  pelagic bacteria 98
uYX 167
uYY 162
uYZ 163

uZX   167
uZZ   163

Validation procedure   13, 172
Vaucheria   32
Vertical distribution of
    microphytobenthos   108–109
Viscosia rustica   33
VOL   45

W30A   208, 212
W50   70
Waste water discharges   7, 42, 182, 206, 263
Water circulation   21
Watershed   19–20
Wave action   19, 70
WIND   70
Wind
  impact   19
  speed   19, 70, 74–75, 243, 246
Workshop approach   11–12

xfd   162
xfdI   95
xfdX   167
xoxY   158
xoxZ   158
xsalZ   158
xtZ   157

YBILOS   164
YUPLOS   164

ZBILOS   164
Zooplankton (CARN, PCOP, PMIC)
    6, 86–96, 176, 180–182
  carnivorous, see macroplankton
  state variables   86
ZUPLOS   164
ZZSILT   81